INTRODUCTION TO STATISTICS
IN A BIOLOGICAL CONTEXT

Edith Seier and Karl H. Joplin

East Tennessee State University

ii

Cover Photo: Llamas in Huayllay, Pasco, Peru. Huayllay is in the central highlands of Peru at an approximated altitude of 4,300 m (14,108 ft). [Photo credit: Federico Helfgott]

The symbol * indicates that the material can be considered optional in a one-semester introductory course

Contents

Preface

This is a textbook for an introductory course in statistics for undergraduate biology majors, students in pre-professional programs in the health sciences, and anybody interested in learning the basics of statistics in a scientific context. It covers basic ideas and concepts in statistics and probability taking into account the future needs of the students. We have selected the topics considering that:

Some characteristics of the book:

Early introduction to inference

Biologists need to analyze the data they produce and should therefore understand the basic ideas of probability and statistics.

Many biology majors intend to pursue a career in medicine or other health-related professions.

Randomization methods are included

Bioinformatics, which makes extensive use of probability and statistics, is becoming an important tool for biologists and medical researchers. It is not our intention to teach bioinformatics but to include some simple exercises in the context of DNA genetics.

Relevant role of the binomial distribution

Five important features of the book are:

Real data

- An early introduction to probability and inference

- Inclusion of randomization methods

- The important role given to the binomial distribution

- Real data in most examples and exercises

- Commands in **R** at the end of most sections

Why introduce inference early?

Introducing inference early on in the course gives the students more time to get used to the ideas associated with hypothesis testing and estimation. The early introduction of inference requires the use of methods that do not need a large quantity of prerequisites, such as randomization methods. The binomial distribution can also be studied with a minimum background and provides the basis for testing hypotheses involving a population proportion.

Several of the examples and data sets come either from our own statistical consulting or our work in biology. The statistical software to be used in class should be the instructor's choice. However, we have included some instructions in MINITAB and complete instructions for **R**. MINITAB is a user-friendly software and is a favorite of many students. **R** is a free software that is extensively used in statistics and bioinformatics. We think students should be at least exposed to the use of **R**. The commands in **R** are provided **Why R?** not only in the book but also in text files available from our website. Students can copy and paste the commands into **R** and adapt them to work with their own data.

When helpful, we include conceptual maps to explain the relationship between important concepts. These conceptual maps can be used by the instructor to present an overview of a topic or by the students to review the connections between different concepts. In some chapters we include a section called *Beyond Biology* to inform the students about applications of statistics to other fields.

Sets of exercises have been included at the end of each chapter. A first set includes short questions, which students can use to review the topics included in the chapter. A second set of exercises is included for extra practice; some of these exercises can be discussed in class, and others can be assigned as homework. Chapters 2, 10, 11, and 12 include labs that can be worked using a computer. Chapters 2, 3, 5, and 10 include ideas for small data analysis projects to be used at the discretion of the instructor. The extent of each type of exercise varies because each topic has different needs. A list of bibliographical references is provided at the end of every chapter.

This book originated in the statistical component of a one-semester integrated Biology/Statistics course taught by the authors. That course was the first one of a sequence of three integrated Biology/Mathematics/Statistics courses that were created and implemented as part of the SYMBIOSIS project with funding from the Howard Hughes Medical Institute. The SYMBIOSIS project was the response of the Department of Mathematics and Statistics and the Department of Biological Sciences at East Tennessee State University to the initiative BIO2010 from the National Research Council of the National Academies. We also have taken into consideration the learning goals in the Scientific Foundations for Future Physicians report from AAMC-HHMI 2009. However, a statistician should feel comfortable teaching from this book because the biological context is explained clearly.

We have designed the book to be used as textbook for a stand-alone introductory statistics course. It has three main units:

1. Chapters 1-5: The basics of data collection, probability and inference.

2. Chapters 6-9: More on probability and its applications.

3. Chapters 10-12: Classical topics of inference and modeling.

Some chapters need to be covered before others as shown in the flow chart.

Are there some topics that need to be covered before others?

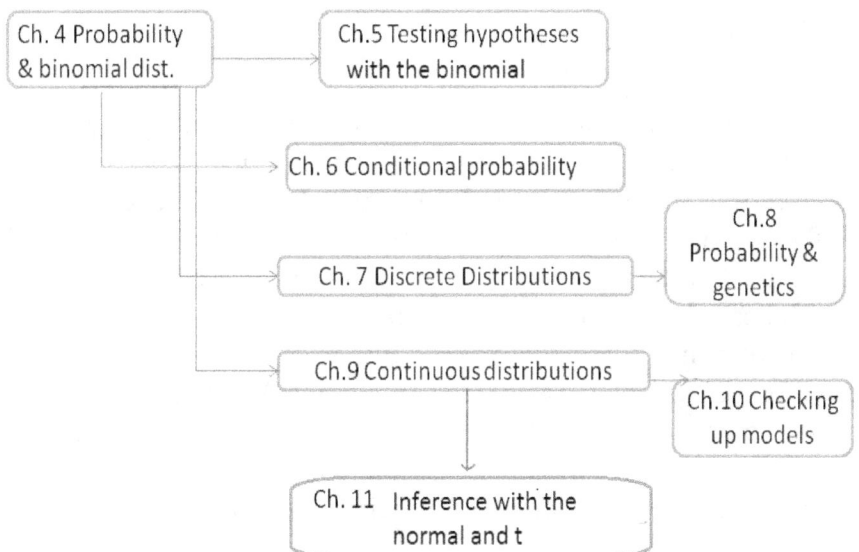

Chapter 1 has a brief description of the scientific method and the role that statistics plays in it. We include the definition of concepts such as statistical hypotheses, variables, population and samples, parameters and statistics, and the basics of data production through experimentation and observation. The p-value is defined as the probability of getting the results shown by the experiment or the survey, or more extreme ones, when the null statistical hypothesis is true. Students are told that a distribution is needed in order to calculate that probability. The distribution describes what can happen when the null hypothesis is true, and there are several ways of obtaining such distribution. After the first chapter, the instructor can select which chapters to cover, and in what order, depending on the nature of the course and the time available. We explain statistical graphs and descriptive statistics in Chapter 2. Side-by-side boxplots, scatterplots and correlation are included because many of the questions that arise in science allude to associations between variables and comparison of groups. In addition to time series plots, survival plots are included. We introduce statistical inference using randomization in Chapter 3. The methods included are the randomization test, the bootstrap and Fisher's exact test for two-way tables. The randomization test to compare two populations and for paired data, and the bootstrap to build confidence intervals are introduced using hands-on activities and later applied with programs in **R** that mimic the tactile simulations. Chapters 4 and 5 need to be kept together. Chapter 4 is an introduction to probability and the binomial distribution. In Chapter 5 the binomial distribution is used to test hypotheses about a population proportion. The notion of power of a test is introduced in the context of the exact binomial test for proportions.

Brief description of each chapter

Chapter 6 explains conditional probability in the context of medical diagnosis using the information about sensitivity and specificity of the test and the prevalence rate of the disease in the population of interest. The application of Bayes' theorem is presented in two alternative ways: using probability trees and applying the formula directly. In class, we use probability trees. The uniform, binomial and geometric distributions are included in Chapter 7. Chapter 8 is optional, it does not contain new probability concepts but applications of probability to both Mendelian and DNA genetics. We include a brief introduction to the vocabulary of genetics (genotype and phenotype) and DNA sequences. Chapter 9 is about continuous distributions; it includes the uniform, normal, chi-square and exponential distributions.

Chapter 10 discusses how to check for assumptions and includes the following chi-square tests: goodness of fit, independence and homogeneity. It also includes tests of normality. Chapter 11 covers the methods of inference based on the normal and t distributions. The sample size issue is discussed for both cases: confidence interval estimation and hypothesis testing. Chapter 12 introduces simple linear regression. It includes the case in which both variables have been previously transformed using logarithms to discuss allometric relationships in biology. Multiple regression and non-linear estimation are briefly discussed.

To facilitate the organization of the material for a one-semester course, we prepared a list of eighteen topics. The topics marked with a (*) in the table of contents of the book are not included in this list (except for selected topics from Chapter 8) and can be omitted or used as enrichment notes for the more motivated students.

1. Hypotheses, variables, parameters and statistics (Ch. 1, Sec. 1.1-1.4).

2. Producing data and preparing data files (Ch. 1, Sec. 1.5-1.8).

3. Graphs and statistics for one quantitative variable (Ch. 2, Sec. 2.1-2.11).

4. Graph and statistics for two quantitative variables: scatterplots and correlation (Ch. 2, Sec. 2.12-2.13).

5. Graphs and statistics for categorical variables (Ch. 2, Sec. 2.14-2.15).

6. Randomization test (Ch. 3, Sec. 3.1).

A list of 18 topics suggested for a one-semester course

7. Bootstrapping to build confidence intervals (Ch. 3, Sec. 3.2).

8. Introduction to probability and the binomial distribution (Ch. 4).

9. Testing hypotheses using the binomial distribution (Ch. 5).

10. Conditional probability and Bayes Rule (Ch. 6, Sec. 6.1-6.4.1).

11. More on discrete distributions (Ch. 7, Sec. 7.1-7.4, 7.6).

12. Applications of probability to Mendelian and DNA genetics (Selected topics from Ch. 8).

13. Normal and Chi-square distributions (Ch. 9, Sec. 9.1-9.3, 9.5).

14. Checking models and assumptions (Ch. 10).

15. Sampling distributions and confidence intervals (Ch. 11, Sec. 11.1-11.2).

16. Tests for means (Ch. 11, Sec. 11.3, 11.5).

17. Large sample inference for proportions (Ch. 11, Sec 11.4).

18. Introduction to linear regression and the allometric model (Ch. 12).

Estimated time to cover each topic

The estimated time to cover topics 1&2 is one week, two weeks for topics 3-5 and one week for topics 6 & 7. Topics 8-14 and 18 require one week each. It might take from 2 to 3 weeks to cover topics 15-17. This makes for a total of 14 or 15 weeks depending on how much time is devoted to do exercises in class. The instructor should refer to the previous flowchart of prerequisites in case he/she wants to omit some topics.

We thank the Howard Hughes Medical Institute for the funding of the SYMBIOSIS project (HHMI grant # 52005962) and the Chairs of the departments of Mathematics & Statistics and Biological Sciences at ETSU for their initiative and continuous support of the project and for motivating us to write this book. We also thank our CO-PIs and collaborators with whom we lived through the SYMBIOSIS experience. The first cohort of freshmen from the integrated version of the course, for whom we wrote the initial version of the class notes in the Fall of 2007, were a strong and joyful motivation to start this work. Among them were Lindsey, Jeremy, Lauren, McKayla, Amanda, Joe and Brandy, who went on to complete the sequence of the three SYMBIOSIS courses. A draft of the current version of the book was used for the first time as a textbook of a stand-alone statistics course in the Spring of 2010. Thanks to Rachel Matheson, a student in that course, who helped to review the current version. Over the years we have benefited from collaborations and conversations with educators and researchers in different fields, some of whom have generously contributed data. We are grateful to our respective families for their support, encouragement, and valuable feedback.

Edith Seier   Karl H. Joplin
Department of Mathematics   Department of Biological Sciences
and Statistics

East Tennessee State University
Johnson City, TN
August 2011

Data files and R-commands are available from
http://faculty.etsu.edu/seier/IntroStatsBioBook.htm

Chapter 1

Hypotheses and data

The concepts of hypothesis and theory are explored within the context of the scientific method. At the same time, basic definitions in statistics such as statistical hypotheses, individual, population, sample, variable, parameter and statistic are introduced. Guidelines for the production of new data, either by experimentation or observation, and for the preparation of data files, are provided.

1.1 The Scientific Method: A way of knowing

The scientific method is an approach to understanding and describing events that occur in the natural world; its main characteristic is the production of testable and falsifiable hypotheses.

Theories and hypotheses

A hypothesis is a statement about a process that describes what the scientist thinks will happen under specified conditions. A hypothesis is based on observations or previous experiments and is never put in the form of a question. Hypotheses cannot be 'proved', since we never know all of the conditions that may affect the outcomes; they can only be either supported or disproved. Data are collected from replicated, controlled experiments and if they constitute strong evidence against the hypothesis or against a prediction based on the hypothesis, we conclude that the hypothesis is false.

Scientific hypotheses are never proved, only disproved

A theory is generally formed by a group of hypotheses or statements that have been extensively tested and that are supported by data. Theories are used to explain natural phenomena or make predictions. A theory can be replaced when data show evidence that contradicts the predictions or

Scientific theories

explanations based on the theory. However, it must be replaced with a theory that can explain everything the previous theory explained as well as the new findings. Theories are never 'proven to be true,' they can only be 'proven to be false.' Thus, to a scientist, a theory is not a guess, and it is not the first thing that is produced. It is instead the result of a long process of hypothesis testing and is the strongest statement that can be made about a natural event.

An interesting example occurred early in the history of scientific inquiry. Jan Baptist van Helmont (Belgium, circa 1580 − 1644) was a medical doctor who did a number of studies on basic physiology of humans. He also investigated the accepted belief, first stated by Aristotle, that plants derived their growth from the soil. He planted a sapling (2.3 kg (5 lbs)) in a large container of soil (90.9 kg (200 lbs)) that he had weighed, and grew it for 5 years. He watered the plant, prevented leaves from falling into the container, and weighed the tree (76.8 kg (169 lbs)) and the soil (90.8 kg (199.8 lbs)) at the end of the experiment. The plant added nearly 74.5 kg of material, and the soil lost only 0.1 kg of mass. From this he concluded that Aristotle was wrong and that the weight gain of the tree had come from the water (hypothesis). As it turned out, van Helmont was incorrect with this hypothesis because he did not know that photosynthesis makes sugar from CO_2 in the air and water, in the presence of light. This shows not only the power of hypothesis testing, but also the importance of testing your own hypotheses. The power of the scientific method is that when scientists do an experiment, they are attempting to do everything they can to disprove a hypothesis, not to 'prove' it. If they cannot disprove the hypothesis, then it becomes an increasingly stronger statement.

1.2   Variables, parameters and statistics

In Statistics, the **individual** is the basic entity about which we collect information. An individual can be a person, organism, cell, plant, institution or object. The whole collection of individuals of a given type form a **population**, which can be delimited geographically or with respect to time. Some examples include the population of maple trees in the Cherokee National Forest, the population of spiders in a cave, the population of elementary schools in the state, the population of visitors to Yellowstone National Park during the year 2008, or the population of cells in the roots of an onion.

Variables represent characteristics of the individuals. Variables can be either **quantitative** (naturally expressed in numbers such as height, weight, length, altitude of a site) or **categorical** (naturally expressed in words such as gender, color, blood type). Assigning codes to the categories or classes of a categorical variable does not make it quantitative. For example, using the code '1' for green and '0' for yellow does not make the variable, color, quantitative. **Quantitative variables** can be either **discrete** if they can only take integer values (such as the number of eggs in a nest that hatch) or **continuous** if they can take any value in an interval (such as width in centimeters of a leaf). **Categorical variables** can be either **nominal** if there is no natural ordering in the categories, such as in color (*yellow* or *green*), or **ordinal** if there is a certain natural ordering such as in location of the nest in the tree (*low*, *medium*, or *high*).

Variables represent characteristics of individuals

Variables can be quantitative or categorical

Parameters are quantities that describe a population in relation to the variables of interest. For categorical variables, the parameter of interest is usually the proportion (p) of individuals that fall into a given category or exhibit a certain trait, for example, the population of maple trees in the Cherokee National Forest that are infected by a disease. In the case of quantitative variables, a parameter can be the mean value (μ) of the variable in the population, for example, the mean canopy volume of all maple trees in the Cherokee National Forest.

A parameter summarizes the values of a variable in the population

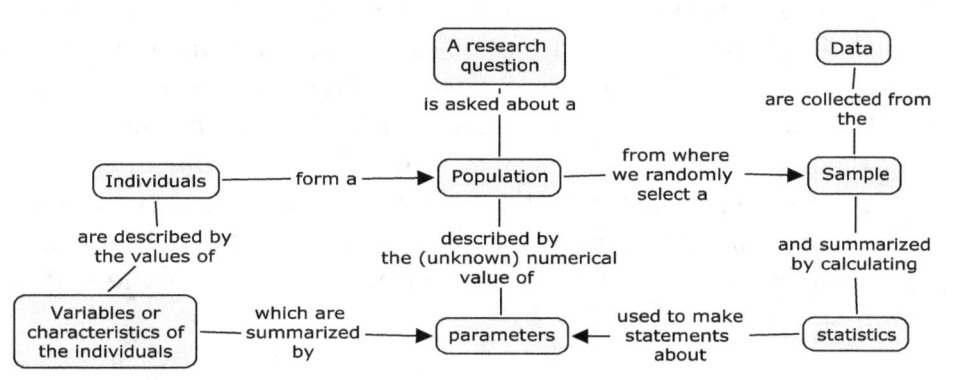

Figure 1.1: Population, sample, variables, parameters, statistics

The population mean (μ) and the population proportion (p) are parameters

Data for the entire population is seldom available. Data are usually collected from randomly selected elements of the population; those selected elements form a **sample**. The quantities calculated from the **sample** in relation to the variables of interest are called **statistics** and are used to estimate or make statements about the parameters, whose numerical value we do not know. The relationships among these concepts are summarized in the conceptual map in Figure 1.1. The symbols used for the sample mean and the sample proportion are $\bar{x}$ and $\hat{p}$, respectively.

A statistic summarizes data from the sample

The sample mean ($\bar{x}$) and the sample proportion ($\hat{p}$) are statistics

Example 1.3.1 What is the yield of a tomato plant?

**For a
quantitative
variable
a parameter
could be
the mean (μ) of
a population**

A farmer would like to estimate the yield in pounds of a given variety of tomatoes under certain conditions of altitude, amount of water, and fertilizer. The farmer plants 20 tomato plants, harvests them, and weighs the yield per plant. The average yield of the 20 plants is 2.3 pounds. In this case the **population** is formed by all the tomato plants of this type that could be planted under the specified conditions. The **variable** is yield in pounds, the **parameter** is the mean yield of all the tomato plants that could eventually be planted, the **statistic** is the average yield of the 20 tomato plants that were actually planted, and the **value of the statistic** is 2.3 pounds ($\bar{x}=2.3$).

Example 1.3.2 What percent of the Tennessee population are current smokers?

**For a
categorical
variable the
usual parameter
is the population
proportion (p)**

Given that tobacco has been an important crop in Tennessee, public health researchers are interested in knowing what percent of the population 18 years or older are current smokers. A random sample of 2466 individuals 18 years or older is selected, and using several questions, each individual is classified as *current smoker*, *former smoker* or *never smoker*. Six hundred and eleven individuals in the sample are current smokers. The **population** is formed by all the residents of Tennessee 18 years or older, the **sample** is formed by the 2466 individuals who were interviewed, the **variable** of interest is whether the individual is a current smoker or not, and the **parameter** is the proportion of current smokers in the population (unfortunately, we do not know the numerical value of that proportion). The **statistic** is the proportion of current smokers in the sample and the **numerical value of the statistic** is 0.248 ($\hat{p} = \frac{611}{2466}=0.2477697 \sim 0.248$).

1.3   Statistical hypotheses

Statistical inference is used to arrive at conclusions about one or more populations based on the study of samples drawn from those populations. It is also used to arrive at conclusions about treatments based on carefully designed experiments. Either a very specific research question, or a hypothesis posed by the scientist, is translated into two concrete opposing statements that we call the **null** (H_o) and **alternative** (H_a) **statistical hypotheses**, which are written in terms of population parameters. Historically, the null hypothesis was introduced first.

Example 1.4.1 Which medicine works faster?

The research question is whether it takes a shorter time on average for a new medicine to relieve headaches. The 'null' and 'alternative' statistical hypotheses are:

Null hypothesis or Ho: It takes, on average, the same amount of minutes to make the headache disappear with either the old or the new medicine.

Alternative hypothesis or Ha: The new medicine works faster.

These statistical hypotheses can be written using mathematical notation:

$$H_o : \mu_{old} = \mu_{new}$$
$$H_a : \mu_{old} > \mu_{new}$$

The null hypothesis H_o

Statistical hypotheses are written in terms of parameters

where μ represents the mean time in minutes that it takes the headache to disappear. Of course, to be able to make a decision about the null statistical hypothesis (reject or not reject it) and answer the research question, we need DATA. An experiment needs to be conducted in which both medicines are compared, for several individuals, in terms of the time that it takes for the headache to disappear. Time is a quantitative variable, that is why the parameter of interest is *mean time* (μ). The statistic to be calculated from the subjects in each treatment will be the average time, or sample mean ($\bar{x}$). The final decision is expressed in terms of the null statistical hypothesis: either reject or not reject H_o.

Example 1.4.2 Are female mallards attracted to the green color in food?

Male mallards have green heads whereas females do not. It might be that female mallards are attracted to the green color. A researcher in animal behavior elaborates the hypothesis that female mallards are attracted to the green color in food. The statistical hypotheses are written in terms of p, the proportion of the female mallard population that would pick the green color when offered two pieces of bread. The pieces of bread are identical except that one has been colored with a green dye and the other has not. The null and alternative statistical hypotheses are:

$$H_o : p = 0.5$$
$$H_a : p > 0.5$$

One-sided alternative hypothesis H_a

Now consider the possibility that the research question is whether female mallards have equal preference for the green and plain bread. The negation of equal preference includes two possibilities: that the female mallards either

Two-sided alternative hypothesis H_a

prefer the green color in food or that they do not like the green color in food. In that case the null and alternative statistical hypotheses are:

$$H_o : p = 0.5$$
$$H_a : p \neq 0.5$$

Statistical hypotheses are always written in terms of parameters whose numerical values are not known, NOT in terms of statistics. Once data are available, statistics can be calculated. In Example 1.4.2, the parameter is the population proportion p. In order to decide to reject or not reject the null hypothesis, we need data. A study needs to be conducted in which several female mallards are offered green and plain bread to see which one they pick first. Which statistic needs to be calculated is suggested by the type of parameter in the hypotheses. If H_o is written in terms of the population proportion as in Example 1.4.2, the sample proportion ($\hat{p}$: proportion of mallards in the sample that pick the green bread first) is used to summarize the data. If H_o is written in terms of population means as in Example 1.4.1,

Write the statistical hypotheses in terms of the parameters; get data and calculate statistics using the data

then sample means ($\bar{x}_{old}$ and $\bar{x}_{new}$), are calculated.

In Statistics, as in other disciplines, there are different ways of doing things. In the approach followed here and in most introductory statistics textbooks, we do not 'accept' H_o, we say 'there is not enough evidence to reject H_o.' The decision is based on how likely it is for the statistic to take the observed value (or a more extreme one) when H_o is true. If that probability or **p-value** is very small, the data constitute 'evidence' against the null hypothesis and the null hypothesis is rejected. For example, assume that in the mallards case, 10 female mallards are offered 2 pieces of bread each, one plain and one dyed green, and all the 10 mallards go for the green bread first. How likely is this to happen ($\hat{p} = \frac{10}{10} = 1$) if there was no preference for the green color in the female mallard population ($p = 0.5$)? Not very likely, it would be equivalent to getting 10 heads in a row when tossing a fair coin 10 times. If the 10 mallards pick the green bread first we would feel like

If it is unlikely to get such results just by chance when the null hypothesis is true, then reject H_o

rejecting $H_o : p = 0.5$.

Hypothesis testing will be seen several times in this book. Most cases and examples will have in common the following steps, also summarized in Figure 1.2:

1. Identify the parameter that describes the population(s) of interest.

2. Write the statistical hypotheses in terms of the parameters and based on the research question or a scientific hypothesis.

3. Collect data through observation or experimentation.

4. Summarize the data with quantities called *statistics*.

5. Calculate the probability that the statistic takes the numerical value actually obtained, or a more extreme value, when the null hypothesis is true.

6. Decide, based on that probability, if the null hypothesis is rejected or not (if the p-value is small, reject H_o).

The probability or p-value is calculated by comparing the value obtained for the statistic, based on the data, with a distribution of reference. The distribution of reference describes the values (and how likely they are to happen) that the statistic can take when the null hypothesis is true. How that distribution under the null hypothesis is defined or prepared depends on the null hypothesis and the method we decide to use. Chapters 3, 5, 10 and 11 cover different cases of hypothesis testing. All have in common that the actual data are compared with what would happen if the null hypothesis were true. Based on that comparison, we gauge how likely to happen are the obtained observations, or more extreme ones, when the null hypothesis is true. That help us to decide if we should reject or not the null hypothesis.

Decisions are written in terms of the null hypothesis: either 'reject H_o' or 'don't reject H_o'

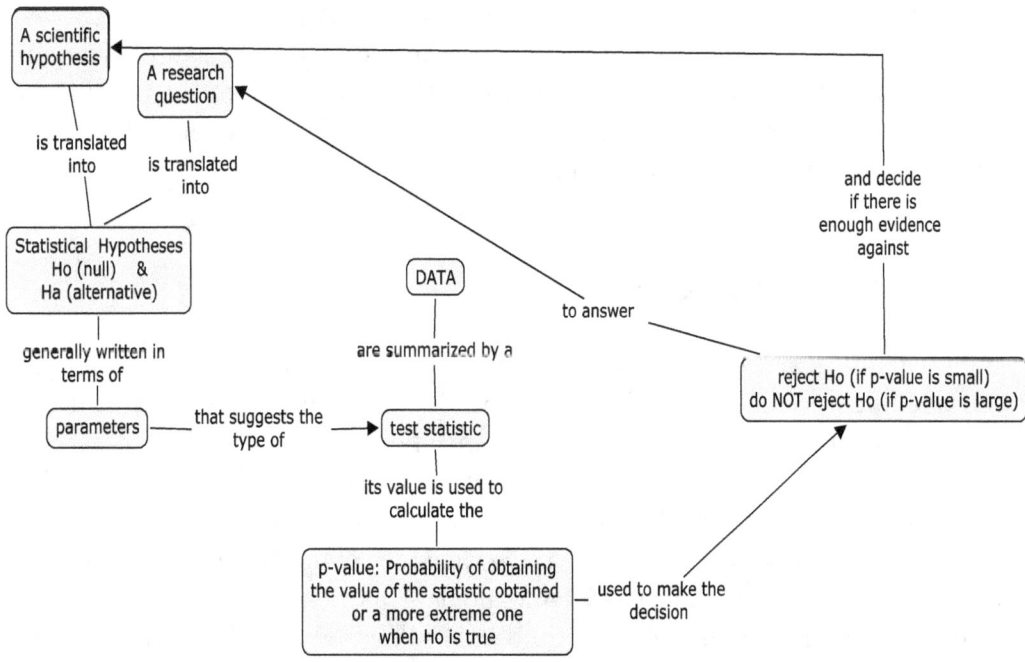

Figure 1.2: Testing statistical hypotheses

1.4   Statistical inference

Statistical inference is a process in which statements are made about populations, based on observations from randomly selected samples. Conclusions can also be drawn about treatments, based on carefully designed experiments. An inference problem might be posed not as a hypothesis testing case but as an estimation case. In estimation problems we are interested in finding a possible interval of values for a parameter of the population based on the data from the sample. Estimating a parameter is sort of making an 'intelligent guess' for its value. For example, think of the population 18 years or older of a state. If we are interested in knowing what percent of the population are current smokers, the problem is an estimation problem (we are trying to estimate the value of the parameter p in the population). There is more than one way of estimating parameters. Chapters 3 and 11 describe two different approaches. On the other hand, if we are interested in knowing specifically if the proportion of people who are smokers in the state is similar to the 22% of a neighboring state or if it is above that quantity, then it is a hypothesis testing problem and we are trying to test the hypothesis $H_o : p = 0.22$ vs. $H_a = p > 0.22$. To solve both, estimation and hypothesis testing problems, we need to collect data. In this example, a survey needs to be done in which the participants are asked several questions to determine if they are current smokers or not. Based on the proportion of individuals in the sample who are current smokers, we will be able to estimate the proportion of the population of the state who are current smokers, and will be able to decide if the null hypothesis should be rejected or not.

Statistical inference: to arrive at conclusions about a population based on data from a sample

We can do ESTIMATION or HYPOTHESIS TESTING

1.5   Producing data

Data are produced by observation or experimentation

A lot can be learned from data collected by others when the studies are well documented, and the information produced is reliable. However, it is a common situation that a research is so specific, or new, that the researcher needs to produce her/his own data. There are two main ways of producing data: **observational studies (surveys)** and **experiments**.

In observational studies, we measure variables of interest or ask questions, while trying to be as unobtrusive and neutral as possible. In experiments, experimental units or subjects are randomly allocated to different treatments and the values of one or more 'response' variables are recorded. If the purpose of the study involves the establishment of cause-effect relationships between variables, then experiments are preferred because in them we can better control for other variables which might interfere with the response variables.

There are some cases in which experiments cannot be performed due to ethical reasons. For example, if the question is whether an allele of the gene BRCA in humans actually causes breast cancer, a random breeding of specific humans to study the occurrence of the disease in a large number of offspring cannot be done. In those cases the researcher needs to find a way of attaining some control over the variables that might influence the response. Sometimes the results of several studies are analyzed together ('meta-analysis'). A classical example of not being able to conduct experiments with humans is that of studying the cause-effect relationship between cigarette smoking and lung cancer (Doll, 2002): a group of people cannot be randomly divided in two subgroups in order to force one of the subgroups to smoke and later be compared with the ones forced to not smoke. An example of an observational study, in which some control over other variables was gained by studying a self-controlled population, is the study of Alzheimer's disease in a population of nuns (Snowdown, 2002). Some complex medical studies have an observational part, and an experimental part, for example the study of the effects of hormone replacement therapy for post-menopausal women (Prentice, 2005). While analyzing the data, models can be used that allow us to examine the effect of one variable 'controlling' for the other variables.

1.5.1 Deciding what type of data to produce

When planning a survey or an experiment, it is very important to decide:

Questions about the production of data:

- Why are the data being produced? We should have a clear idea of the research question or the scientific hypothesis.

  Why?

- Who (or what) is the individual or element of study? (cell? cubic unit of blood? fly? tree? species? person?) From whom will the information be collected? The whole population? A random sample?

  Who?

- What characteristics of the individual will be observed or measured? The 'variables' to be studied have to be defined.

  What?

- How will the data be produced? Observational study or experiment? How will the measurements be done? What units will be used? Will any instrument be used? What criterion will be used to classify the individuals into groups?

  How?

  When?

- When will the study be conducted? Does time of the year or time of the day have an effect?

1.5.2   Designing an observational study

Difference between a census and a sample survey: ALL or just a RANDOM SAMPLE of individuals

In **observational studies**, measurements are made or questions are asked. It is very important that all the information necessary to answer the research questions is obtained in a reliable and consistent way. If the objectives of the study are not clearly stated from the beginning, some important information might be missed or unnecessary data might be collected. An observational study is called a **census** when information is collected from all the individuals in the population. An observational study is called a **sample survey** if information is collected only from a portion of the population called the sample.

Samples need to be randomly selected in order to use the sample observations to arrive at conclusions about the whole population. Common examples of sample surveys are opinion polls. It is very important to respect the privacy of the individuals so that no identifiable information is released.

First, the general objective of the survey is stated, then a list of specific objectives is crafted to make sure that all the necessary questions or measurements are made without including unnecessary ones. It is very important to design a questionnaire that is not too long. Questions should be completely clear, in a language appropriate for the population of interest. It is recommended that the questionnaire be tried out first with a small group of people, to correct it if necessary, before the final printing. The questionnaire is applied to the individuals and the answers or measurements carefully recorded onto a data file. Persons conducting surveys should try to be as unobtrusive as possible in order not to influence the answers of the respondents. Based on the data file, tables and graphs are prepared to display the answers to each question. Reports generally include, at the beginning, the tables for the demographic variables so that the reader gets a description of the sample. Tables and graphs can also be prepared to explore the possible association between answers to different questions. Figure 1.3 displays the connections between some important concepts in surveys.

Samples should be randomly selected

Example: sampling frame of the cells in an onion root

In order to select the individuals who will be part of the random sample, the first step is to create a document that allows us to locate all the individuals in the population. That document is called **sampling frame**. For example, if the population has 100 individuals ($N=100$), the sampling frame can be a list of the 100 names numbered from 1 to 100. A random mechanism is used to select the n individuals which will be in the sample. The random mechanism could be something as simple as writing the numbers from 1 to N in pieces of paper and putting the folded pieces of paper in a box in order to pick n of them at random. However, the most common random mechanism currently used is a computer, and most statistical software have an option to select random samples. The following commands in **R** will create the labels 1 to 100, and select a sample of 6 individuals:

R: random sample

```
i<-seq(1,100,by=1)
sample(i,6)
```

The output is a list of 6 labels indicating which individuals will be in the sample, for example: 40, 37, 5, 44, 6, 53. This method of sampling is called **simple random sampling**. When using simple random sampling, all the possible subsets of size n of the N elements in the population have the same probability of being chosen as the sample. Simple random sampling is applied when the sampling frame enumerates all the individuals in the population and the conclusions are wanted for the population as a whole.

Simple random sampling (SRS)

Sometimes it is necessary to apply a more complex sampling method due to either operational reasons or the fact that the sampling frame lists groups of individuals instead of the individuals themselves. Some of those methods are 'stratified sampling' (the population is divided in sections or 'strata' and a random sample is selected from each one because estimates are desired for each stratum), 'cluster sampling' (groups or 'clusters' of individuals are selected because it is easier to access them in that way), 'two-stage sampling' (when the sampling is done in two stages, first groups are selected, and then some individuals are selected from each group). It is important to be aware that when one of those more complex sampling methods is used, the analysis of the data is also more complex and beyond the scope of this book.

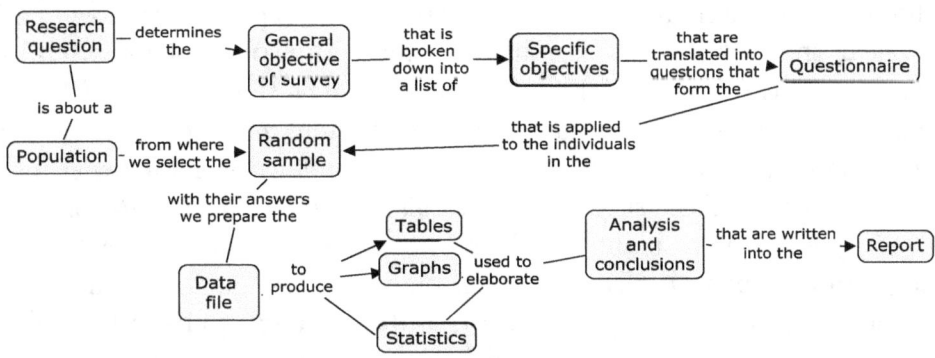

Figure 1.3: Conducting a survey

1.5.3   Designing an experiment

Completely
randomized
design

Experiments are conducted to observe the effect of one or various **factors** in one or more **response variables**. For example, an experiment can be conducted to learn how the intensity of light and the frequency of watering affect the growth of a particular type of plant. The levels of those factors determine the **treatments**. For example, if the levels of light are *low*, *medium* and *high*, and the frequency of watering is **once** or **twice** a week, there will be 6 treatments that are the combination of the levels of each factor as shown in Figure 1.4. When each plant is assigned to one of the treatments using a random mechanism, it is said that a **completely randomized design** is being applied.

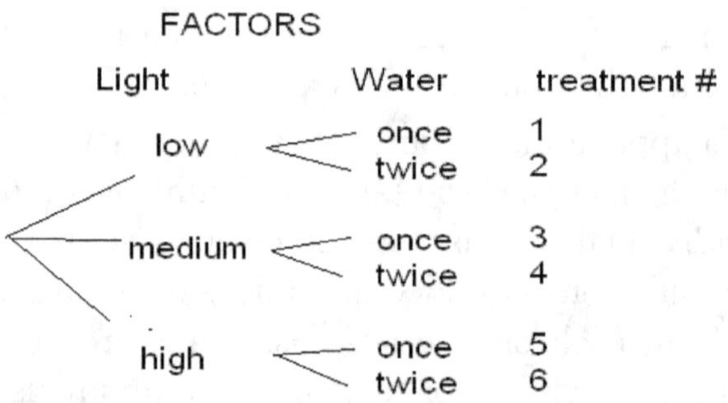

```
FACTORS

Light              Water         treatment #

      low    <     once             1
                   twice            2


<         medium   <    once        3
                        twice       4


      high   <     once             5
                   twice            6
```

Figure 1.4: Treatments as combinations of the levels of two factors

Principles of
experimental
design:
replication,
randomization
and control of
other variables

The treatments are applied to the **subjects** or **experimental units** and the values of the response variable(s) are recorded for each individual. The values of the response variable are summarized for each treatment and, based on those values, the treatments are compared. Several plants are grown under each one of the 6 treatments, keeping constant (**control of other variables principle**) all the other variables that can influence the growth of the plant such as temperature, type of plant, soil composition, fertilizer, etc. The ability to control for other variables is the main reason for preferring experiments when the purpose of the study is to establish a *cause-effect relationship*. At least two plants (more plants would be better) are needed for each treatment (**replication principle**). Which plants are assigned to which treatment has to be determined at random (**randomization principle**). In case experiments have to be performed one by one, the randomization principle is applied to the order in which they are performed.

Controls are always necessary. Consider the example in which the objective of the experiment is to learn about the effect of two different types of fertilizer on the yield of potato plants. If the two fertilizers of interest are A and B, the potato plants available will be randomly divided in three groups. One group will be planted using fertilizer A, another one using fertilizer B and one group (the **control group**) will be planted without using fertilizer.

Do not forget the control group

Human subjects might act differently because of the presence of the researcher. In the social sciences this is known as the Hawthorne effect. The name originated in a study on the effect of low and high lighting on productivity, conducted in a factory named Hawthorne Works. It happened that productivity increased because of the attention workers received, independently of the light level. In medical studies, patients might report they feel better because of the attention they receive from health professionals, not because of the treatment itself. This is the reason that in medical experiments with human subjects, the control group receives a placebo (something that looks like medicine but is not) in order to observe the so-called **placebo effect** (the result of both the attention received and the thought that they are taking a medicine). Having a control group allows the researcher to distinguish the **placebo effect** from the effect of the medicine. For example, in experiments with anti-depressants, children and younger adolescents tend to be very responsive to the placebo (Bridge et al. 2009). Experiments in which the subjects do not know if they are receiving the real treatment or the placebo are called **blind experiments**. Double blind experiments are those in which neither the subject nor the specialist who evaluates them know if they are receiving the treatment or the placebo. Figure 1.5 displays the connections among several important concepts in experimental design.

Placebo effect

Blind and double blind experiments

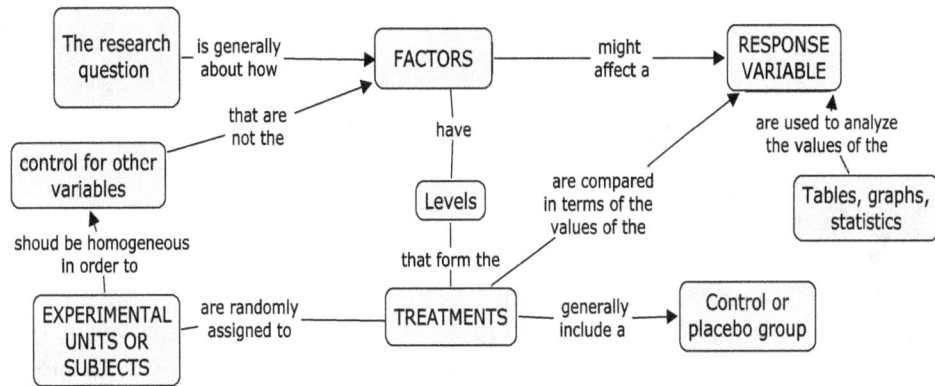

Figure 1.5: Designing experiments

There are of course several special experimental designs. In some experiments the subjects are paired (matched-pairs design), such as is the case on

Matched
pairs
design

experiments with twins. One element of the pair receives one treatment and the other receives the other treatment or a placebo. In that case the analysis is focused on the difference in the response variable between the two elements of each pair. The null hypothesis is written in terms of the mean difference μ_d. In some other experiments, each individual receives all the treatments with a washout period in between the treatments. Sometimes there are quite different groups or blocks of experimental subjects, and all the treatments are applied in each one of the blocks. The analysis of the data depends on the type of design used. Unless otherwise stated, it will be assumed in the book that a completely randomized design has been used in the experiments.

1.6   Data files

Preparing
a data file:
one row for
each individual,
one column for
each variable

Cell ID	Root	Region name	Length
1	1	apical	0.023
2	1	apical	0.030
3	1	apical	0.029
4	1	elongated	0.105
5	1	elongated	0.113
6	1	elongated	0.058
7	2	apical	0.033
8	2	apical	0.021
9	2	apical	0.031
10	2	elongated	0.107
11	2	elongated	0.052
12	2	elongated	0.127

Figure 1.6: Example of data file

After data are collected, a data file is prepared and checked for errors. If the data file was prepared by somebody else, we need to fully understand its structure, the information contained in it and the coding system used. Usually numerical codes are used to represent the categories of a categorical variable, e.g. 1 for YES, 0 for NO. Data are frequently typed into spreadsheets to be analyzed later using statistical software. A dot, '*' , NA, or a numerical code such as 99 might be used for missing data or 'no response,' depending on the software to be used. A data file has one row for each individual and one column for each variable. Some columns are created to indicate which treatment or group (if any) the individuals belong to. The data file in Figure 1.6 was created to store the lengths (in millimeters) of 12 onion root cells. There are 2 roots and 6 cells in each root (3 from the apical region and 3 from the elongation region). The first column is for the label or ID of the cell (from 1 to 12), the second column for the ID of the root. There is a column to indicate the region where that cell is located and finally there is a column with the lengths of the cells. This particular design of the data file facilitates later analyses by allowing to compare the values of the variable (length) between regions or between roots, or between regions within roots. It is important to keep each individual observation in the data file and not only the means by group or treatment, because the variability among the individual observations plays an important role in the analysis.

Keep the individual observations, not only the means per group

Data files for this book and information about software

Data files (ANSI) have been prepared with the data used in the book. They are text type files with one or several columns of numbers and can be read by **R** and other statistical software. Brief instructions to work with MINITAB are included in most sections. **R** is a free software extensively used in statistics and bioinformatics that can be downloaded from http://www.r-project.org. Detailed instructions to work with **R** are included in the book. Data files and text files with the commands in **R** (to be copied and pasted into the **R** session) are available from the website http://faculty.etsu.edu/seier/IntroStatBioBook.htm.
Note: User friendly interfaces to **R** do exist, such as *R-commander* available from http://socserv.mcmaster.ca/jfox/Misc/Rcmdr/ (Fox, 2005) and *RStudio* available from http://www.rstudio.org/.

Data files for this book

Using statistical software:
In MINITAB, use FILE>Open Worksheet to open a MINITAB data file (.mtw). To open ANSI files, use FILE>Other files, or simply open the data file with Notepad and copy and paste the values into the MINITAB worksheet. When working with **R***, data need to be typed into the program using:*
x< −c(, , ,)
or can be read from a (ANSI) data file. Assuming that the name of the data file is datafile and that it is located in drive **e:***, use the command 'scan' if the data file has only one column of data:*
x<-scan('e:datafile.dat')
If there is more than one column of data, use the command 'read.table':
x< −read.table('e:datafile.dat')

MINITAB & R: Reading data

1.7   What is Statistics?

Statistics is a science that creates, develops, analyzes, and compares methods for collecting, organizing, and analyzing data in order to arrive at conclusions that will in turn provide useful information for the advancement of scientific knowledge and decision making. Data has been collected for practical purposes since humans began to organize themselves into states. Probability, the main mathematical tool for statistics, started to develop in the 17th century. In the 19th century life tables were constructed in the insurance business, and statistical data were used to provide public administration with information for making decisions. The fields of application of statistics today are diverse and include economics, biology, public health, medicine, education, psychology, sociology, and engineering, among others. Scientists who developed statistical methods for scientific research at the end of the 19th century and the first part of the 20th century, were not mathematicians isolated from applications. Rather they were mathematicians, or scientists with mathematical training, who did research in genetics, biology, and agriculture. Biological research provided a strong impulse to statistical research.

One of the oldest journals in statistics, Biometrika, was founded in 1901 by Karl Pearson. A milestone in the application of statistics to research in biology was the publication in 1925 of the first edition of the book *Statistical Methods for Research Workers* by R.A. Fisher. In part motivated by the advancements in genetics, medicine and other fields, and in part due to its own dynamics and computational availability, the development of statistics continues in our time. The advent of huge data sets from DNA sequencing pose new needs for statistical analysis. Statistics and probability are major components of the field of Bioinformatics.

Statistics and Biology have a long history together

Statistics is a powerful tool to do research in the experimental sciences. Many statistical methods exist that are beyond the scope of this book. Statistics is also a science that continues to develop, and new advances are published in several statistical journals. Methods of analysis are refined, modified, improved, or designed; theoretical constructs are developed and discussed.

1.8   Beyond biology

Observational studies are frequently conducted by government agencies, research organizations, private companies and individuals. A census is conducted every 10 years to estimate the size of the total population of the country as well as of smaller geographical areas. Results of the census are used for planning purposes, including the allocation of funds and political decisions. All issues related to the census and its results can be found at http://www.census.gov.

The General Social Survey is conducted by the National Opinion Research Center at the University of Chicago. The GSS covers hundreds of variables, from demographic variables to beliefs, and works with random samples of over 3,000 individuals. National and statewide surveys are conducted on specific problems such as tobacco, alcohol and illegal drugs use in order to estimate the proportion of the population who abuse or are dependent on a given substance. Opinion polls are also conducted frequently; the most widely known is probably the weekly Gallup poll (http://www.gallup.com). In the Gallup poll, the proportion of U.S. residents in favor or against a certain issue is estimated based on the opinion of a little over a thousand randomly selected residents. Graduate students working on their theses and faculty doing research also design and conduct surveys on specific topics. A document with guidelines for the conduction of surveys (Scheuren, 2006) is available from the Survey Research Methods section of the American Statistical Association (http://www.whatisasurvey.info/).

It is important to distinguish between reliable results coming from well-organized surveys that work with random samples and non-scientific surveys that some communication media do by asking their listeners or viewers to call and give their opinion. These activities can be a source of entertainment but they do NOT provide reliable information about the whole population because the samples are self-selected and only the people with very strong opinions, one way or the other, tend to call.

Another important issue in relation to surveys, which will be clarified in Chapter 11, is that the size of the sample is not necessarily a pre-determined fraction of the population size. Rather it is determined based on the precision and confidence we want for our estimations, and depends also on how diverse the population is. Estimates with a desirable confidence and precision can be obtained for the opinions of the adult (18 years or older) U.S. population, which is over 200 million people, based on a **random** sample of about 1,500 individuals. However, there are some special surveys, such as those to estimate the proportion of residents of a state who are dependent on illegal drugs, for which larger samples (5,000 individuals) are recommended because the proportion of individuals in the population who are addicted is low. It should be emphasized that the non-negotiable part when working with samples is their 'random' nature: **the individuals who will form the sample need to be selected using a random mechanism.** Currently

Do not trust results from studies with self-selected samples

that random mechanism is usually a random number generator implemented on a computer.

Experiments are frequently planned and conducted as part of scientific research, especially in medicine and agriculture. The area of design of experiments was initiated by Ronald A. Fisher, who worked in agricultural experiments at the Rothamsted Experimental Station in England. Since then, there is a long tradition of experimental design in colleges of agriculture. Experiments are also used in the design of industrial products.

1.9   Exercises

1.9.1   Review questions

1. What is a variable? What types of variables are there?

2. What is a parameter? What is the typical parameter of interest when studying a categorical variable? What is the typical parameter of interest when studying a quantitative variable?

3. What is a statistic? Mention a typical statistic for quantitative variables and a typical statistic for categorical variables.

4. Assume that we are trying to find out what percentage of Pima cotton seeds would germinate in the soil type and weather conditions particular to a region. We select a random sample of 100 seeds, plant them and follow their development; only 78 germinate. Identify population, variable, parameter of interest, sample, statistic and its numerical value for this example.

5. Assume that we are trying to find out what is the mean yield (lb) per plant of a variety of potato, under certain conditions of watering and fertilizer, harvested after a given number of days since planting. We plant 20 potatoes under those conditions and harvest them at the indicated time. The average yield of those 20 potato plants is 4.35 pounds. Identify population, variable, parameter of interest, sample, statistic and its numerical value for this example.

Review questions for Chapter 1

6. Why is it said that cause-effect relationships can be established from experiments but not from observational studies? What are the 3 principles of experimental design?

7. A biologist hypothesizes that adult male flesh flies are more prone to exhibit aggressive behavior than females of the same species and age. He plans to observe a group of 30 male flies and 30 female flies, all of them 4 days old, during 10 hours and record for each one of the flies whether they exhibit aggressive behavior or not. Which of these options is the correct way of writing the null and alternative statistical hypotheses?
   A) $H_o : p_M = p_F$ $\qquad\qquad H_a : p_M > p_F$
   B) $H_o : \mu_M = \mu_F$ $\qquad\qquad H_a : \mu_M > \mu_F$
   C) $H_o : \hat{p}_M = p_F$ $\qquad\qquad H_a : \hat{p}_M > \hat{p}_F$
   D) $H_o : males = females$ $\qquad H_a : males > females$

8. What is the 'p-value'?
   A) The probability that the alternative hypothesis is true
   B) The probability that the null hypothesis is true
   C) The probability that the statistic takes the observed value or a more extreme one when the null hypothesis is true
   D) The probability of making an error
   E) A population proportion

9. When is the data considered enough evidence against a null hypothesis as to reject it? When the p-value is _____
   A) very small $\qquad$ B) very large $\qquad$ C) 0.5

10. Identify the factors and their levels, treatments and response variable in the following story: We want to know how the teaching environment and the time of the day at which classes are held affect the performance of students. Two teaching environments (large section without computers, small section in a computer lab) and three times of the day (early morning, noon, evening) will be considered. The same teachers will teach the same material during two weeks and afterwards a common exam will be administered to all the students. Draw a diagram similar to that of Figure 1.4 to list the treatments.

1.9.2   Exercises for discussion or homework

1. In 1954 an experiment was conducted to test the first vaccine against poliomyelitis (Meldrum, 1998; Meier, 2000). The research question was whether the polio vaccine reduces the proportion of children who get the disease. The experiment involved 400,000 children, half of them were randomly selected to receive the vaccine while the other half received a

placebo. What treatments are being compared? What is the parameter of interest? Write the null and alternative hypotheses. Out of the children who received the placebo, 142 got polio. Only 56 children, among those who received the vaccine, contracted polio. What type of statistic would you use in this case? Calculate the numerical value of the statistic for each group.

Exercises for class discussion or homework in Chapter 1

2. The Woman Health Initiative study (Prentice et al. 2005) had an observational component and an experimental component. One of the research questions was whether hormone replacement therapy for post-menopausal women reduces the risk of coronary heart disease. The experiment enlisted 16,608 post-menopausal women, 8,506 of whom received a therapy of estrogen+progestin, while 8,102 women received a placebo. From the treatment (estrogen+progestin) group 188 women presented coronary heart disease, whereas 147 women from the placebo group presented coronary heart disease. What are the two populations to be compared in this case? What is the parameter of interest? Is the value of the parameter known for each population? What are the sample sizes? What statistic would you use? What is the value of the statistic for each group?

3. Darwin (1876) knew that inbreeding is not good for humans and wondered whether something similar happens with plants. He conducted some experiments with corn plants (*Zea mays*). He produced 15 plants by self-fertilization and 15 plants by crossed-fertilization and focused on how tall the plants were after a given number of weeks. He wondered if plants produced by self-fertilization would be smaller. What is the response variable? Is that variable categorical or quantitative? Would you work with means or proportions? What are the treatments being compared?

   Next assume two different scenarios (neither of them describes exactly the way in which Darwin conducted his experiment). In the first scenario, assume that Darwin planted the plants in pairs using 15 rectangular pots, on one side of a pot was a plant produced by self-fertilization and on the other was a plant produced by cross-fertilization. Assume that which plant to plant at the right and which to plant at the left in each pot was determined by flipping a fair coin. Write the null and alternative hypotheses. Assume that the height for each plant is recorded after a given number of weeks. What variable would you focus on? What statistic would you calculate?.

In the second scenario, assume that Darwin randomly planted the 30 plants in a plot of land, the location of each plant was determined using a random mechanism and no pairing of plants was involved. Write the null and alternative hypotheses. What statistic would you calculate for the plants in each treatment?

4. A researcher in the health sciences conducts an experiment with mice (McAmis, 2011). She wants to study the effect of a chemical substance and the effect of a specific type of stress on glucose. The chemical substance can be administered at the levels of 0, 25 or 50 mg. Half of the mice will be subject to stress and half of them will not. Before applying the treatments, the amount of glucose (mg/100 ml) is measured for each mouse. The glucose is measured again a week into the treatment, and the change in glucose for each mouse is calculated. What are the factors? What are the levels for each factor? How many treatments are there? Draw a diagram similar to the one in Figure 1.4 to list the treatments. Which is the control group? What is the experimental unit? What is the response variable? How many mice are needed in total if we want to assign 10 mice to each treatment? Discuss the 3 principles of experimental design in the context of this experiment.

5. Assume that you are organizing a survey on tobacco consumption and that you will interview 400 randomly selected individuals using the following questionnaire:

   (a) Gender?

   (b) Age?

   (c) Have you smoked more than 100 cigarettes in your lifetime?

   (d) Have you smoked cigarettes during the past 30 days? (if the answer is yes, you will ask the remaining questions)

   (e) How many cigarettes per day do you smoke on average?

   (f) Do you plan to quit smoking in the next six months?

   (g) In your last visit to the doctor, did she or he advise you to quit smoking?

   Sketch a data file in which to type in the answers to these questions. What codes would you use for the categorical variables?

References

Bridge, J.A., Birmaher, M.D., Ivengar, S., Barbe, R.P. and Brent, D.A. (2009) Placebo Response in Randomized Controlled Trials of Antidepressants for Pediatric Major Depressive Disorder. *American Journal of Psychiatry* **166**: 42-49.

Darwin, C. R. (1876) *The effects of cross and self fertilization in the vegetable kingdom.* London: John Murray.
http://www.gutenberg.org/ebooks/4346

Doll, Richard (2002) Proof of causality: deduction from epidemiological observation. *Perspectives in Biology and Medicine* **45**: 499515.

Fisher, R. (1970) *Statistical Methods for Research Workers.* Edinburgh: Oliver and Boyd.

Fox, J. (2005) The R commander- a Basic-Statistics Graphical User Interface to R. *Journal of Statistical Software* Vol 9 (4)
http://www.jstatsoft.org/v14/i09/paper

McAmis, L. (2011) *Role of Stress in the Onset of Diabetes Mellitus in Mice.* Honors in Discipline B.S. Thesis, East Tennessee State University.

Meier, P. (1972) The Biggest Public Health Experiment Ever: The 1954 Field Trial of the Salk Poliomyelitis Vaccine, in *Statistics a guide to the unknown* ed. Judith Tanur et al. pp 2-13. San Francisco: Holden Day.

Meldrum, M. (1998) A Calculated Risk: the Salk Polio Vaccine Field Trials of 1954. *British Medical Journal* **317**: 1233-1236.

Prentice, R.L. et al. for the Women's Health Initiative Investigators (2005) Combined Postmenopausal Hormone Therapy and Cardiovascular Disease: Toward Resolving the Discrepancy between Observational Studies and the Women's Health Initiative Clinical Trial. *American Journal of Epidemiology* **162**: 404-414.

Snowdown, D. (2002) *Aging With Grace: What the Nun Study Teaches Us About Leading Longer, Healthier, and More Meaningful Lives.* New York: Bantam

Scheuren, F. (2006) *What is a survey?* American Statistical Association
http://www.whatisasurvey.info

Chapter 2

What are the data telling us?

Nature is full of information. It is just a matter of asking the appropriate questions, collecting reliable data and knowing how to look at the data in order to extract the relevant information. The theme of this chapter is the display and summarization of data. Which type of graph or statistics to use depends on the nature of the variable (categorical or quantitative). The material is organized as follows:

- *One quantitative variable for one group of individuals (Sections 2.2 - 2.7)*

- *Comparing two or more groups in terms of one quantitative variable (Sections 2.8 - 2.9)*

- *One quantitative variable studied through time (Sections 2.10 - 2.11)*

- *Two or more quantitative variables for the same individuals (Sections 2.12-2.13)*

- *Categorical variables (Sections 2.14 - 2.15)*

Statistical graphs and numerical summaries will be presented simultaneously. The effect of transformations such as change of units or logarithms will be be discussed. In addition to review questions, exercises and projects, this chapter includes a lab in two versions (MINITAB and **R***).*

2.1   Variables and statistical graphs

In Statistics, the object, organism, person, or organization **about whom information is collected** is called an **individual**, **statistical unit** or **subject**. An individual can be, for example, a person, an animal, a plant, a colony of bees, or a square mile of forest. The characteristics of the individuals are called **variables**. Variables are of two types:

nominal and ordinal categorical variables

discrete and continuous quantitative variables

Categorical: Variables are said to be categorical when they are naturally expressed in words instead of numbers, such as 'red blood cell shape' (*normal* or *sickle*), or 'individual status' (*infected* and *not infected*). If there is no natural ordering in the categories of the variable, they are called **nominal categorical variables**. If there is a natural ordering in the categories such as *small*, *medium*, and *large*, they are called **ordinal categorical variables**.

Quantitative: Variables are said to be quantitative when they are naturally expressed in numbers. The units (centimeters, millimeters, kilograms, etc.) should be mentioned. There are two types of quantitative variables: **discrete** and **continuous**. If the variable can take only integer values (0,1,2,3,...) like when we count, it is called discrete. A variable is called continuous if it can take any value in a given interval. For example, the length of a cell in the tip of an onion root can take any value between 0 and 0.04 millimeters.

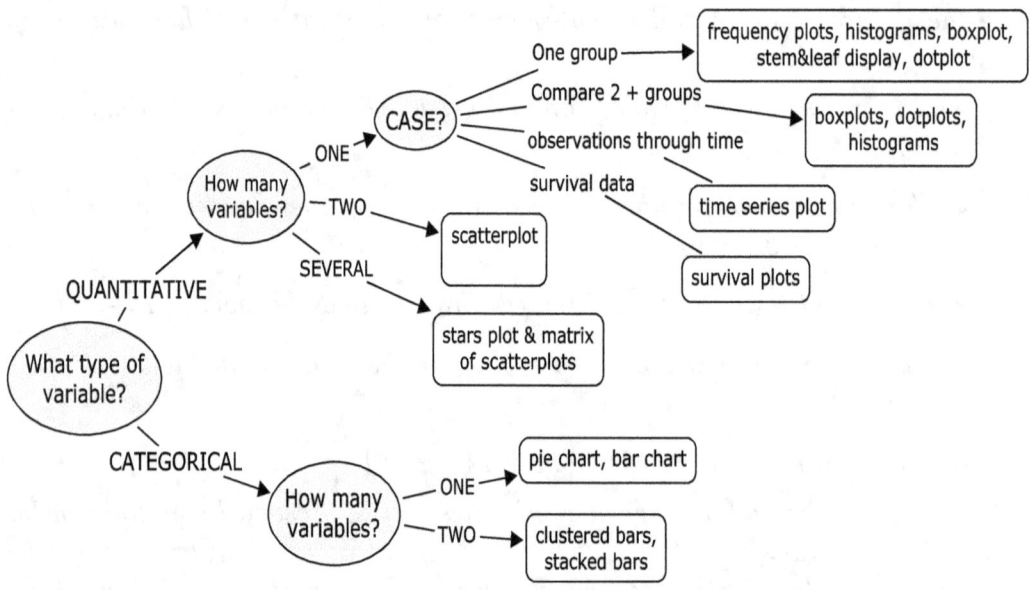

Figure 2.1: Statistical graphs and how to decide which one to use

It is very important to identify the type of variable because the graphs and methods of analysis to be used depend on the nature of the variable. Different statistics are calculated for categorical and quantitative variables. The answers to the following questions determine which statistical graph to use to display the data:

1. Is the variable categorical or quantitative?

2. Are we describing a single group of individuals or are we comparing two or more groups?

3. Are we analyzing just one variable at a time or are we interested in associations between variables?

4. Has the variable been measured for one place or individual throughout time? Is the variable of interest the number of individuals from a cohort who will die or 'fail' each day?

Figure 2.1 describes which statistical graphs to use depending on the answers to those questions. Figure 2.2 can serve as a guide to decide which statistics to use in a given situation.

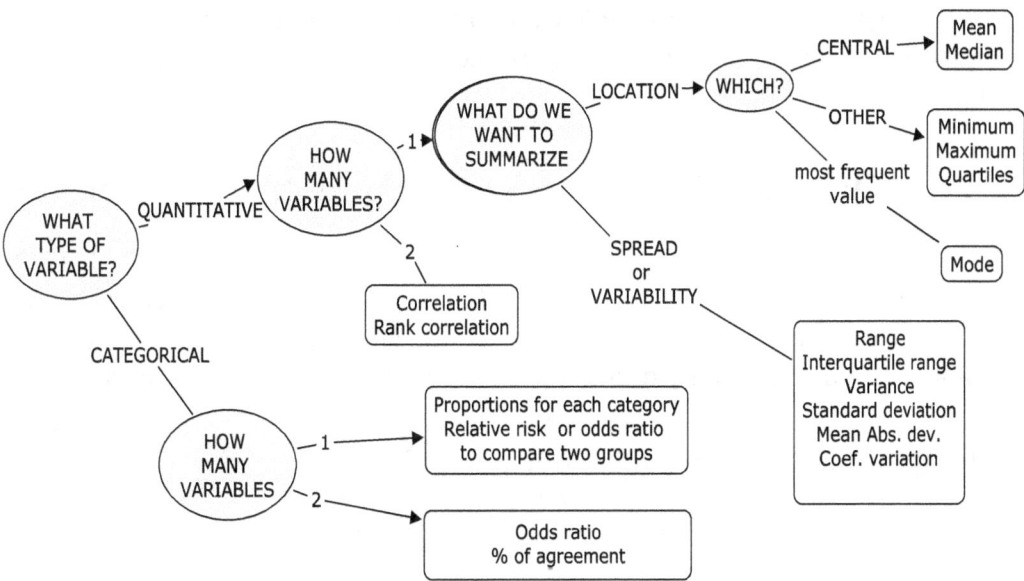

Figure 2.2: Descriptive statistics and their purpose

2.2 Case of one quantitative variable

The statistical graphs most frequently used to display observations for one quantitative variable are frequency plots, histograms, dotplots, and stem-and-leaf displays (or 'stemplots'). Some questions of interest that we can answer by looking at those displays are:

- What values does the variable take? (Min , Max)

- How frequent is each value or interval of values?

- Is there a typical or more frequent value?

- Is there some observation that looks unusually large or small? (outliers)

Things to look for in data displays

- Does it seem that different groups of data exist?

- How would you describe the shape of the distribution?

Variability is ubiquitous in nature. Variability is one of the hallmarks of biology. There is variability among cells of the same type in the same organism, variability among individuals of the same species, variability among species of the same family, and there might be even variability in different measurements done to the same individual due to the imprecision of measuring tools. The analysis of variability is one of the main concerns in statistical analysis.

2.2.1   Quantitative variables that only take integer values

When you count things, the variable 'X: number of ...' can take the value 0, 1, 2, 3,... These variables are called *discrete quantitative variables*. To tabulate the observations, frequency tables and frequency plots are used.

Example 2.2.1 Number of abnormal cells

Number of abnormal cells

There are 50 cell cultures, the number of abnormal cells in each culture is counted and the results are:

```
0   2   2   3   3   4   4   0   0   2   0   1   3   2   2
1   4   2   0   1   2   4   2   0   2   2   0   4   0   1
1   1   3   0   1   4   1   4   5   3   0   3   0   0   3
2   2   1   2   1
```

Tables and frequency plots

The different values of the variable are listed and the number of cultures with each number of abnormal cells or **frequency** is written next to the value of the variable. A number and a title are assigned to the table as shown in

Table 2.1. Either Figure 2.3a or Figure 2.3b could be used to display the frequencies in the second column of Table 2.1. Figure 2.3a is a dotplot in which each observation is represented by a dot or other symbol. In Figure 2.3b, lines with height equal to the frequencies are drawn.

Relative frequencies ($frequency/n$) can be included in the table as well. Cumulative relative frequencies, obtained by adding the relative frequencies, have been included in the last column of Table 2.1. These cumulative frequencies are plotted in Figure 2.3c using 'steps;' each step has the height of the frequency corresponding to the value of the variables (the horizontal line at 0.5 is optional and can be useful to locate the value that divides the distribution in the lower and upper 50%). Cumulative relative frequencies might become handy to write comments such as 'it is clear from Table 2.1 that 22/50 or 44% of the cultures have at most 1 abnormal cell.'

Frequency tables and plots answer the question: How many times does each value of the variable appear in the data set?

Table 2.1: Cultures classified by the number of abnormal cells

Number of abnormal cells	frequencies	relative frequencies	cumulative rel. freq.
0	12	0.24	0.24
1	10	0.20	0.44
2	13	0.26	0.70
3	7	0.14	0.84
4	7	0.14	0.98
5	1	0.02	1.00
Total	50	1	

Table and plots for data of one quantitative variable that takes only integer values

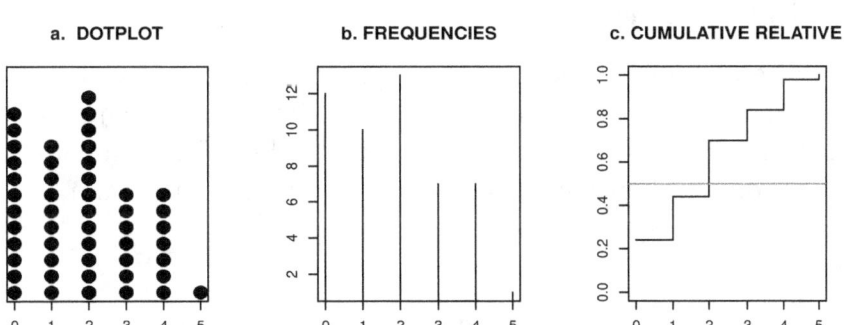

Figure 2.3: Dotplot, frequency plot, and cumulative relative frequencies

Using statistical software:
To read the data and prepare the frequency table using R, type:

R: Reading and tabulating data

```
cells<-c(0,2,2,3,3,4,4,0,0,2,0,1,3,2,2,1,4,2,0,1,2,4,2,0,2,2,
0,4,0,1,1,1,3,0,1,4,1,4,5,3,0,3,0,0,3,2,2,1,2,1)
table(cells)
```

The output is:

```
cells
 0  1  2  3  4  5
12 10 13  7  7  1
```

To prepare Figure 2.3a, after reading the data, use:

```
stripchart(cells,method="stack", offset=1, at=0, pch=19,cex=2,
main='DOTPLOT')
```

To prepare Figure 2.3b, enter the values of the variable and the frequencies and then ask for the frequency plot:

```
abnormalcells<-c(0,1,2,3,4,5)
frequencies<-c(12,10,13,7,7,1)
plot(abnormalcells,frequencies,'h',main='FREQUENCIES')
```

MINITAB: Frequency plots

In MINITAB, to tabulate the data use the option STAT>Tables>Tally individual variables. To prepare dotplots, use GRAPH>Dotplot. To plot the frequencies type the values 0,1,2,3,4,5 in one column, the frequencies in another column and use GRAPH>Scatterplot>simple and in 'Data view' choose 'project lines.'

Numerical summaries

The data can be summarized by calculating some statistics such as the mean, median, and mode. A statistic is an expression that is defined in terms of the elements of the sample. The numerical value of the statistic will vary depending on which elements of the population are in the sample.

Mode: the most frequent value

Mode

The mode is the most frequent value. In the example, the mode is 2 because there are 13 cultures that have 2 abnormal cells, and there is no other value with a higher or equal frequency. In other examples, there might not be a most frequent value or there might be more than one. Thus, the mode does not always exist, and if it exists, it may not be unique.

Mean

The mean is the average value. To calculate the mean add all the observations and divide the sum by the number of observations. If the observations are 0, 1, 8, 12 and 19, the sum of those values is 40; thus, the mean is $40/5 = 8$. **Mean: the average** In the case of the abnormal cells example, the sum of the 50 observations is 90, which divided by 50 gives 1.8. Note that the mean of integer valued data might not be an integer. The mean can be calculated from the already tabulated values $(0 \times 12 + 1 \times 10 + 2 \times 13 + 3 \times 7 + 4 \times 7 + 5 \times 1)/50 = 1.8$. The symbol used to represent the sample mean is $\bar{x}$. The mathematical expressions or formulas that describe the calculation of the sample mean from raw and tabulated data are:

$$\bar{x} = \frac{\sum_{i=1}^{n} x_i}{n} \qquad and \qquad \bar{x} = \frac{\sum_{i=1}^{n} x_i f_i}{n} \qquad (2.1)$$

where $\sum$ is the symbol for 'sum,' n is the sample size and f_i are the frequencies.

Median

The median is the value that occupies the central position after the observations have been ordered. If the observations are 0, 1, 8, 12, 19, the median is **Median: the center of the distribution** 8 because it is in the central position. In the abnormal cells example, since there are 50 observations, the median is the average of the two in the center (25th and 26th positions). First we sort the values:

$$0\ 0\ 0\ 0\ 0\ 0\ 0\ 0\ 0\ 0\ 0\ 0\ 1\ 1\ 1\ 1\ 1\ 1\ 1\ 1\ 1\ 2\ 2\ 2$$
$$2\ 2\ 2\ 2\ 2\ 2\ 2\ 3\ 3\ 3\ 3\ 3\ 3\ 3\ 4\ 4\ 4\ 4\ 4\ 4\ 5$$

The two observations in the center are equal to 2, thus the median is 2. The median can be visualized in the plot for cumulative relative frequencies if we trace a line at the 0.5 level in the Y axis and, when we encounter the step plot, come down to the X axis (Figure 2.3c).

Summary for Example 2.1 Abnormal cells

Twenty four percent of the cultures do not have any abnormal cell at all; the most common value is 2 cells per culture, and the maximum number of abnormal cells per culture is 5. Most of the cultures have a low number of abnormal cells, at least 50% (actually 70%) of the cultures have at most 2 abnormal cells (2 is the median), and the average value is 1.8 abnormal cells per culture.

Using statistical software:
In **R***, once the data set for the variable 'cells' has been read, the command*
'mean(cells)' calculates the mean. The command 'median(cells)' calculates

R & MINITAB: *the median.*
calculate   *In MINITAB, statistics such as mean and median are calculated using STATS >*
mean and   *Basic Statistics > Display basic statistics*
median

2.2.2   Quantitative variables in general

There are several displays and plots that can be used to display the values
of a continuous variable or a discrete variable with a long range of values.

Example 2.2.2 Red blood cells of parrots

RBC of   Consider the data set in Table 2.2: the values correspond to the typical dry
Psittacidae   cell area (in μm^2) of erythrocytes (red blood cells, RBC) for 34 bird species
                of the family Psittacidae. There is variability in the size of RBC for species
                of the same family of birds. The information contained in Table 2.2 can be
                better appreciated in a stem-and-leaf display, as the one below, or tabulated
                as in Table 2.3 or in one of the plots or statistical graphs in Figure 2.4. The
                data can also be summarized using statistics such as the mean or median.

Stem-and-leaf display

The stem-and-leaf display or 'stemplot' presents all the observations sorted
from smallest to largest. Each observation is split in two parts, the *stem* and
Stem-and-leaf   the *leaf*. Since several observations might share the same stem value, the
or stem-plot: stem is written only once in each line.
displays all the
observations,
in order

```
The decimal point is 1 digit(s) to the right of the |
   5 | 667889999
   6 | 0001112223334
   6 | 556677
   7 | 133
   7 | 58
   8 | 2
```

The shape of the distribution (skewed with a tail to the right) can be per-
ceived from the display. The stem-and-leaf display was introduced into statistical practice by John Tukey (circa 1970) as part of the exploratory data
analysis methods (EDA). The way to construct the stem-and-leaf display or

Table 2.2: Dry red blood cell area of Psittacidae

ID	Species		area (μm^2)
1	*Agapornis*	*cana*	57.78
2	*Agapornis*	*pullaria*	57.89
3	*Alisterus*	*amboinensis*	59.95
4	*Alisterus*	*scapularis*	62.68
5	*Amazona*	*albifrons*	71.07
6	*Amazona*	*amazonica*	73.46
7	*Amazona*	*autumnalis*	82.25
8	*Amazona*	*dufresniana*	65.93
9	*Amazona*	*imperialis*	67.39
10	*Amazona*	*leucocephala*	66.32
11	*Ara*	*ararauna*	62.60
12	*Ara*	*macao*	55.94
13	*Ara*	*severa*	61.57
14	*Aratinga*	*solstitialis*	59.39
15	*Brotogeris*	*jugularis*	73.18
16	*Coracopsis*	*nigra*	61.04
17	*Coracopsis*	*vasa*	63.66
18	*Cyanoliseus*	*patagonus*	59.12
19	*Cyanoramphus*	*novaezelandiae*	57.32
20	*Enicognathus*	*leptorhynchus*	62.36
21	*Graydidascalus*	*brachyurus*	59.60
22	*Myiopsitta*	*monachus*	58.93
23	*Pionites*	*melanocephala*	64.93
24	*Pionus*	*menstruus*	64.61
25	*Pionus*	*senilis*	77.60
26	*Platycercus*	*caledonicus*	61.47
27	*Platycercus*	*elegans*	61.21
28	*Platycercus*	*eximius*	59.37
29	*Psittacula*	*alexandri*	61.56
30	*Psittacula*	*cyanocephala*	55.61
31	*Psittacula*	*krameri*	59.89
32	*Psittacus*	*erithacus*	66.74
33	*Pyrrhura*	*hoffmanni*	74.64
34	*Tanygnathus*	*megalorhynchos*	62.84

Source: Gregory (2005) http://www.genomesize.com/cellsize

stemplot is flexible. Each value of the 'stem' can have one, two, or 5 lines, depending on the range of the variable. In this case there are observations in the 50's, 60's, 70's, and one in the 80's. There are two lines for the 60's. There is a line for observations from 60 to 64, and another for observations from 65 to 69, so that the stemplot does not end up having few but very long lines. The same is done for the other values of the stem. The values 55.94 and 55.61 have been rounded to 56 because the leaf can have only one digit. Notice that the stem 5 is written just once, and the leaf 6 appears twice because there are two observations with the value 56. Extreme distant observations can be displayed at the beginning or the end of the stem-and-leaf display, separately from the other observations and using the label *low* or *high*. If the raw data are no longer available and you are reading an already prepared stemplot, make sure to pay attention to the information about the location of the decimal point (if using R) or the units that the numbers in the leaf part represent (if using MINITAB).

Using statistical software:
In MINITAB, stem-and-leaf displays are found under STAT > EDA.
The stem-and-leaf display or stemplot for the data in Table 2.2 was obtained with R by reading the data for the variable area and using the command stem.

MINITAB & R
Stemplots

```
area<-c(57.78, 57.89, 59.95, 62.68, 71.07, 73.46, 82.25, 65.93,
67.39, 66.32, 62.60, 55.94, 61.57, 59.39, 73.18, 61.04, 63.66,
59.12, 57.32, 62.36, 59.60, 58.93, 64.93, 64.61, 77.60, 61.47,
61.21, 59.37, 61.56, 55.61, 59.89, 66.74, 74.64, 62.84)
stem(area)
```

Frequency tables

Frequency tables with intervals

Table 2.1 displays isolated values of the variable because the variable can take only integer values and the range of values (0 to 5) is rather short. The variable dry cell area (data in Table 2.2) is continuous, the minimum value of the variable is 55.81 and the maximum is 82.25. The difference between the extremes is 26.44, and out of convenience it could be rounded up to 30 because $30 = 6 \times 5$. Six intervals, each one of width 5, are defined starting at 55 and ending at 85, in order to produce Table 2.3. The frequency table reports how many individuals (species in this case) fall in each interval. The value 60 appears listed in the first two intervals; if there was a species with exactly 60 μm^2 it would be included in the one with the bracket at 60.

Table 2.3: Typical dry area of red blood cells for 34 species of the family Psittacidae

Dry cell area μm^2	Frequencies
[55 − 60)	12
[60 − 65)	12
[65 − 70)	4
[70 − 75)	4
[75 − 80)	2
[80 − 85)	2
Total	34

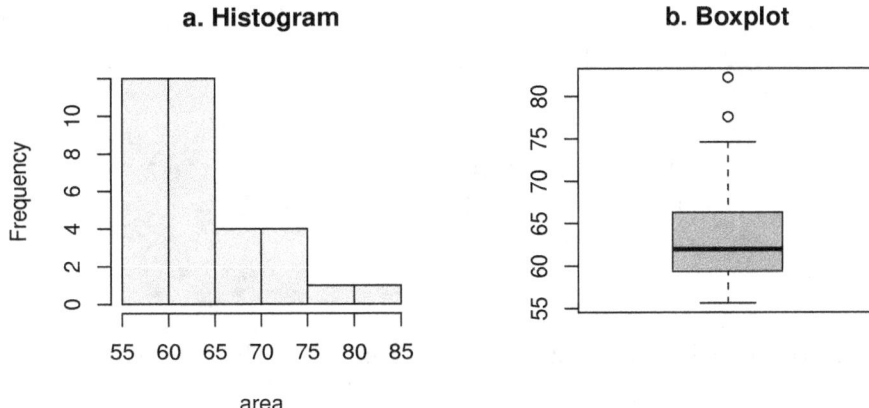

Figure 2.4: Histogram and boxplot of dry RBC area for Psittacidae

Histogram

The histogram in Figure 2.4a displays the same information as Table 2.3. Intervals of values of the variable are defined, a rectangle is drawn for each interval, the base of the rectangle is equal to the width of the interval and the area of the rectangle is proportional to the frequency of the interval. The simplest case is when all the intervals are of the same width, as in Table 2.3. In that case the height is made equal to the frequency and the proportionality will be maintained. The first rectangle in the histogram in Figure 2.4a indicates that there are 12 species for which the typical red blood cell area is between 55 and 60 μm^2. If the widths are not all the same, since $frequency = width \times height$ then $height = frequency/width$.

Using statistical software:
In R, enter the data for the variable 'area' and type the command 'hist(area)'
to produce the histogram in Figure 2.4a.

R: Histograms

Five-number summary

five-number summary: Minimum Q_L, Median Q_U Maximum

The five-number summary consists of:

minimum, lower quartile, median, upper quartile and maximum

of a set of observations. The median divides the data set in two halves, in turn the quartiles divide each half in two halves. Below the lower or first quartile (Q_L or Q_1) are the 25% of the species that have smaller red blood cells. Above the upper or third quartile (Q_U or Q_3) are the 25% of the species with larger red blood cells. The five-number summary for the cell area example is:

Min.	Q_L	Median	Q_U	Max.
55.61	59.44	61.97	66.22	82.25

Outliers

Outliers are observations that are very different from the others. In the case of values for a single quantitative variable, outliers are values that are either too small or too large. A practical rule to identify outliers is to calculate the 'fences':

A rule for finding outliers

$$\text{Lower fence: } Q_L - 1.5 \times IQR = 59.44 - 1.5 \times 6.78 = 49.27$$
$$\text{Upper fence: } Q_U + 1.5 \times IQR = 66.22 + 1.5 \times 6.78 = 76.39$$

IQR = 66.22-59.44 = 6.78 is the 'interquartile range' or difference between the two quartiles. Any observation beyond the fences, that is smaller than 49.27 or larger than 76.39, is considered an outlier.

The boxplot displays the five-number summary

Boxplot

The boxplot can be considered the plot of the five-number summary. A box is drawn from the lower quartile to the upper quartile with a line for the median. The whiskers are drawn from the box to the lowest and highest observations that are not outliers. The outliers are marked either with X, O, or $*$. Figure 2.4b is the boxplot for the dry cell area example.

Summary for Example 2.2.2, dry cell area of RBC of Psittacidae

The observations correspond to the typical dry area (μm^2) of the red blood cells of 34 species of the family Psittacidae. The smallest area is 55.61 and the maximum 82.25. Half of the species have red blood cell area of at most 61.97 (median). The upper 25% of species have red blood cell area of at least 66.22 and 25% of the species have red blood cells of at most 59.44. The distribution is skewed with a tail to the right indicating that most species have smaller cells and only a few have larger cells. There are two species that can be considered outliers with regard to the size of their red blood cells, they are *Pionus senilis* (76.6) and *Amazona autumnalis* (82.25). Outliers are NOT a nuisance, sometimes they lead us to new questions (Why do those species have larger RBC?), further research and even to the formation of biological hypotheses.

Using statistical software:
The **R** *commands to read the cell area data, and to produce Figure 2.4 are listed below. The command* **summary** *produces not only the five-number summary but the mean as well.*

R: Histograms, Boxplots, stemplot five-number summary

```
## to read the data
area<-c(57.78, 57.89, 59.95, 62.68, 71.07, 73.46, 82.25, 65.93,
67.39, 66.32, 62.60, 55.94, 61.57, 59.39, 73.18, 61.04, 63.66,
59.12, 57.32, 62.36, 59.60, 58.93, 64.93, 64.61, 77.60, 61.47,
61.21, 59.37, 61.56, 55.61, 59.89, 66.74, 74.64, 62.84)
stem(area)  ## to get the stem-and-leaf display
summary(area) ## calculates five-number summary and mean
par(mfcol=c(1,2))## 1 row and 2 columns of plots
hist(area,col='khaki', main='a. Histogram')
boxplot(area,col='coral',main='b. Boxplot')
```

In MINITAB histograms and boxplots are obtained by choosing the appropriate options in GRAPHS. The descriptive statistics can be calculated with STAT > Basic Statistics > Display basic statistics and choosing the statistics you want to calculate from the list that appears after clicking in 'statistics.'

MINITAB: Histograms and boxplots

2.3 More on numerical summaries

The values that a variable takes for all the individuals in a **population** are summarized by a **parameter**. The values that the variable takes for the individuals in the **sample** are summarized by a **statistic**. The two characteristics of a distribution that are usually summarized are location and spread. We have already introduced some location and spread statistics, now we will learn some more details about them. We will also learn a few additional statistics. The number of observations in a sample is represented by n while the number of elements in a population is represented by N.

n and N

2.3.1   Location

Mean

The mean diameter at breast height (1.3 m, 1.37 m or 1.4 m from the ground, depending on the system used) of all the trees in a national forest would be a population mean. That population mean is a parameter that describes the population of trees in the forest, its numerical value is generally not known because information for each single element of the population is not available. The mean diameter at breast height of 100 trees, randomly selected from that national forest, would be a sample mean. The sample mean is a statistic that is calculated with the sample data. Once the 100 trees have been randomly selected and their diameter at breast height measured, the sample mean will have a numerical value. The symbol for the population mean is μ, the symbol for the sample mean is $\bar{x}$. Statistics are used as **estimators** of parameters. Once the sample is drawn and the numerical value of the statistic is calculated, that value is called an **estimate** of the unknown value of the parameter. The population and sample means are:

Estimators and estimates

The mean in populations and samples

$$\mu = \frac{\sum_{i=1}^{N} x_i}{N} \qquad and \qquad \bar{x} = \frac{\sum_{i=1}^{n} x_i}{n}. \qquad (2.2)$$

A single observation can have a lot of influence over the mean, specially when there is only a small number of observations. Compare the two data sets in Figure 2.5; they only differ in one observation. However, the mean of the second data set is larger than the mean of the first data set.

The mean is sensitive to outliers

Now look at the data set in Figure 2.6, where the distances from each observation to the mean are represented with line segments; notice that the distances from the observations to the mean add up to zero, which is true for any data set. The mean can be considered to be a point of equilibrium.

Median

The median is the center of the distribution, it divides the distribution in the lower and upper 50%. The sorted or ordered observations are called

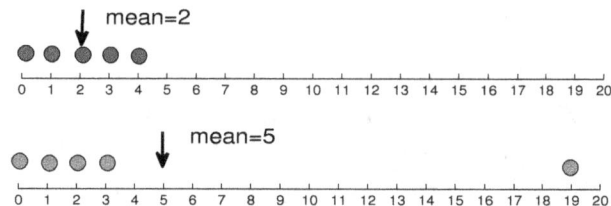

Figure 2.5: Means of two data sets

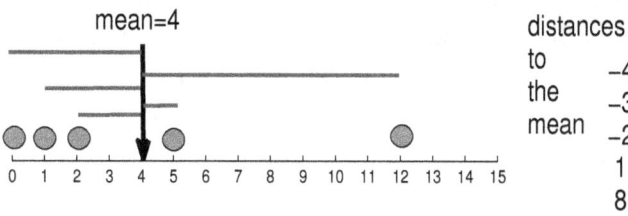

Figure 2.6: Distances to the mean

order statistics and the observation in the ith position is represented by $x_{(i)}$. Thus, the median is $x_{(n+1)/2}$ in the case of an odd number of observations and $(x_{(n/2)}+x_{(n/2+1)})/2$ when n is even. **The median is said to be resistant to outliers**; the median is equal to 2 for both data sets in Figure 2.5. Because the mean gets affected by extreme observations and the median does not, it is likely that the mean takes a larger value than the median in situations such as the second data set in Figure 2.5. If the outlier were to be to the other side (an extremely low value), the mean would be smaller than the median.

Comparing mean and median

Another situation in which the median is preferable to the mean is the case of 'censored' data. One example would be a race in which there are 100 participants who will run 5 miles but time is kept only during the first 40 minutes, and the variable 'time to complete the 5K' is registered only for those participants who finished in 40 minutes or less and the rest are marked as *40+'* or *more than 40*. Only 93 of the 100 participants finished the race in 40 minutes or less. The mean requires to add all the observations and we do not know really how long it took for the remaining 7 participants to complete the distance. The median can be calculated just by ordering the data: the 7 observations 40+ will be at one extreme, and the median will be the average of the observations in the 50th and 51th positions.

If the data are censored, calculate the median instead of the mean

Quartiles and Percentiles

Some special quantiles: quartiles percentiles

In a similar way that the median divides the distribution in two parts, **the quartiles divide the distribution in four parts**. If the amount of observations is large enough we might also be interested in the percentiles that divide the distribution in 100 parts. It is common to report the results of exams taken by a large number of students, such as the SAT or the GRE, as percentiles. When a person is in the 68th percentile, it means that 68% of the students scored equal or less than that person. The generic term is **quantile**, and can be represented by q. Given a value p, between 0 and 1, $q(p)$ is the value that separates the distribution in the lower $p \times 100\%$ and the upper $(1 - p) \times 100\%$. The lower quartile, the median and the upper quartile are the quantiles: $q(0.25)$, $q(0.5)$ and $q(0.75)$.

2.3.2   Spread or variability

It is important to know how the value of a given variable differs among the individuals of a population or sample. There are different ways of measuring the spread or variability in a data set.

Range

Range = max − min. For the first data set in Figure 2.5 the range is 4, for the second data set the range is 19. One single observation can make the range vary greatly.

Summarizing variability: Range, IQR, MAD, Variance, Standard-deviation

Interquartile range

$IQR = Q_U - Q_L$ represents the length of the interval in which the central 50% of the observations are contained. It can be easily visualized in the boxplot as the length of the box.

Mean Absolute Deviation, Variance, and Standard Deviation

Consider a set of values for a variable as in Figure 2.6, and think of the distances from the values to the mean. The larger the distances the more disperse the observations are. The sum of those distances is always zero. A way of transforming those distances is needed so that the negative and the positive values do not cancel each other out. One option is to consider the absolute values of those distances and calculate the average absolute distance, which is called the 'mean absolute deviation':

$$MAD = \frac{\sum_{i=1}^{n} |x_i - \bar{x}|}{n} \qquad (2.3)$$

The mean absolute deviation could also be calculated with respect to the median instead of the mean. For the values in Figure 2.6, $MAD = 18/5 = 3.6$. Since the median is 2, the mean absolute deviation with respect to the median would be $(2 + 1 + 0 + 3 + 10)/5 = 16/5 = 3.2$.

Another option is to consider the squares of the distances. The average of the squares of the distances from the observations to the mean is called the **variance**. The square root of the variance is called **standard deviation**. The symbol σ^2 is used for the population variance and s^2 for the sample variance:

$$\sigma^2 = \frac{\sum_{i=1}^{N}(x_i - \mu)^2}{N}, \qquad s^2 = \frac{\sum_{i=1}^{n}(x_i - \bar{x})^2}{n-1}. \qquad (2.4)$$

> Variance is the average of the squares of the distances from the observations to the mean

Consequently, the standard deviation is defined (for the population and sample case respectively) as:

$$\sigma = \sqrt{\frac{\sum_{i=1}^{N}(x_i - \mu)^2}{N}} \qquad and \qquad s = \sqrt{\frac{\sum_{i=1}^{n}(x_i - \bar{x})^2}{n-1}}. \qquad (2.5)$$

> Standard deviation is the square root of the variance

Mean absolute deviations, variances and standard deviations can never take a negative value because they are calculated based on either absolute values or squares of distances, and absolute values and squares are always equal to or greater than zero. Variability statistics only take value 0 in cases where all the observations have exactly the same value. The variance and the standard deviation tend to be more sensitive to extreme observations or outliers than the mean absolute deviation (MAD) because they consider the squares of the distances.

> The variance, standard deviation, and MAD can never be negative

Statistics of relative variability

These statistics compare the variability to the average value or central value of the variable and are specially useful when comparing groups of data. The most commonly used one is the **coefficient of variation**:

$$CV = \frac{Standard\ deviation}{Mean} \qquad (2.6)$$

> The coefficient of variation is used to compare the variability of groups with different means

When comparing two data sets with different means, it is considered that the one with higher CV has larger variability or heterogeneity, and the one with lower CV shows more homogeneity.

Another statistic of relative variability is the **coefficient of dispersion**. COD is equal to the mean absolute deviation (with respect to the median) divided by the median.

R: Variance and more

Using statistical software:
In MINITAB use STATS>Basic Statistics>Display basic statistics, click on Statistics and select the ones you want to calculate.
The following commands in **R** *will read the data in Example 2.2.2 and calculate the spread statistics described in this section.*

```
## to read the data
x<-c(57.78, 57.89, 59.95, 62.68, 71.07, 73.46, 82.25, 65.93,
67.39, 66.32, 62.60, 55.94, 61.57, 59.39, 73.18, 61.04, 63.66,
59.12, 57.32, 62.36, 59.60, 58.93, 64.93, 64.61, 77.60, 61.47,
61.21, 59.37, 61.56, 55.61, 59.89, 66.74, 74.64, 62.84)
sd(x)  ## Calculates standard deviation
var(x)  ## Calculates the variance
sd(x)/mean(x) ## Calculates the coefficient of variation
## Calculates mean absolute deviation with respect to the mean
dev<-abs(x-mean(x)); mean(dev)
```

2.4   More on outliers

A practical rule to identify outliers, in data for one quantitative variable, was shown in Section 2.2.2. First, the 'fences' are calculated; any observation beyond the fences is considered an outlier.

Lower fence: $LowerQuartile - 1.5 \times InterquartileRange$
Upper fence: $UpperQuartile + 1.5 \times InterquartileRange$

If you see an outlier, check first if the observation is correct

Boxplots automatically display outliers as isolated dots or some other symbol. The first thing to do when encountering outliers is to check if they are valid observations and not the result of a typo or error in measurement. If the value is correct, the first question that comes to mind is: Why is this observation so different from the others?

Example 2.4.1 Red blood cells of mammals

Red blood cells of mammals

Table 2.4 and Figure 2.7 display the number of red blood cells (erythrocytes) by cubic centimeter of blood for 13 species of mammals. The data come from

Table 2.4: Typical number of red blood cells for 13 mammal species

Species	million RBC
human	5.00
opposum	4.90
skunk	7.70
raccoon	8.70
weasel	7.40
fox squirrel	8.50
red squirrel	8.90
rabbit	6.30
llama	14.10
dog	7.15
cat	9.00
capuchin monkey	5.50
elephant	3.04

several sources, mainly from Youatt et al. (1961). Of course, it would have been better to include information about more species of mammals. The outlier is the llama from the order Artiodactyla, family Camelidae. The llama has a much higher number of red blood cells than the other twelve species of mammals in Table 2.4. The same happens for other species closely related to the llama such as vicuña, guanaco, and alpaca. What makes those species different from others? They live at high altitudes in the South American Andes. The presence of the outlier in Figure 2.7 brings us to ask the question: Is there a potential association between altitude and number of red blood cells? This question will be addressed for humans in Section 2.12 where we will learn how to find out if two quantitative variables are associated.

An outlier can be a starting point for further research

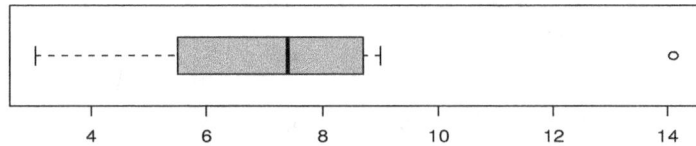

Figure 2.7: RBC (million) for 13 species of mammals

Another question about outliers is: How is the outlier going to affect the statistics to be calculated? Figure 2.21c displays the lengths of a large number of cells taken from the elongation region of several onion roots. There are outliers in that data set. The mean length of the cells is 0.04706 mm and the median is 0.041 mm. The mean is higher than the median, as usually happens in skewed distributions with a longer tail to the right side. If the two more distant outliers had not been there, the mean would be 0.04472 and the median would be 0.0405. The impact of the two distant outliers on the mean is not large because there are 270 observations. The impact of an outlier is more dramatic when the sample is small.

2.5   More on shapes

In addition to location and dispersion or spread, the shape of a distribution contains more information about the data. In Figure 2.8 there are six examples of different shapes of distributions. The three in the top row are symmetric; however, the distribution in **a)** is 'unimodal,' **b)** is 'U-shaped,' and **c)** is 'uniform.' A symmetric unimodal shape shows that the values in

'Read'
the shape!

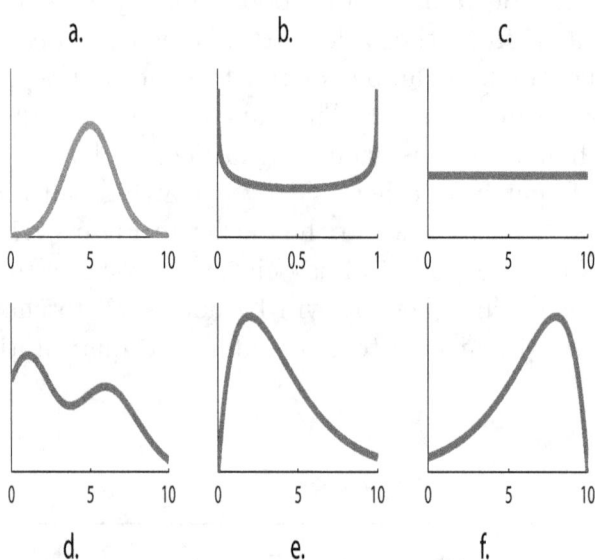

Figure 2.8: Examples of shapes of distributions

the center happen more frequently and that lower or higher values occur less frequently. The unimodal symmetric shape in **a)** is common when the individuals belong to a single population with natural individual variability around a central value. The U-shaped distribution in **b)** indicates that the values in the center occur less frequently and that the smaller and larger values are more frequent. The uniform shape in **c)** indicates that all the values of the variable are equally likely to happen. The distribution in **d)** is 'bimodal;' it indicates that the population is likely to be a mixture of two subpopulations with different means. The distributions in **e)** and **f)** are skewed distributions. Distribution **e)** has a longer tail to the right indicating that smaller values of the variable are more frequent and higher values are less frequent. The distribution in **f)** is skewed to the left, indicating that high values of the variable are more frequent and the lower values happen less frequently. The symmetry (or lack of symmetry) of a distribution says **Symmetric** a lot about the data. Histograms and stem-and-leaf displays are usually ac- **or skewed?** companied by a comment about the shape of the distribution, whether it is symmetric or skewed. Two histograms are shown in Figure 2.9. The distribution in the first one is fairly symmetric while the second one is skewed with a longer tail to the right.

Pulse rate
of students

Example 2.5.1 Pulse rate

Figure 2.9a) displays the self-reported number of heart beats per minute of 210 college students while sitting in class (data file *pulserate*). The minimum and maximum pulse rates are 40 and 112 per minute, the mean is $\bar{x} = 75.729$ and the standard deviation $s = 12.191$. The distribution is fairly symmetric, the most frequent values are those around the mean, with very low and very high pulse rates being infrequent.

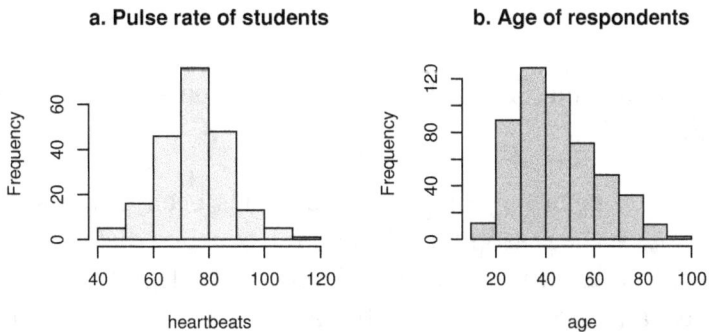

Figure 2.9: Symmetric and skewed frequency distributions

Example 2.5.2 Age of respondents in a survey

Age of
respondents

The distribution of the age of 503 randomly selected individuals (18 years or older) who responded to a drug and alcohol survey is shown in Figure 2.9.b (data file *age.dat*). The households were randomly selected using the random digit dialing method for a phone interview. The distribution is skewed because there are fewer old people in the population. Symmetric distributions are said to have skewness zero. Skewed distributions with a longer tail to the right have positive skewness.

Unimodal distributions such as the two histograms in Figure 2.9 indicate, in each case, a single population with variability among its members with an interval of values that are more frequent.

Example 2.5.3 Age at time of death

Age at time
of death

**Bimodality
usually indicates
a mixed
population**

The histogram in Figure 2.10a was prepared with the values of X: *age at time of death* for a sample of 135 tombstones from the Greenhill Cemetery in Laramie, WY (data file: *greenhill*). The distribution is clearly bimodal.

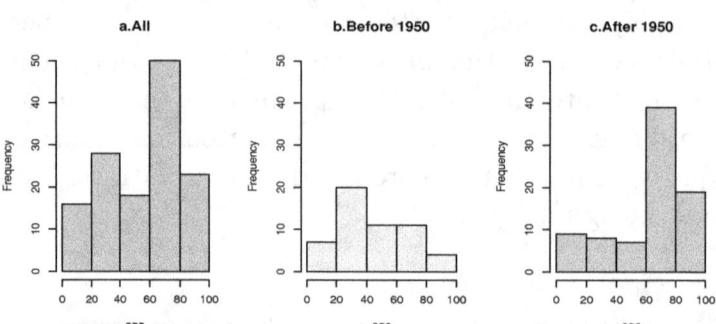

Figure 2.10: Age at time of death for 135 individuals, 53 died before 1950 and 82 died after 1950

Greenhill is the only cemetery in Laramie and has been in use since the late 1800's. What can be the reason for a bimodal frequency distribution? People live longer now, thanks in part to the existence of antibiotics, which started to be developed and used in the 1930's and 1940's. Life expectancy has changed dramatically from the first part of the 20th century to the present day; childbirth and tuberculosis were common causes of death in the USA in the early part of the 20th century. Infections took the lives of young people. The 135 observations were grouped according to the year of death (before and after 1950). The histogram in Figure 2.10b was prepared with the values of the variable 'age at time of death' for 53 individuals who died before 1950. The histogram in Figure 2.10c corresponds to 82 individuals who died after 1950. Each distribution has a peak on a different location, which explains the bimodal nature of the frequency distribution for the whole sample. This is an example of a 'hidden' variable (year of death) that makes the distribution of 'age at time of death' bimodal.

Another characteristic of the shape of a distribution, not as well known as symmetry, is **kurtosis**. A formal explanation of kurtosis is beyond the scope of introductory statistics, but an intuitive explanation of the concept of kurtosis can be given. Kurtosis is associated with the concentration of mass toward the center and the tails of the distribution. Distributions with considerable mass toward the center and heavy tails are said to have high kurtosis. When talking about kurtosis, distributions are traditionally compared with the normal distribution (to be studied in Chapter 7). However, one simple way of visualizing peak and tails is to compare the distribution with a uniform distribution with the same median and variance, as in Figure 2.11 (Kotz & Seier, 2008). Each of the distributions in Figure 2.11 has more mass toward the center and heavier tails than the uniform distribution.

Peak and tails

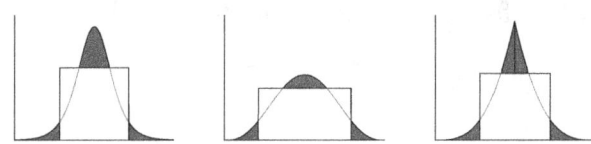

Figure 2.11: Peak and tails of distributions as compared with the uniform

2.6 The effect of transforming the data

In case of change of units, statistics do not need to be re-calculated from scratch

Assume that a data set consisting of the lengths in inches of a sample of leaves has been analyzed. The histogram is ready; the mean and standard deviation have been calculated. Assume that a change of units from inches into centimeters is needed to compare the results with other studies. Is the shape of the distribution going to change? Are the mean and other statistics going to change? That is the question to be answered in the first part of this section. The second part is dedicated to the 'logarithmic' transformation, which is quite important in biology.

2.6.1   Change of units

Transformations that involve only the four basic operations $(+, -, \times, \div)$ are called **'linear transformations'** and the general expression for them is:

$$Y = aX + k$$

A new variable Y is created by multiplying a variable X by a constant a and adding a constant k.

Adding a constant

Adding or subtracting a constant will change the location statistics but not the spread statistics since the distance among the observations (and thus the variability or spread) does not change.

Adding the same number to all the observations affects the location statistics but not the spread statistics

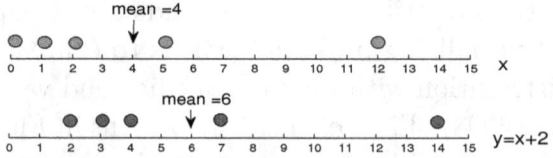

Figure 2.12: Effect of adding a constant to all the observations

If a constant is added to all the observations, as in Figure 2.12, all the location statistics will increase by the same number, but the standard deviation will remain the same. In general, if $y = x + k$ then $\bar{y} = \bar{x} + k$.

Multiplying by a constant

Multiplying all the observations by the same number affects both location and spread statistics

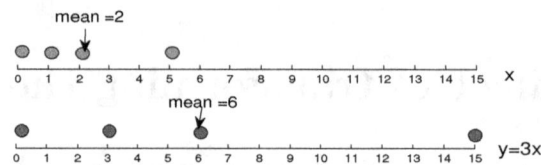

Figure 2.13: Effect of multiplying all the observations by a constant

When all the observations are multiplied (or divided) by the same number, as in Figure 2.13, the distances among the observations change. Thus, not only the location statistics but also the spread statistics will change. If $y = ax$, then $\bar{y} = a \times \bar{x}$ and $s_y = a \times s_x$.

An inch has 2.5 centimeters. When changing from inches to centimeters, each length is multiplied by 2.5. The mean, median and standard deviation will get multiplied by 2.5.

A kilo has 2.2 pounds. When changing from pounds to kilos, each weight is divided by 2.2. The mean, median and standard deviation will get divided by 2.2.

inches to cm, pounds to kilos, F to C^o

The formula to convert temperatures from Fahrenheit to Celsius is $C = (F - 32) \times \frac{5}{9} = F \times \frac{5}{9} - 32 \times \frac{5}{9}$. From each observation for temperature (F), the value 32 is subtracted and the result multiplied by 5/9. Subtracting 32 affects the mean and the median but not the standard deviation. However, multiplying by 5/9 affects the spread statistics as well.

The shape of the distribution, in terms of skewness or kurtosis, does not change under linear transformations such as the change of units. The distribution might shrink or expand (changing the spread) and the location will change, but the overall shape will not.

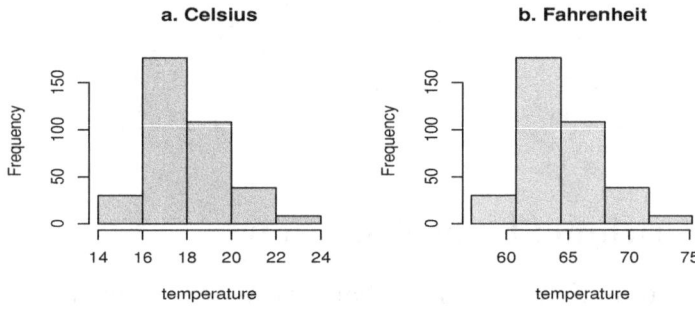

Figure 2.14: Histogram of sea surface temperature

Example 2.6.1 Sea surface temperature

Consider a data set (file *tempsea*) that contains the monthly average temperature of the sea surface in Callao (a harbor in Peru, South America) during 30 years (360 observations). The data set was recorded in the Celsius scale (Figure 2.14.a) and it has been converted into Fahrenheit (Figure 2.14.b). Notice that the scale of the horizontal axis is different. Since $F = \frac{9}{5} \times C + 32$, the multiplication by the constant 9/5 affects the mean and other location statistics AND the standard deviation as well. The mean and the spread are different in the two histograms in Figure 2.14, but the shape of the distribution is the same.

Sea surface temperature

Temperature	Min.	Q_L	Median	Q_U	Max.	Mean	StDev.
Celsius	14.60	16.80	17.60	19.10	23.70	18.03	1.695108
Fahrenheit	58.28	62.24	63.68	64.46	66.38	74.66	3.051195

**Celsius
to Fahrenheit**

*Using statistical software:
In MINITAB use CALC > Calculator to calculate linear transformations
of the variables, such as the change of units from Celsius to Fahrenheit. The
command $MTB > let\ C2 = (C1 * 9/5) + 32$ serves the same purpose.
The following* **R** *commands read the temperature data, transform from Fahrenheit into Celsius and produce the histograms in Figure 2.14. Notice that one
more option for the histogram has been included: 'breaks' indicates the endpoints of the intervals.*

**MINITAB & R:
Change of units**

```
par(mfcol=c(1,2))  ## puts both histograms in one figure
Celsius<-scan('e:tempsea.dat') ## reads data file in Celsius
Fahrenheit<-(Celsius*9/5)+32    ## transforms to Farenheit
a<-c(14,16,18,20,22,24)         ## endpoints for intervals in C
b<-(a*9/5)+32                   ## endpoints for intervals in F
hist(Celsius,breaks=a,col='palegreen3')   # histogram
hist(Fahrenheit,breaks=b,col='lightcyan3') # histogram
```

Logarithms

2.6.2 Logarithmic transformation

Non-linear transformations are those that involve expressions such as square
roots ($\sqrt{x}$), radicals, inverses $1/x$, powers such as x^2, i.e., those transformations of the data that are not obtained by just adding/subtracting and/or
multiplying/dividing by a number. A very important non-linear transformation is the logarithmic transformation. In order to understand logarithms, it
is convenient to see the exponential function first.

The exponential function

To define a function is to assign to each number another unique number
following some rule. The exponential function assigns, to the value x, the
value 2.71828^x, also called $exp(x)$. For example, to 1, the value $2.71828^1 = 2.71828$ is assigned; to 2, the value $2.71828^2 = 7.389056$ is assigned and
so on. Figure 2.15a displays the values of x in the interval $(-2, 3)$ and
the corresponding values of $exp(x)$. Some special numbers have a name in
mathematics. The number 2.71828 is called e, thus another way of writing
$exp(x)$ is e^x.

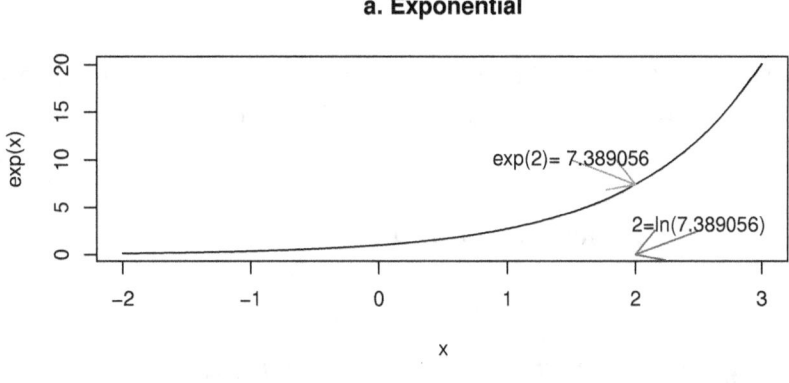

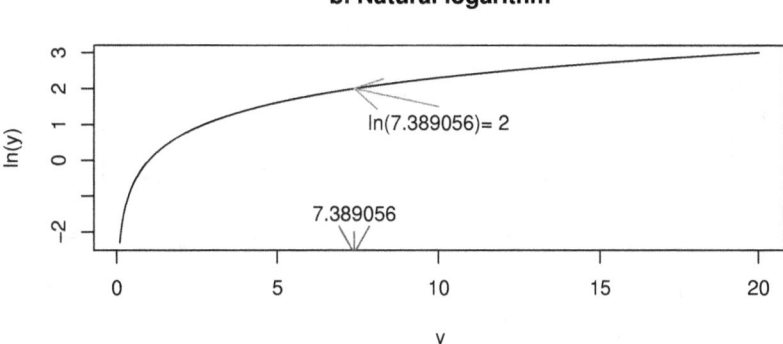

Figure 2.15: Exponential and Logarithmic functions

Logarithms

Remember that e^2 or $exp(2)$ is equal to 7.389056. Thus, the natural logarithm of 7.389056 is 2. If $y = e^x$, the natural logarithm of y is x, or $ln(y) = x$. In mathematics, the logarithmic function is said to be the 'inverse' of the exponential function.

Looking at the vertical axis in Figure 2.15a, it is clear that when x is in the interval $(-2, 3)$, the values e^x are in the interval $(0, 20)$. In Figure 2.15b, to each value in the interval $(0, 20)$, the corresponding value of its natural logarithm is given. Notice that the values of $ln(y)$ tend to be smaller than the values of y; the horizontal axis goes from 0 to 20 but the vertical axis only goes up to 3. This helps to explain the effect of logarithms on the shapes of distributions.

The effect of the logarithmic transformation on the shapes of distributions

Non-linear transformations change not only the values of the variable and the corresponding values of the statistics but the shapes of the distribution as well.

Example 2.6.2 Brain size of mammals

The effect of
the logarithmic
transformation
in the shape of
the distribution

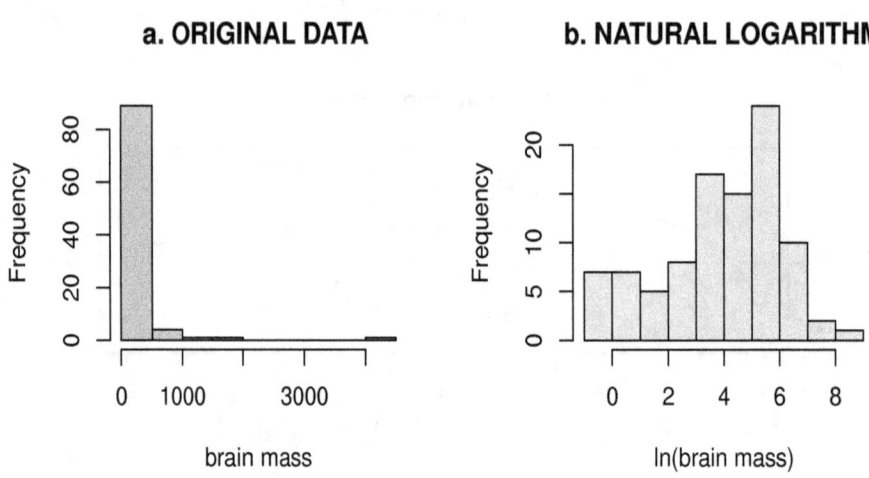

Figure 2.16: Brain mass for 96 species of mammals

Brain size
of mammals

Data on the typical brain weight of 96 species of mammals is found in Ramsey & Schaffer (1997) and in Shacher & Staffeld (1974). The distribution of the variable 'brain mass' is very skewed with a long tail to the right because most mammal species have a small brain and only a few (among them humans with typical brain weight of 1300g) have a large brain. Figure 2.16 displays the histogram of the original data and the natural logarithm of the data. The logarithms of a severely skewed right variable can sometimes have a fairly symmetric distribution. In biology, it is not uncommon to work with base 2 logarithms instead of natural logarithms. Logarithms with base 2 use the number 2 instead of the number 2.71828. If $y = 2^x$, then $log_2(y) = x$.
Using statistical software:
To calculate logarithms using MINITAB either use CALC > Calculator and then select 'natural logs' from the menu, or at the MTB > prompt type loge(x) if the data are in x.

To calculate the natural logarithm of x using R, it is enough to write 'log(x)'. For example ln(5) is calculated by typing log(5). For the brain size example, if the data for brain size are already stored as 'brain', the natural logarithms are calculated by typing 'log(brain).'

R & MINITAB: Logarithms

2.7 Paired data

Sometimes a variable is measured twice for the same individual, as in pre/post tests or before/after studies. Other times, the same variable is measured to each member of a pair, as in studies involving identical twins. Other examples of paired observations are measuring the reaction time for the left and right hand of a group of individuals or the length of the left and right limb of salamanders in a study of symmetry. In health related studies it is common to measure a variable before ('baseline value') and after a treatment. Paired data can also come from experiments in which a matched-pairs design has been used. In matched-pairs designs, there are several pairs of near-identical individuals. In each pair, one individual receives the treatment and the other one forms part of the control group. One example of matched-pairs design is an experiment to compare the yield of two varieties of potato using experimental plots located in different places. Each plot is divided into two; the variety to be planted in each half-plot is randomly determined. In all these examples the focus is on the difference of the two values of the response variable. The variable 'difference' is analyzed using the same plots and statistics learned in previous sections for the one variable case.

Paired data: we measure the same variable twice to each individual or individuals come in pairs

The variable to be analyzed is the difference between the two measurements

Example 2.7.1 Cherry juice and arthritis

An experiment (Delk, 2009) was conducted to observe if drinking cherry juice can help to reduce pain in elderly people suffering from arthritis. The data set on Table 2.5 corresponds to 10 males over 65 years of age suffering from arthritis. The variable observed is the necessary pressure on the knee for them to feel pain. For each individual two values were recorded: one ('baseline') at the beginning of the study and the other one ('final') after 4 weeks of drinking cherry juice. This is a typical 'before/after' case. The differences 'final-baseline' are in the last column of Table 2.5 and Figure 2.17. It is clear from Table 2.5 that all the 10 male patients in the sample were able to tolerate more pressure before feeling pain at the end of the four weeks of treatment. Patients were able to endure between 1.5 and 2.4 units more of pressure. The mean and median are very similar, 1.94 and 1.95 respectively.

Table 2.5: Pressure needed to produce pain in 10 arthritis patients

Individual	baseline	final	final-baseline
1	2.3	4.3	2.0
2	2.6	4.6	2.0
3	2.5	4.9	2.4
4	2.0	3.8	1.8
5	2.4	4.3	1.9
6	2.4	4.2	1.8
7	2.1	4.1	2.0
8	2.5	4.0	1.5
9	2.0	3.9	1.9
10	2.2	4.3	2.1

Source: Delk (2009)

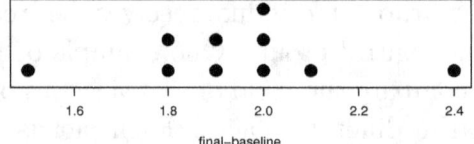

Figure 2.17: Difference in pressure needed to feel pain

Example 2.7.2 Wings of Australasian robins

Wings of male and female robins

Not all the examples of paired observations are of the 'before/after' type. Table 2.6 displays the typical length (mm) of the wing of males and females for the species of the family Petroicidae (Australasian robins). The research question is whether or not there is 'sexual dimorphism.' That is, if there is difference between males and females with regard to the length of the wing. The difference in wing length between males and females of the same species (last column of the table) becomes the variable of interest. Figure 2.18 displays the difference between males and females of the same species in three different plots: boxplot, dotplot, and histogram. Notice that all the differences are larger than 0. The typical wing length for adult males is larger than the typical wing length for the females of the same species. However, the difference is under a centimeter (10 mm) for all species.

Table 2.6: Typical wing length (mm) for male and female Australasian robins

ID	English name of species	Male	Female	Difference male-female
2	Jacky-winter	89.4	87.9	1.5
3	Lemon-bellied Flyrobin	73.6	70.5	3.1
4	Yellow-legged Flyrobin	70.6	65.0	5.6
8	Scarlet Robin	73.4	71.4	2.0
9	Tomtit	69.4	67.0	2.4
10	Red-capped Robin	62.9	61.3	1.6
11	Flame Robin	78.9	74.3	4.6
12	Rose Robin	65.7	63.7	2.0
13	Pink Robin	68.2	65.7	2.5
14	New Zealand Robin	100.0	97.0	3.0
15	Chatham Islands Robin	83.6	81.7	1.9
16	Hooded Robin	97.1	92.0	5.1
17	Dusky Robin	88.9	86.8	2.1
18	White-faced Robin	79.4	72.0	7.4
19	Pale-yellow Robin	75.7	71.6	4.1
20	Yellow Robin	84.6	81.7	2.9
21	Grey-breasted Robin	90.6	85.5	5.1
22	White-breasted Robin	78.0	74.7	3.3
23	Mangrove Robin	82.1	77.1	5.0
25	White-browed Robin	87.8	81.2	6.6
28	Ashy Robin	111.0	106.1	4.9
29	Northern Scrub-Robin	105.7	105.5	0.2
30	Southern Scrub-Robin	96.7	87.5	9.2

Source: Lislevand, Figuerola and Szekely (2007)
Seven species of the family are not included because data are not available.

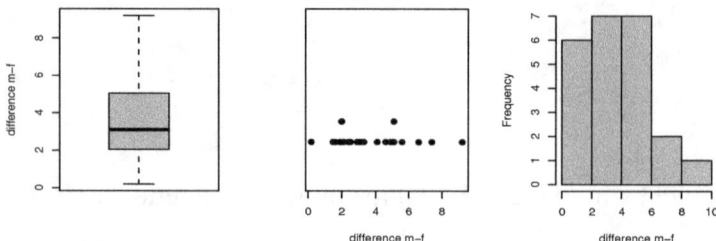

Figure 2.18: Differences between typical male and female wing length (mm) for Australasian robins

2.8　Comparing two or more groups

Consider the situation in which a random sample is selected from each one of two populations, for example a random sample of sugar maple trees and a random sample of sycamore trees. The same quantitative variable, such as height, is measured for each tree. The two groups of data can be compared in terms of location (means, medians), variability, and the shape of the distributions. Two different regions of onion roots are compared in the next example.

Onion cells

Example 2.8.1 Length of onion root cells

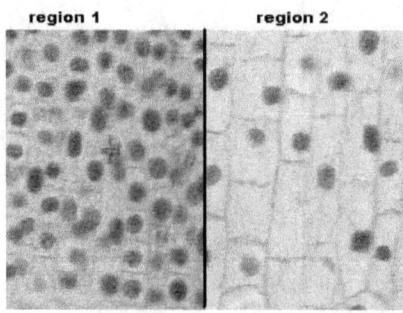

Figure 2.19: Cells in two regions of an onion root

Figure 2.19 shows the location of two regions in the onion root and pictures for the cells in each region. A random sample of 10 cells was drawn from each one of the two regions and their lengths in millimeters were measured. The dotplots in Figure 2.20 indicate variability in length for each region and also that cells in the elongation region tend to be longer and more diverse in length than the cells in the apical region. Some descriptive statistics have been calculated in Table 2.7 for this data set.

Table 2.7: Descriptive statistics for the length of cells from two different regions of onion roots

	Apical	Elongation
Sample size (n)	10	10
Mean	0.0252	0.0612
Median	0.0255	0.052
Standard deviation	0.007420692	0.03082495
Coef. of variation	0.2944719	0.5036757

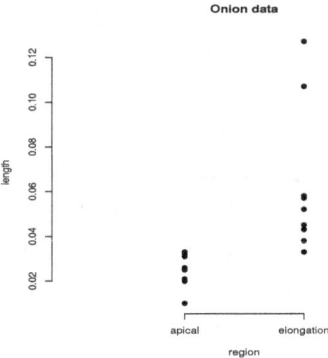

Figure 2.20: Lengths of 10 cells in each one of two regions of an onion root

The cells in the elongation region are on average approximately twice as long as cells in the apical region. In the sample from the elongation region there are two cells that look much longer than the other cells in the same region. Since the means of the two samples are different, the coefficient of variation is used to compare variability. There is more relative variability in the elongation region than in the apical region. It is difficult to comment about the shape of the distribution based on only 10 observations.

Comparing the values of the same variable in two samples

Figure 2.21 displays the lengths of 270 cells in each region, randomly selected from 27 different roots. The boxplot indicates that the cells in the apical region are smaller. The cells in the elongation region tend to be longer in average, and there is considerable variability among them. The boxplot also indicates that the distribution of length is fairly symmetric in the apical region but skewed (with outliers) in the elongation region. Notice that the scales in the two histograms are different. Plotting each distribution with

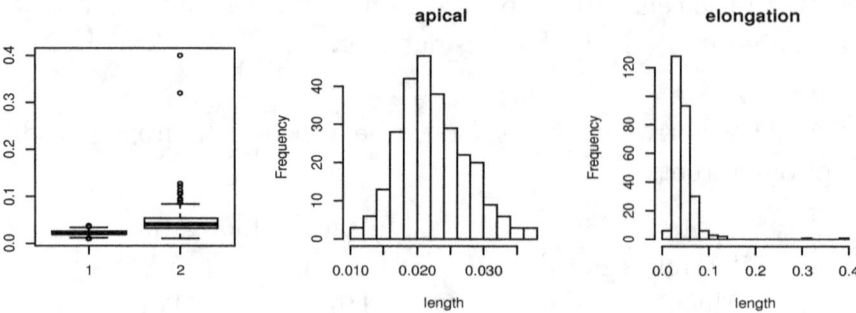

Figure 2.21: Lengths of 270 cells in each one of two regions of onion roots

Comparing shapes

its own scale allows us to better appreciate the shapes. The shape of the first histogram indicates that the cells in the apical region show individual variability around a central value; the values near the median are more frequent, the shorter lengths are as infrequent as the longer cells. The shape of the second histogram is skewed with a longer right tail, it indicates that a few cells are unusually large, approximately 0.3 or 0.4 millimeters. The cells divide in the apical region and new cells emerge; the new cells are small and there is little variability in terms of length. Once even newer cells emerge, the older cells are displaced to the elongation region and start growing; evidently, there is variability in how much they grow.

Example 2.8.2 Lengths of leaves of trees

Side-by-side boxplots

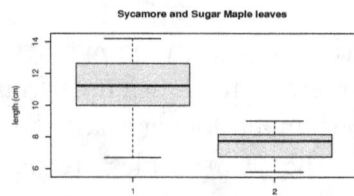

Figure 2.22: Lengths of twenty sycamore and sugar maple leaves

Figure 2.22 displays side-by-side boxplots to compare the length (in centimeters) of leaves of sycamore and sugar maple trees. There are twenty leaves of each type (datafile *leaves*). The boxplots indicate that sycamore leaves tend to be longer on average than sugar maple leaves. There is also more diversity in the length of the sycamore leaves, both in absolute terms (compare the standard deviations) and in relation to the mean (compare the

Leaves of trees

coefficients of variation). Boxplots are particularly useful in comparing two or more groups of data because they display a comparative view, not only of medians, but of minimum, maximum, spread, and symmetry as well.

Statistic	Sycamore	Sugar maple
Mean	11.080	7.555
Median	11.250	7.750
Standard deviation	1.997	0.944
Coefficient of variation	0.180	0.125

Example 2.8.3 Intensity of expression for one gene

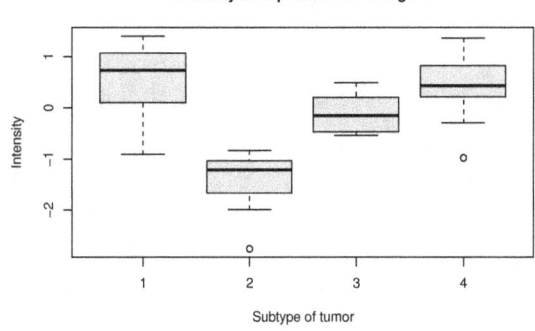

Figure 2.23: Intensity of expression of one gene for 4 different subtypes of tumor

Microarrays are used to observe the intensity in the expression of genes. Figure 2.23 displays side-by-side boxplots for the intensity of the expression of a particular gene for 83 patients with one of four subtypes of blue cell tumors. The data file *onegene* was prepared with values from the well-known Khan data set (Desper, Khan & Schaffer, 2004). Small, round, blue-cell tumors (SRBCTs) appear as four varieties: neuroblastomas (NB), rhabdomyosarcomas (RM), Burkitt's lymphomas (BL), and the Ewing's (EW) family of tumors. It is difficult to distinguish between the four varieties at the clinical level and patterns of expression of genes are valuable in the diagnosis. From the 83 patients, 29 have subtype EW (1), 11 have subtype BL (2), 18 have subtype NB (3) and 25 have subtype RM(4). The data file *onegene* has two columns, one for intensity of the gene and the other for the subtype of tumor (1 to 4). Figure 2.23 shows that the level of expression of this gene is lower for subtype BL than for the other subtypes.

Expression
of one gene

Using statistical software:

In MINITAB, boxplots are obtained using GRAPHS>Boxplot; there are 'simple' and 'with groups' options. To prepare side-by-side boxplots two columns of data are needed, one for the values of the variable, the other for the variable that indicates to which group each individual belongs. The statistics are calculated using STATS>Basic Statistics >Display Basic Statistics. The statistics can be calculated for groups of data using the option 'by' where the name of the indicator variable for groups is selected.

MINITAB & R:
Side by side
boxplots

The R commands to produce Figure 2.22 and to calculate mean, median and standard deviation by type of tree appear below. Notice that all the lengths are in one variable and 'type' is the indicator variable for type of tree. The symbol ~ is used to prepare boxplots by groups. The command by(variable, groups, mean) is used to calculate the mean by groups. The other statistics are calculated in a similar way.

```
length<-c(11.2,13.0,12.5,10.0,11.3, 7.0,10.0,10.2,10.6,6.7,9.3,
11.0,12.8,9.5,12.2,13.5,13.1,11.3,12.2,14.2,7.6,5.8,7.7, 9.0,
8.1,8.6,6.3,7.5,8.6,8.7,7.8,6.5,6.1,8.0,7.5,8.0,7.0,6.2,7.9,8.2)
type<-c(1,1,1,1,1,1,1,1,1,1,1,1,1,1,1,1,1,1,1,1,2,2,2,2,
2,2,2,2,2,2,2,2,2,2,2,2,2,2,2,2)
boxplot(length~type,main='Sycamore and Sugar Maple leaves ',
ylab='length (cm)', col='lightgreen')
## Calculating statistics by group
by(length,type,mean); by(length,type,median)
by(length,type,sd)
```

In R, side-by-side boxplots can also be prepared if the data have been entered separately; 'boxplot(x,y)' will do side-by-side boxplots for the objects 'x' and 'y.' In the case of Figure 2.23, the data file 'onegene' has two columns, one for the variable and one to indicate the groups. The following commands are used to read the data file and prepare the boxplots:

R: Reading
a data file with
several variables

```
## read a data file with 2 or more columns using read.table
## columns have no names thus header=FALSE
gene<-read.table('e:onegene.dat',header=FALSE)
## objects are formed with each column
intensity<-gene[,1]   ## all rows of column 1
subtype<-gene[,2]     ## all rows of column 2
## produce side by side boxplots for the gene expression example
boxplot(intensity~subtype, main='Intensity of expression of one gene',
xlab='Subtype of tumor', ylab='Intensity' ,col='cadetblue2')
```

2.9 Two groups of paired observations

Sometimes two groups or treatments need to be compared, but the variable of interest in each group is the difference in a pre/post or other type of study that produces paired observations. For example, the difference in reaction time between the left and right hand is studied for males and females. Females and males are the two groups to be compared, but the variable of interest measured to each individual is the difference between the reaction time of his/her left and right hand.

Example 2.9.1 Cherry juice vs. placebo

Table 2.2 contains data on pressure required to feel pain in the knee at the beginning and end of a four week treatment with cherry juice for a group of 10 individuals. This group is to be compared with another group, called *control* of the same gender, similar age, and status of the disease (arthritis). The control group received a placebo, something that looked like cherry juice but that was not cherry juice and contained no active ingredient. The purpose of such control groups in experiments is to study the placebo effect, the possibility that the patients feel better because they think they are being treated and because of the attention the receive from the researcher.

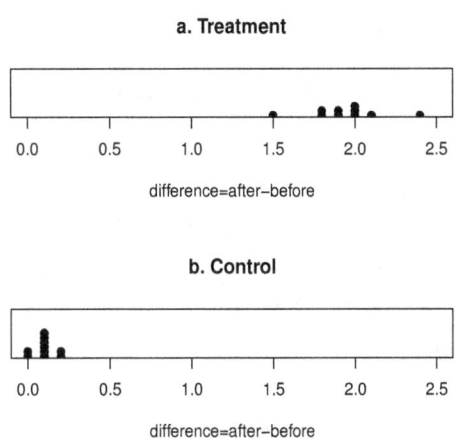

Figure 2.24: Difference final-baseline for treatment and control groups

Table 2.8 displays the difference final-baseline for the *treatment* and *control* groups. The differences are plotted in Figure 2.24. The difference between the two groups is quite clear, the control group did not show practically any improvement and everybody in the treatment group showed some improvement. If more observations were available, boxplots could have been used instead of dotplots, but for small data sets dotplots are preferable.

Table 2.8: Treatment and control group in an experiment with cherry juice

	Cherry juice		Placebo
Individual	final-baseline	Individual	final-baseline
1	2.0	11	0.1
2	2.0	12	0.0
3	2.4	13	0.1
4	1.8	14	0.1
5	1.9	15	0.0
6	1.8	16	0.2
7	2.0	17	0.2
8	1.5	18	0.1
9	1.9	19	0.1
10	2.1	20	0.1

Source: Delk, K.L. (2009)

MINITAB & R-
Dotplots by groups

Using statistical software:
In MINITAB dotplots by groups can be produced with GRAPHS>Dotplot by choosing the option 'with groups.'
The R commands used to produce Figure 2.24 appear next. Notice that that the same scale has been for both dotplots to facilitate the comparison of the groups.

```
treatment<-c(2,2,2.4,1.8,1.9,1.8,2,1.5,1.9,2.1)
control<-c(0.1,0,0.1,0.1,0,0.2,0.2,0.1,0.1,0.1)
a<-c(0,2.5) ## entering the endpoints of the x axis
par(mfcol=c(2,1)) ## 2 plots in the same figure
## cex controls the size of symbols
## offset controls the vertical space between symbols
## pch=19 indicates dots instead of squares
## xlim controls the endpoints of the x axis
## stripchart prepares the dotplot
stripchart(treatment,method='stack',offset=0.5,at=0,cex=1,
pch=19,xlim=a, main='a. Treatment',xlab='difference=after-before' )
stripchart(control,method='stack',offset=0.5,at=0,cex=1,
pch=19,xlim=a,  main='b. Control', xlab='difference=after-before')
```

2.10 Plotting time series

A sequence of values observed for a variable over time is called a **time series**. The corresponding plot is called a time series plot or 'time-plot' and it is formed by connecting the dots corresponding to the consecutive observations. There are several methods to analyze time series, which are studied in more advanced courses. At a descriptive level, the questions about a time series that can be addressed are:

Plotting values of a variable over time

1. Is the average value more or less constant or is there an upward, downward or irregular trend?

2. Is the variability more or less constant or are there portions of the time series with higher variability?

3. Are there any noticeable cycles?

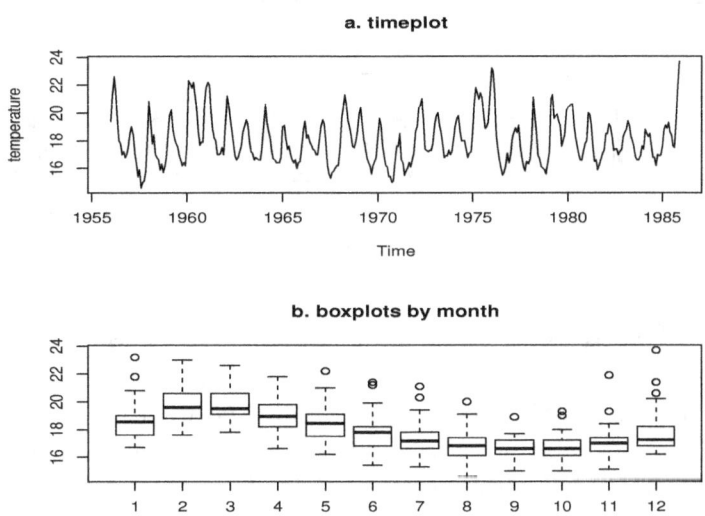

Figure 2.25: Monthly average of sea surface temperature in Callao, Peru

Sea surface temperature

Example 2.10.1 Sea surface temperature

Figure 2.25a displays a time series plot or 'timeplot' for the monthly average temperature (Celsius) of the sea in Callao, Peru, from 1956 to 1985 (Data from IMARPE). There are 12 observations for each one of the 30 years. The same data set was used to prepare the histogram in Figure 2.14. The histogram gave an idea of the range of values for the monthly average temperature and which values were more frequent. However, plotting the temperature through time as in Figure 2.25a gives an insight into the behavior of temperature of the sea that the histogram does not give. During these 30 years no upward or downward trend is noticeable but cycles are present. Figure 2.25b displays boxplots for the 30 observations for each one of the 12 months. The purpose of the boxplots is to compare the twelve months to see if there is a seasonal pattern. In the case of the temperature of the sea surface, a seasonal pattern is clear: higher temperatures in the summer (January-March, since the location is in the southern hemisphere) and lower temperatures in winter (July-September). Longer cycles that last several years also exist. Those longer cycles might be connected to 'El Niño' and 'La Niña,' names given to oscillations in the temperature of the surface of the Pacific Ocean.

Activity
of a bee

Example 2.10.2 Activity of a bee in constant darkness

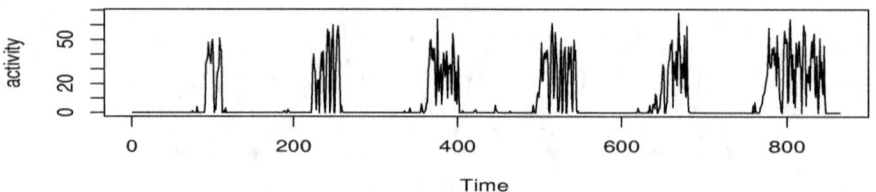

Figure 2.26: Activity of a bee in constant darkness for 6 days

Figure 2.26 displays 864 observations collected on the activity of a bee (Source: Dr. Moore's laboratory at ETSU). An adult honey bee (*Apis mellifera*) is placed in a pexiglass chamber with an infrared light-emitting and detection diode pair. A computer records the number of times that the bee interrupts a beam of infrared light during each one of many consecutive 10-minute intervals; the values recorded represent the activity of the bee. The bee is trained during 3 days, alternating 12 hours of light and 12 hours of darkness and then the lights are turned off. The time series plot displays the activity of the bee during 6 days in constant darkness. Cycles of activity and rest are clearly present, those cycles are approximately one day long. Since the bee does not receive an external cue of light or darkness, the cycles of activity and rest are determined by the internal clock of the bee. Thus, the cycles reveal the presence of a *circadian rhythm.*

Biologists use a special type of graph called an 'actogram' to display time series consisting of activity data. Actograms display the same information than the regular time series plot but using bars instead of connecting lines and separate plots for each day. The purpose of having several plots is to facilitate the comparison of the different days, especially the onset of activity in each day. Separate time series plots could also be obtained for each day. The actogram in Figure 2.27 was prepared with the same data set as the time series plot in Figure 2.26.

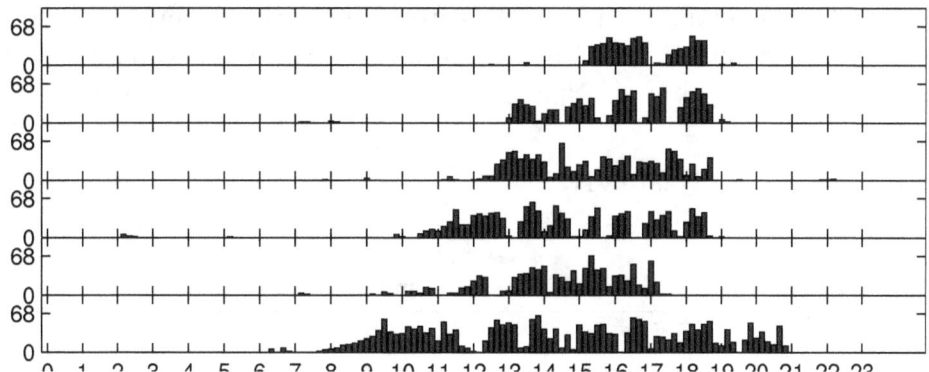

Figure 2.27: Actogram for the activity of a bee in constant darkness for 6 days

Using statistical software:
MINITAB has the option GRAPH>Time series plot. To prepare the boxplots in Figure 2.25b, first a column with numbers from 1 to 12 is created to label the month for each observation using CALC>Make pattern data. Then, the option GRAPHS>Boxplots>with groups is used.
Only two commands were needed in R to prepare Figure 2.26. Figure 2.25 requires more commands in order to put the dates and prepare the boxplots:

MINITAB &R:
Time series plots

```
## commands for Figure 2.26
activity<-scan('e:onebeesu.dat') ## reads the data
ts.plot(activity)    ## plots the time series
## commands for Figure 2.25
temp<-scan('e:tempsea.dat')    ## reads data
temperature<- ts(temp, start = c(1956, 1), frequency = 12) ##dates
t<-rep(seq(1,12,by=1),30)   ## creates labels for months
par(mfrow=c(2,1))   ## to plot two figures together
ts.plot(temperature, main='a. timeplot')
boxplot(temperature~t, main='b. boxplots by month')
```

2.11   Plotting survival data

Plotting the
proportion
of the
cohort
that is still
alive at
each point
of time

Fly survival

A very special type of data collected over time is that of survival data. The
study starts with an initial cohort of individuals and the observation recorded
for each period (day, week, or month) is the number of individuals who die
or 'fail' in a more general health or industrial context (such as going out
of remission, going back to using drugs, failure of an electronic component,
etc.) during that period. Instead of plotting the data as a time series for
the variable X: *number of individuals who died or failed each day*, the plot
displays the proportion of individuals from the initial cohort who are survivors at each point of time. Figure 2.28 has survival plots for two groups of
individuals (male and female flies).

Example 2.11.1 Fly survival

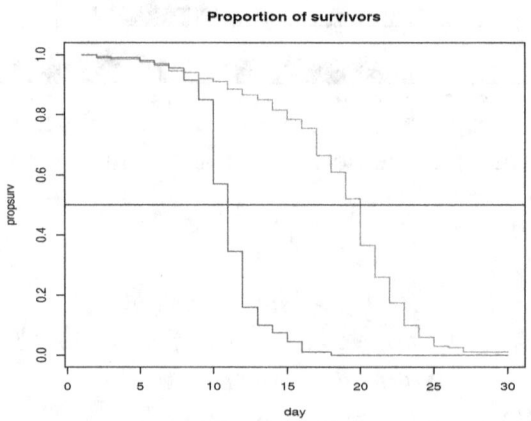

Figure 2.28: Survival plots for male and female flies

This data set was produced by a group of high school students in their
AP Biology class. They placed flies (*Sarcophaga crasipalpis*) in cages under
different conditions (males alone, females alone, mixed cages). Each day
the students counted how many new dead flies were lying at the bottom of
the cage. Table 2.9 displays the data for one cage of males and one cage
of females, starting with 200 flies in each cage. The plot obtained is called
the **estimated survivor curve** for just one group or for two groups as in
Figure 2.28. They are not, in fact, 'curves' but step functions; however, the
theoretical version is a curve. To find the median life length of the flies, trace
a horizontal line at the value of 0.5 in the vertical axis and check where the

Table 2.9: Survival data for flies (*Sarcophaga crasipalpis*)

	MALES				FEMALES			
Day	Dead	Cum.	Alive	% alive	Dead	Cum.	Alive	%alive
1	0	0	200	1.000	0	0	200	1.000
2	2	2	198	0.990	1	1	199	0.995
3	0	2	198	0.990	2	3	197	0.985
4	0	2	198	0.990	0	3	197	0.985
5	2	4	196	0.980	2	5	195	0.975
6	3	7	193	0.965	1	6	194	0.970
7	2	9	191	0.955	5	11	189	0.945
8	8	17	183	0.915	1	12	188	0.940
9	13	30	170	0.850	4	16	184	0.920
10	56	86	114	0.570	2	18	182	0.910
11	45	131	69	0.345	5	23	177	0.885
12	37	168	32	0.160	4	27	173	0.865
13	12	180	20	0.100	3	30	170	0.850
14	5	185	15	0.075	7	37	163	0.815
15	6	191	9	0.045	6	43	157	0.785
16	7	198	2	0.010	6	49	151	0.755
17	0	198	2	0.010	18	67	133	0.665
18	2	200	0	0.000	11	78	122	0.610
19	0	200	0	0.000	18	96	104	0.520
20	0	200	0	0.000	31	127	73	0.365
21	0	200	0	0.000	21	148	52	0.260
22	0	200	0	0.000	17	165	35	0.175
23	0	200	0	0.000	0	165	35	0.175
24	0	200	0	0.000	23	188	12	0.060
25	0	200	0	0.000	6	194	6	0.030
26	0	200	0	0.000	1	195	5	0.025
27	0	200	0	0.000	3	198	2	0.010
28	0	200	0	0.000	0	198	2	0.010
29	0	200	0	0.000	0	198	2	0.010
30	0	200	0	0.000	0	198	2	0.010

Source: Mr. Rollins AP Biology class, University School, Johnson City, TN

horizontal line encounters the survival curves. Figure 2.28 indicates that the median life for males is 11 days and the median life for females is 20 days. When living in separate cages, females have a longer median life than males. It is a different scenario when they are put in the same cage.

It is common for survival data to not come in the form of Table 2.9 (number of dead flies per day) but rather reporting how long each individual lives. There is an entire area of statistics called Survival Analysis that analyzes the data on percent of survivors in relation to other variables that might be associated with the survival of individuals.

R: survival
plots

Using statistical software:
These commands were used to obtain the plot for female flies in Figure 2.28:

```
nindi<-200   ## how many flies we start with
## we type in the data of number of dead flies each day
x<-c(0,1,2,0,2,1,5,1,4,2,5,4,3,7,6,
     6,18,11,18,31,21,17,0,23,6,1,3,0,0,0)
## NO MORE INPUT IS NEEDED
n=length(x)  ## calculates the number of days
day<-seq(1,n,by=1)      ## creates the values for 'day'
cum<-cumsum(x)          ## accumulates the values
surv<-nindi-cum         ## calculates the number of survivors
propsurv<-surv/nindi    ## calculates proportion of survivors
plot(day,propsurv,'s',col='red',ylim=c(0,1))
abline(0.5,0)           ## adds the horizontal line
## END HERE if there is only one group
```

2.12   Analyzing two quantitative variables

Generally, observational studies or experiments are not limited to the study of one single variable. It is common to work in a bivariate (two variables) or a multivariate environment. One question that frequently arises is whether there is an association or relationship between two given variables. This section focuses on the case in which both variables are quantitative.

2.12.1   Scatterplots

A scatterplot is the statistical graph used to display the values of two quantitative variables simultaneously, with the purpose of exploring if a relationship exists between them. Each individual is represented by a dot, whose coordinates are the values for the two variables.

Studying two quantitative variables simultaneously

Example 2.12.1 Altitude and number of red blood cells in humans

It was observed in Section 2.4 that South-American camelids living at high altitudes in the Andes, such as llamas, alpacas, vicuñas and guanacos, have a high number of red blood cells compared to some other mammals. We wonder if the number of red blood cells in humans is related to the altitude at which they live. The data displayed in the scatterplot in Figure 2.29a, and in Table 2.10, come from page 274 of Spector (1956). Only long time residents at each altitude were considered. Each one of 17 individuals is represented by a dot according to the value of *X:altitude* and *Y:# (million) of red blood cells.*

Altitude and red blood cells

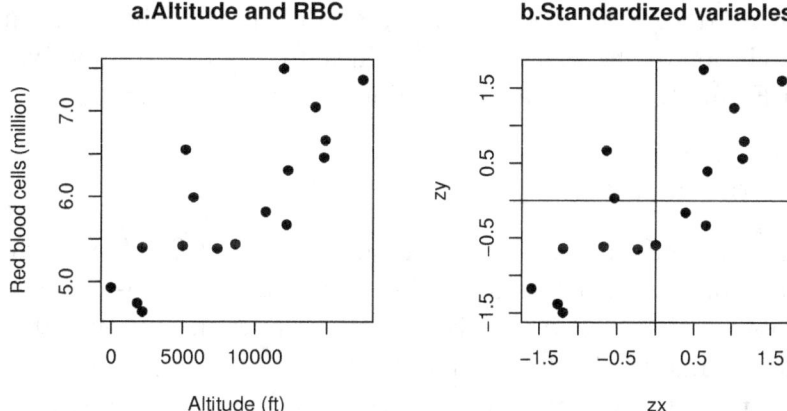

Figure 2.29: Scatterplot for altitude (X) and red blood cells (Y) for 17 individuals

The scatterplot suggests a positive relationship in the sense that, the higher the altitude, the more red blood cells humans tend to have. Biologically that is a sign of adaptation since red blood cells transport oxygen and at high altitudes there is less oxygen. Individuals living at high altitudes adapt by having more erythrocytes (also called red blood cells or RBC) to transport oxygen. In the USA this is put into practice by athletes who frequently train in the Rocky Mountains region in Colorado. In some countries in South America, the more successful runners competing at sea level are those who normally live at higher elevations in the Andes.

Notice that the scatterplot in Figure 2.29b looks very similar to the one in Figure 2.29a, but the location of the origin and the scales in the axes are different. To **standardize** the values of a variable, the mean is subtracted and that difference is divided by the standard deviation:

$$z_x = \frac{x - \bar{x}}{s_x} \qquad and \qquad z_y = \frac{y - \bar{y}}{s_y}. \qquad (2.7)$$

Before standardization, the variable *altitude (ft)* took values in the interval $(0, 17500)$. The variable *red blood cells (million)* took values in the interval $(4.65, 7.5)$. The mean and standard deviation of the data for each variable are:

	Mean	Standard deviation
Altitude	8640	5403.269
RBC	5.962	0.877

The standardized version of a variable is free of units. The mean and the standard deviation of the standardized values are 0 and 1, respectively. Notice that in Figure 2.29b the range of values for z_x and z_y are similar.

2.12.2   Correlation

The correlation coefficient r is a statistic that quantifies how strong the linear association between two quantitative variables is, based on a set of n pairs of observations. The idea is to observe how the two variables vary together, after adjusting for location and spread. In Figure 2.29b notice that most points are on two quadrants that form a diagonal, i.e., individuals with higher values for z_x tend to have higher values of z_y. Individuals with lower values of z_x tend to have lower values of z_y. There are few points outside the diagonal quadrants; this is a rough indicator of a fairly strong positive correlation. Formula (2.8) defines Pearson's correlation coefficient r in terms of the standardized version of the variables.

$$r = \frac{1}{n-1}\sum z_x z_y \qquad where \qquad z_x = \frac{x - \bar{x}}{s_x} \qquad z_y = \frac{y - \bar{y}}{s_y} \qquad (2.8)$$

The process of calculating the correlation coefficient r between *altitude of residence* and *million of erythrocytes per* cm^3 *of blood* for the 17 individuals is shown in Table 2.10 (we do not use now the last two columns of Table 2.10, they are to be used in the next section). Notice that the sum of the products of the standardized values at the bottom of Table 2.10 is 12.845, thus

Table 2.10: Altitude of residence (X) and red blood cells (million) (Y)

ID	Variables		standardized vars.		product	ranks	
#	X	Y	z_x	z_y	$z_x \times z_y$	for X	for Y
1	0	4.93	-1.59903	-1.17722	1.88241	1	3
2	1840	4.75	-1.25850	-1.38247	1.73984	2	2
3	2200	5.40	-1.19187	-0.64126	0.76430	3.5	5
4	2200	4.65	-1.19187	-1.49651	1.78364	3.5	1
5	5000	5.42	-0.67367	-0.61846	0.41663	5	6
6	5200	6.55	-0.63665	0.67011	-0.42663	6	13
7	5750	5.99	-0.53486	0.03153	-0.01686	7	10
8	7400	5.39	-0.22949	-0.65267	0.14978	8	4
9	8650	5.44	0.00185	-0.59565	-0.00110	9	7
10	10740	5.82	0.38865	-0.16233	-0.06309	10	9
11	12000	7.50	0.62185	1.75341	1.09035	11	17
12	12200	5.67	0.65886	-0.33338	-0.21965	12	8
13	12300	6.31	0.67737	0.39643	0.26853	13	11
14	14200	7.05	1.02901	1.24027	1.27625	14	15
15	14800	6.46	1.14005	0.56748	0.64695	15	12
16	14900	6.66	1.15856	0.79554	0.92168	16	14
17	17500	7.37	1.63975	1.60517	2.63208	17	16
				TOTAL	12.845		

Source for values of X and Y: Spector (1956) p. 274

$r = \frac{12.845}{16} = 0.8028$. It can be said that there is a strong linear correlation between the altitude at which individuals in the sample live and the number of red blood cells they have. Figure 2.30 summarizes the interpretation of the correlation coefficient: values near zero indicate a weak or non-existent **r measures** linear relationship, values close to 1 and -1 indicate a strong relationship **the strength** (positive or negative). The correlation coefficient is generally calculated using **and** software. A change of units in one or both variables will not affect the **direction** standardized values, and thus, it will not affect r because it is calculated **of linear** with the standardized version of the observations. **relationships**

The correlation coefficient r takes values between -1 and 1. Figure 2.31 shows six scatterplots with their corresponding value for the correlation. The **r does not** scatterplots depict different cases of strength and direction of the linear as- **change when** sociation between two quantitative variables. In Figure 2.31a) and d) the **the units** relationship is perfect, positive in one and negative in the other. The value **of the variables** of the correlation is 1 and -1 respectively. When there is no association be- **are changed**

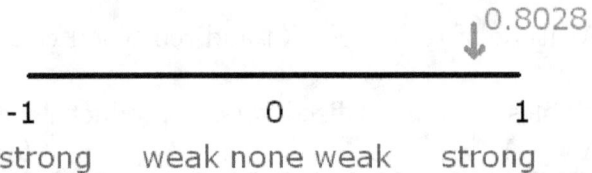

Figure 2.30: Interpretation of the value of the correlation coefficient **r**

tween the variables such as in Figure 2.31b, r is 0 or close to 0. A negative correlation indicates that, as one variable increases its value, the other one goes down. Figures 2.31c and 2.31e represent situations of strong but not perfect relationships. In Figure 2.31f the relationship is perfect but not linear, thus, **r** is not 1. Pearson's correlation r only measures the strength of 'linear' relationships.

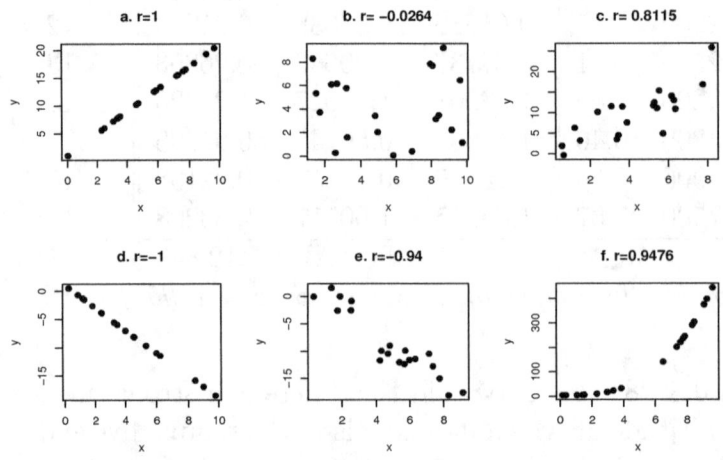

Figure 2.31: Six scatterplots with corresponding correlations

MINITAB & R :
Scatterplots,
and correlation

Using statistical software:
In MINITAB, use the option GRAPH > scatterplot. To calculate the correlation select STAT > Basic statistics > Correlation.
The following commands will enter the data in **R**, *obtain the plot and calculate the correlation:*

```
Altitude<-c(0,1840,2200,2200,5000,5200,5750,7400,8650,10740,12000,
12200,12300,14200,14800,14900,17500)
RBC<-c(4.93,4.75,5.40,4.65,5.42,6.55,5.99,5.39,5.44,5.82,7.50,5.67,
6.31,7.05,6.46,6.66,7.37)
```

```
plot(Altitude,RBC) ## gets scatterplot
cor(Altitude,RBC)## calculates Pearson's correlation r
```

2.12.3 Rank correlation *

When measurements are not exact, it might be preferable to calculate the correlation between the ranks of the observations instead of the correlation between the observations themselves. This correlation is called Spearman's rank correlation. Rank correlation indicates whether high values of one variable are associated with high (or low) values of the other variable. If the measurements have been done in a precise form, rank correlation loses some information by replacing the observed values by their ranks. However, if the measurements are not precise (such as in the self-evaluation of pain), the replacement might be desirable because the attention is not put in the precise numerical value but how it ranks overall with respect to the other observations. Rank correlation is popular in applications to the social sciences.

Rank correlation: Correlation between the relative positions of the observations

    The last two columns of Table 2.10 contain the ranks for the data in the altitude/erythrocytes example. If two observations have the same value, the rank assigned to them is the average of the two positions. The rank correlation between between altitude and RBC is 0.7995097 ≈ 0.8.

Using statistical software:
These **R** *commands calculate the ranks and their correlation, once the data for altitude and RBC have been read.*

R : Rank correlation

```
ralt<-rank(Altitude)
rRBC<-rank(RBC)
cor(ralt,rRBC)
```

2.13 Plotting several quantitative variables

Frequently, data are available for several quantitative variables for a group of individuals. Two questions of interest are:

Data for several quantitative variables

- Which two variables are more highly correlated?

- Are there groups (clusters) of individuals that can be formed based on the values of those variables?

Example 2.13.1 Typical values of five variables for the males of 20 species of birds

20 bird species

The last five columns of Table 2.11 contain data for the typical mass (V1), wing length (V2), tarsus length (V3), bill length (V4) and tail length (V5), for the males of 20 species of birds (data file *birds20*) from different families. Those 20 species are a small sample of all the species in the data base by Lislevand et al. (2007). If the objective is to examine which variables are related, scatterplots for each pair of variables can be prepared as in Figure 2.32. The scatterplots reveal, among other things, that the relationship between mass and wing length, and between mass and tarsus length, are not linear.

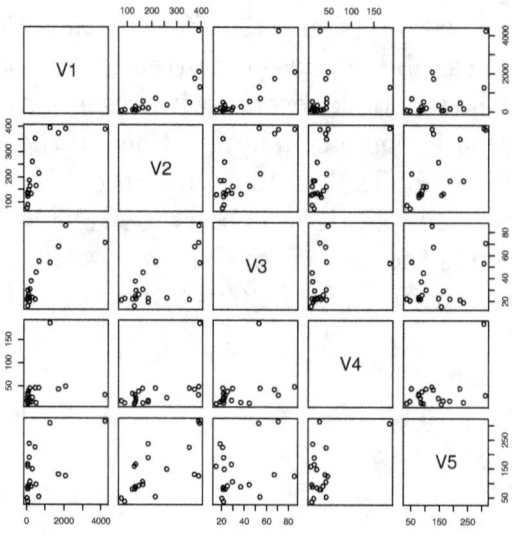

Figure 2.32: Scatterplot matrix for typical mass (V1), wing (V2), tarsus (V3), bill (V4) and tail (V5) length for the males of 20 species of birds

If the objective is to form groups or 'clusters' of species according to the typical values of the variables, then a plot like the one in Figure 2.33 is suggested. The graph in Figure 2.33 is called a 'stars' plot and is used in multivariate statistical analysis. A star or another icon is created to represent each individual, species in this case. To form the star, several axes are traced (one for each variable) coming out from a dot in the center. Over each axis the value of the variable is marked and the dots are connected. The idea is that similar shapes would indicate similar proportions between the values of the different variables. In each star, the variables are clockwise ordered.

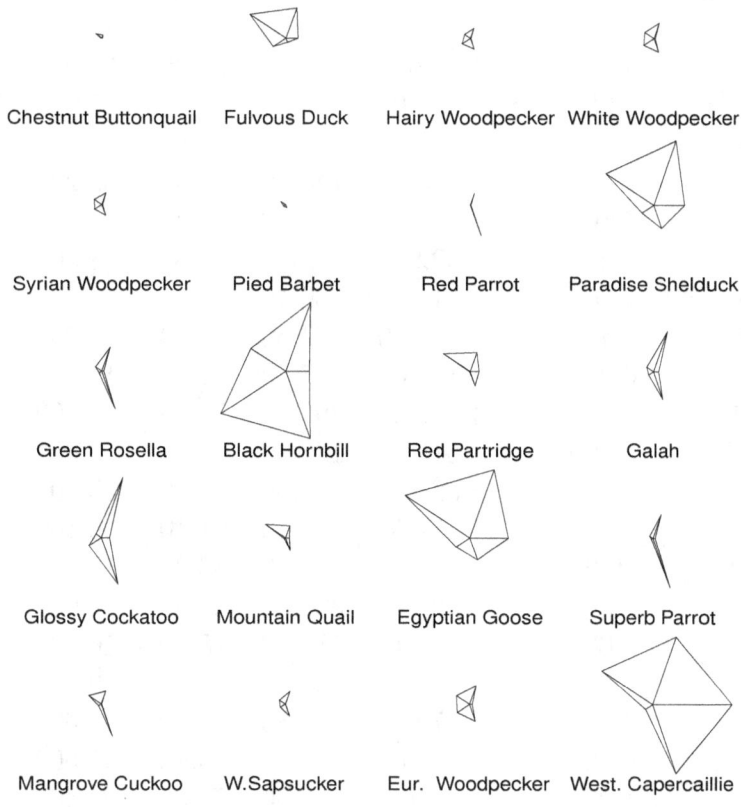

Figure 2.33: Stars plot of 20 species of birds based on typical mass, wing, tarsus, bill,and tail length

Notice that the stars, corresponding to species coming from the same family, look similar. The following groups could be formed:

1. Egyptian goose and paradise shelduck

2. Red-rumped parrot , green rosella and superb parrot

3. All the woodpeckers

4. Galah and glossy black-cockatoo (one is larger than the other but the proportions between variables are similar)

5. Red-legged partridge and mountain quail

Table 2.11: Egg mass, clutch size and typical body measurements for males
of 20 species of birds

ID	Egg mass	Clutch size	mass V1	wing V2	tarsus V3	bill V4	tail V5
1	6.60	4.00	68.9	90.60	21.80	14.00	35.0
2	51.00	13.34	675.0	216.00	54.20	46.20	51.5
3	4.30	3.93	69.8	122.82	21.63	34.33	77.4
4	6.70	4.00	108.5	148.00	25.30	39.20	88.6
5	5.10	5.10	79.5	132.00	23.30	31.70	77.6
6	3.00	2.50	32.9	77.50	20.40	18.50	46.2
7	4.50	4.80	61.4	130.40	15.10	12.20	156.6
8	91.00	9.40	1712.0	374.50	67.10	43.70	128.2
9	8.80	6.00	148.6	187.20	21.90	19.50	186.5
10	44.70	1.50	1264.0	395.00	53.00	186.00	306.0
11	20.10	13.00	516.0	165.00	44.50	13.90	93.2
12	14.40	3.70	324.2	261.80	22.40	25.40	146.1
13	25.50	1.00	462.5	353.10	21.10	46.10	222.2
14	10.41	10.90	244.7	133.70	36.90	15.30	86.1
15	97.00	8.50	2075.0	392.00	85.50	49.60	123.0
16	8.10	5.00	153.1	186.80	18.80	16.30	234.4
17	10.10	3.65	64.3	136.50	28.75	19.54	163.4
18	3.70	4.77	47.6	136.80	21.50	25.60	83.2
19	8.90	6.00	189.0	164.00	29.70	45.50	103.0
20	53.00	8.00	4240.0	390.00	70.60	31.40	312.0

Source: Lislevand, Figuerola and Szekely (2007) Ecological Archives
http://esapubs.org/archive/ecol/E088/096/metadata.htm

Using software:
In MINITAB use GRAPH>Matrix plot.

MINITAB & R:
matrices of
scatterplots
R: *Stars plot*

Working with **R**, *once the data file has been read with the command read.table,*
the matrix of scatterplots is produced with the command **pairs( )** *and the stars*
plots with **stars***. The data file birds20wn.dat has the names of the species in*
the first column. The other 5 columns are V1-V5 from Table 2.11

```
## reads the data file with several columns
birdswn<-read.table('e:/DATdatafiles/birds20wn2.dat')
birds<-birdswn[,2:6] ## creates object birds with numerical values
stars(birds,labels=birdswn[,1]) ## makes stars, puts labels
pairs(birds) ## produces matrix of scatterplots
```

2.14 One categorical variable

Categorical variables are those that are naturally expressed in words, not in numbers. Numerical codes can be used to represent the categories, but that does not make the variable quantitative. Some common codes are 1 for *yes* and 0 for *no*, 1 for *male* and 5 for *female*. Categorical variables can be ordinal (if there is a natural ordering among its categories, for example: *low, medium, high*) and nominal.

To tally data for one categorical variable, the two questions asked are:

- What are the categories? (there can be two or more categories)

- How many individuals or elements fall in each category?

Example 2.14.1: Round and sickle red blood cells

Normal red blood cells in humans have a round dimpled shape. In a blood disorder called 'sickle cell anemia,' some red blood cells adopt a 'sickle' shape. Assume that you take a blood sample from an individual and in a small unit of volume count how many red blood cells of each type you find. In this example the 'statistical unit' or 'individual' is the erythrocyte or red blood cell. The variable 'shape' is a categorical variable with two categories: *round* and *sickle*. Assume that there are 160 cells, 60 have the regular shape and **Table** 100 have the sickle shape (Table 2.12). **for one categorical variable**

Table 2.12: Red blood cells classified by shape

Shape	Frequency	Relative Frequency	Percent
round	60	0.375	37.5%
sickle	100	0.635	62.5%

A bar graph or a pie chart can be used to graphically display this type of data as shown in Figure 2.34. In the bar chart a bar is drawn for each category, with height proportional to the frequency of each category. We should not attempt to interpret 'the shape' of a bar chart as we did with histograms in an earlier section. This is due to the fact that when the variable **Never use** is nominal, the order in which the categories are displayed is arbitrary. **pie-charts**

In the pie chart, the idea is to divide the 360 degrees proportionally to **for quantitative** the frequencies of the categories. To draw a pie chart it is necessary to know **variables** the total number of individuals and how many fall in each category.

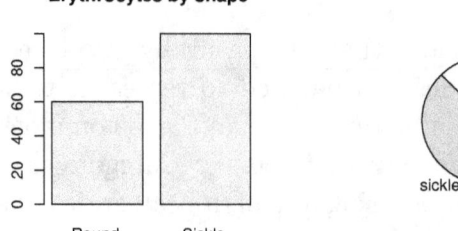

Figure 2.34: Bar and pie charts for the shape of red blood cells

R- Bar charts
Pie charts

Using statistical software:
The commands in **R** *to produce Figure 2.34 are listed below:*

```
par(mfcol=c(1,2))    ## 1 row and 2 columns of plots
g<-c(60,100)    ## enter frequencies
names(g)<-c("Round ", "Sickle ")
barplot(g, main= "Erythrocytes by shape ", col= "bisque")
pie(g,c("round","sickle"))
```

Example 2.14.2 Papilloma virus 18

A simple frequency table used in the study of DNA counts how many times each one of the nucleotides C, A, T, and G is present in a given sequence. The nucleotide frequency in one of the strands for the Human Papilloma virus 18, associated with cervical cancer, is displayed in Table 2.13. The sequence of nucleotides was obtained from the NCBI (National Center for Biotechnology Information) website (data file *hpv18.doc*).

GC content

Table 2.13: Nucleotide frequency in the HP virus 18

Nucleotide	Frequency	%
G	1680	21.3822
C	1497	19.0531
A	2365	30.1005
T	2315	29.4642

The percentage of G added to the percentage of C is known as the GC content. DNA with a higher GC content is more stable than DNA with higher percentages of A and T. The GC pair is bound by three hydrogen bonds, while AT pairs are bound by two hydrogen bonds. HPV 18 has a GC content of about 40% .

2.15 Two-way tables

The need for a two-way table usually arises in one of the following situations:

- Comparing two or more groups in terms of one categorical variable (Section 2.15.1)

- One group of individuals classified according to two categorical variables (Section 2.15.2)

- Paired categorical data (Section 2.15.3)

2.15.1 Comparing two or more groups

In experiments, such as the one in Example 2.15.1, two or more treatments are compared with regard to a categorical outcome. In observational studies, samples from several existing populations or sub-populations can be compared with respect to a categorical variable. For example, current smokers, former smokers and 'never smokers' (individuals who have smoked less than 100 cigarettes in their lifetime) can be compared with regard to having or not having lung cancer. Example 2.15.2 refers to an observational study.

A two-way table to compare groups with respect to one categorical variable

Example 2.15.1 Poliomyelitis vaccine

In well-planned experiments, each individual is randomly assigned to one of the treatments. The number of individuals in each treatment is known from the beginning. The individuals are then followed through time to observe the outcome. One very well-known example (Table 2.14) is the double-blind experiment with the first vaccine against poliomyelitis in 1954. This was a randomized experiment because children were assigned at random either to the vaccine (treatment) or placebo (control) group.

Poliomyelitis vaccine experiment

Table 2.14: Salk vaccine experiment in 1954

Group	Polio	No polio	Total
Placebo	142	199858	200000
Vaccine	56	199944	200000

Source: Meier(1972)

From the point of view of time this is a **prospective** study because either placebo or vaccine was given to each child and and the child was observed for a period of time to see if he/she developed polio. The total of each row in Table 2.14 was known from the beginning of the study. For more information on the experiment with the Salk vaccine see Meier (1972) or Meldrum (1998).

It is interesting to see what percentage of each group (placebo and vaccine) got polio. We calculate the percentages with respect to the row totals. The risk of getting polio for a child who did not receive the vaccine is:

$$142/200000 = 0.00071.$$

The risk of getting polio for a child who received the vaccine is:

$$56/200000 = 0.00028.$$

Relative risk: to compare two groups in an experiment

The **relative risk** is the ratio of the risk of getting polio for each group:

$$0.00071/0.00028 = 2.535714$$

The risk of contracting poliomyelitis was more than double for children who did not receive the vaccine.

Example 2.15.2 Understory vegetation in birds habitat

A researcher collected information about understory coverage, both in places known to be habitat for a species of bird and at some random places (Clark-Schubert, 2009); the data are in Table 2.15. Understory vegetation is the area of the forest that grows in the shade of the forest canopy. The percentage of area covered is a quantitative variable, but it is time consuming to do an exact measurement. The variable has been 'categorized' in the sense that instead of making an exact measurement of the coverage, intervals have been formed and the researcher writes down in which interval a particular location is instead of a precise value.

Table 2.15: 561 sites classified by type and percentage of understory vegetation cover

Site	0-20%	20-40%	40-60%	60-80%	80-100%	Total
Random	77	49	45	39	45	255
	30.20%	19.22%	17.65%	15.29%	17.65%	100%
Habitat	36	43	101	75	51	306
	11.76%	14.05%	33.01%	24.51%	16.67%	100.00%

Source: Clark-Schubert (2009)

The question is whether both types of sites (the sites used as habitat by this species of birds and other sites) have the same distribution in terms of percentage of understory vegetation cover. From Table 2.15 and Figure 2.35 it is clear that, among the habitat sites, the most common interval for cover is 40−60%; among the random sites, the most common interval is 0−20%. Sites favored by this specie of birds tend to have a higher understory vegetation coverage. Figure 2.35 displays the distribution of percentage of understory vegetation cover for the two types of sites using **clustered bar charts**.

Vegetation coverage

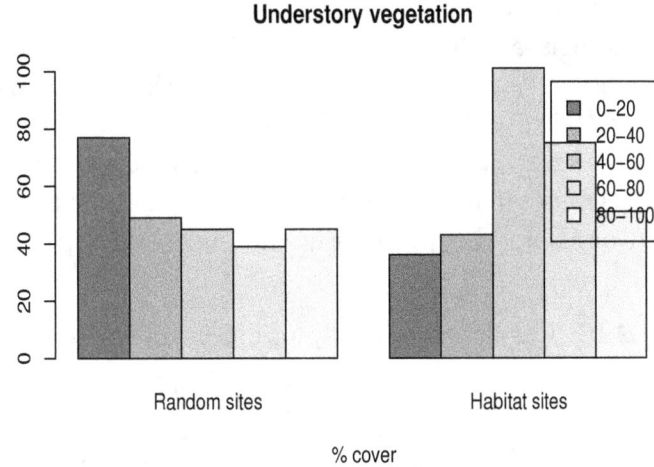

Figure 2.35: Cluster bars to compare random and habitat sites in terms of understory vegetation coverage

Using statistical software:
MINITAB has the option 'cluster' in GRAPH > Bar.
The following commands in **R** *read the data in Table 2.15 and produce the clustered bar charts in Figure 2.35*

MINITAB & R:
Clustered
bars chart

```
## clustered bars charts
## next we enter the rows  of data
A<-matrix(c(77,49,45,39,45,36,43,101,75,51),nrow=5,ncol=2)
## enter the names of the categories
rnames<-c('0-20 ', '20-40 ','40-60 ', '60-80', '80-100')
## this command prepares the plot
barplot(A,beside=TRUE, names=c('Random sites','Habitat sites'),
xlab='% cover',main='Understory vegetation',legend=rnames)
```

Example 2.15.3 Peptic ulcers and blood type

ulcers and blood type

Case-control studies are observational studies in which the groups are formed according to a current condition, and then each group is classified according to another categorical variable. A study was done on 521 patients with peptic ulcers (**case group**) and 680 individuals who do not have peptic ulcers (**control group**) (Clarke et al. 1959; Lange 2002). Both groups were classified with respect to their blood type (A, B, AB or O). Stacked bars charts and clustered bars charts are used in Figure 2.36 to display the data in Table 2.16. The totals of the columns are known from the beginning of the study because it is known how many people are in the case and control groups. Since there is a different number of individuals in the case and control groups, plotting the percentages instead of the frequencies facilitates the comparison of the distribution with regard to blood types within each group. The percentages have been calculated with respect to the totals of each column. It seems that in the 'no ulcers' group there is a larger presence of people with A or B blood type than in the 'ulcers' group.

Table 2.16: Case-control study of ulcers and blood type

Blood type	Ulcers		No Ulcers	
A	186	35.7%	279	41.03%
B	38	7.3%	69	10.15%
AB	13	2.5%	17	2.5%
O	284	54.5%	315	46.32%
Total	521	100%	680	100%

Source: Clarke et al. (1959) & Lange (2002)

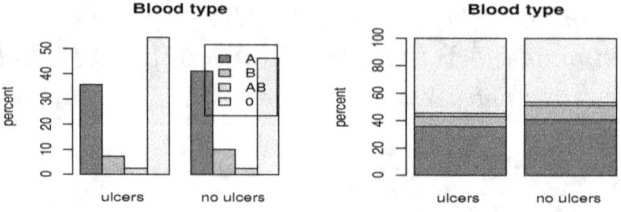

Figure 2.36: Clustered and stacked bar charts for ulcers and blood type

In case-control studies the question asked to the individuals in both groups can refer to something that happened in the past, so this is called a retrospective study. **The relative risk is NOT calculated for case-control or retrospective studies.**

Using statistical software:
MINITAB has the options 'cluster' and 'stack' in GRAPH > Barchart.
Next are the R commands that produced Figure 2.36

```
## clustered and stacked bars charts
par(mfcol=c(1,2))     ## 2 plots appear side by side
## next we enter the 2 columns of data or percentages
A<-matrix(c(35.7,7.3,2.5,54.5,41.03,10.15,2.5,46.32),nrow=4,ncol=2)
rnames<-c('A ', 'B ','AB ', 'O')
## the only difference is in the BESIDE option of the barplot
barplot(A,beside=TRUE, names=c('ulcers ', 'no ulcers '),
main='Blood type',ylab='percent',legend=rnames)
barplot(A,beside=FALSE,names=c('ulcers ', 'no ulcers '),
main='Blood type',ylab='percent')
```

2.15.2   One group and two categorical variables

In this case n individuals are classified according to two categorical variables. The total number of individuals n is known in advance but the totals of rows or columns are not known until the individuals are classified. The research question is whether the variables behave independently or if there is some sort of association between them. This type of two-way tables appears frequently in the analysis of surveys; several questions are asked and individuals are classified with respect to their answers to two of the questions.

A two-way table for one group and two variables

Example 2.15.4. Blood type

Blood type (A, B, AB, O) is determined by the presence or absence of two sugars (A and B) in the surface of red blood cells. Table 2.17 was prepared for a hypothetical sample of 10000 people based on information about the presence of the different types of blood in the U.S. population.

blood type

Table 2.17: Presence of sugars A and B for a hypothetical sample

	B present	B not present	TOTAL
A present	400(AB)	4200(A)	4600
A not present	1000(B)	4400(O)	5400
Total	1400	8600	10000

The odds ratio will be calculated to summarize the information in Table 2.17. Among those who have sugar A, the odds of having sugar B are:

$$400/4200$$

Among those who do not have sugar A, the odds of having sugar B are:

$$1000/4400$$

Next the ratio of those odds is calculated:

$$(400/4200)/(1000/4400) = (400 \times 4400)/(1000 \times 4200) = 0.4190476$$

Thus, the odds of having sugar B are about half for those with sugar A than for those without sugar A (an odds ratio of 1 would mean that the variables behave independently).

Odds ratios are used for cross-sectional and retrospective studies

Studies in which individuals are studied in a given time, neither followed into the future as in prospective studies, nor classified based on their current status and asked for something in their past as in retrospective studies, are called **cross sectional**. In cross sectional studies, a random sample of individuals is selected, and questions are asked or variables are measured at that time. When two categorical variables are being analyzed together, the statistic to summarize the information in the resulting table is the **odds ratio**.

2.15.3   Paired categorical data

Paired data: Each individual is measured twice or individuals come in pairs

A case of paired observations can also happen in the context of categorical variables when the same categorical variable is observed twice (such as 'pre-post' studies) for the same individual or once for each element of a pair of individuals. Some examples are:

- Two different physicians evaluate the same group of patients. Each doctor can classify each patient as having or not having schizophrenia. The doctors may or may not agree in their diagnoses.

- An osteoporosis diagnosis (yes or no) is made for each hip (left and right) for each woman in a sample.

- A quiz is given to a group of students before and after being assigned a reading. They are given a pass/fail grade each time.

Example 2.15.4 Osteoporosis in each hip

Table 2.18 displays the diagnosis of osteoporosis (yes or no) in the left and right hips of a group of 3012 women. The percent of women who have the same diagnosis in both hips can be calculated by adding the numbers in the diagonal:

$$(2340 + 455)/3012 = 0.9279548$$

Thus, 92.8% of the women have the same status (either osteoporosis or no-osteoporosis) in both hips. Also, it could be of interest to observe if the prevalence of osteoporosis is similar for the right hip than for the left hip. This type of analysis is called the 'analysis of symmetry.' With that purpose, the marginal frequencies can be compared: 586/3012 for the left hip and 541/3012 for the right hip.

Table 2.18: 3012 white women 50 years or older classified by osteoporosis of the hips

	Right hip		
Left hip	No	Yes	Total
No	2340	86	2426
Yes	131	455	586
Total	2471	541	3012

In the case of a diagnosis done by two physicians, it might be of interest to calculate the percent of patients for which the diagnoses done by both physicians coincide. This type of analysis is called sometimes **rater's agreement** and the analysis is focused on comparing the sum of the frequencies in the diagonal with the total number of individuals. **Rater's agreement**

2.16 Beyond Biology

The statistical graphs and statistics learned in this chapter are extensively used in practically all the areas to which Statistics is applied. Two-way tables are used in the analysis of survey data to compare the answers to two different questions in social studies. Time series plots are useful in the analysis of financial and meteorological data. Survival data are of interest in industrial studies related to failure of components.

2.17   Exercises

2.17.1   Review questions

1. For data of what type of variables (quantitative or categorical) are boxplots, dotplots, stemplots, and histograms considered appropriate statistical graphs?

2. For data of what type of variables (quantitative or categorical?) are bar-charts and pie-charts considered appropriate?

3. Which statistics form the five-number summary?

4. Comparing the mean and the median:
   A) Which is more sensitive to outliers?
   B) Which is likely to be larger in a skewed to the right distribution?
   C) Which one would you use in the case of censored data?

5. A random sample of 5 cells from the apical region of a single onion root was selected. Their lengths (in millimeters) are:

   0.015, 0.028, 0.023, 0.030, 0.029.

   A) What is the median?
   B) What is the mean?
   C) Which is lower, the mean or the median? Why?
   E) Is there a mode?
   F) Calculate the standard deviation.
   G) Why is the standard deviation not 0?

6. In Figure 2.4b:
   A) Identify the minimum, maximum, median, lower and upper quartile.
   B) Interpret the value of the upper quartile.
   C) How large should an observation be in order to be considered an outlier?

7. What is the first thing you should do when finding an outlier?

8. What might be the reason for obtaining a bimodal distribution if we take a random sample of cells from an entire onion root and measure their length?

9. Assume the mean of a sample is 8, the median is 5 and the standard deviation is 2. What would be the new mean, median, standard deviation and variance if the number 4 is added to each one of the observations?

10. Assume that the mean and standard deviation of the heights of 20 individuals are 68 inches and 10 inches, respectively. What would be the mean and standard deviations in centimeters?(1 inch=2.5 centimeters)

11. Assume that the mean and standard deviation of the daily temperature (F) at noon in a city during the month of January is 42 degrees and 10 degrees respectively (Fahrenheit). What would be the mean and standard deviation in the Celsius scale? (C=(F-32)5/9)

12. The correlation between the mean daily temperature (F) and the daily precipitation in inches during 120 days is -0.28. What would be the correlation if the temperature is expressed in Celsius and the precipitation in centimeters?

Review questions for Chapter 2

13. Somebody is trying to compare the variability in temperatures at noon during the month of March in Bozeman, MT, and Miami, FL, places known for having very different average temperature. Which statistic do you recommend to use in order to do the comparison?

14. Somebody is studying the number of bacterial colonies in each one of 200 samples of frozen fish. Most of the fish is in good condition and has very few bacteria, but a few samples are highly contaminated. Sketch the distribution. Now, think of the distribution of the logarithm of the data (of course, to apply logarithms we have to make sure that none of the data is 0 because ln(0) is not defined). Do you think that the distribution of the logarithm will be more skewed or less skewed?

15. What statistical graph do you typically use to compare two groups with regard to a quantitative variable?

16. What statistical graph do you typically use to compare two groups with regard to a categorical variable?

17. Assume that you are studying the relationship between two quantitative variables (X and Y) and the correlation (Pearson's correlation) is r=-0.92.
A) What is that value telling you?
B) Does that mean that there exists a cause-effect relationship between X and Y?

18. Somebody is analyzing the data of a survey in which two questions are: How many cigarettes per day do you smoke? and What is your gender? (code used: males $=1$, females $=2$) . This person wants to calculate Pearson's correlation coefficient between *number of cigarettes* and *gender*. What would you tell this person? What statistic would you recommend instead?

19. Data has been collected about height, arm span, foot length, hand span and head circumference for a group of 300 ten-year old males. We want to have an overall picture of how these variables relate to one another for this group of children. What statistical graph do you recommend?

20. The monthly precipitation in inches is available from the meteorological station in Morristown, TN, for the past 40 years. Which statistical graphs can be used to display that data set? (mention 2 graphs)

Review questions for Chapter 2

21. Calculate the relative risk of coronary heart disease for women in the treatment group as compared with the women in the placebo group using the data of Exercise 1.2.

22. Two questions in a survey were 'What is your gender?' and 'Have you smoked marijuana at least once in your life?' Based on the two-way table below calculate the odds ratio of having smoked marijuana for males as compared to females.

	marijuana		
	YES	NO	All
MALE	70	130	200
FEMALE	77	226	303
All	356	147	503

23. Explain why we should not describe and interpret the shape of the distribution in Figure 2.36a but why we could allow a description of the shape of the distribution in Figure 2.35a.

2.17.2   Lab on Chapter 2

Lab on Chapter 2- R version

R commands that will be useful for this lab are: **mean(x), boxplot(x), summary(x), hist(x), cor(x,y), plot(x,y), boxplot(x,y)**. The 'x' and the 'y' can be replaced by the name given to the variables. Three different ways of reading data will be used:**x<-c()** to type the data, **x<-scan()** to read data files with a single column and **x<-read.table()** to read files with several columns. To copy figures to your report, right click on the figure and choose the option 'copy as bitmap'

1. **How fast are these students?** In an Intro Stats class, I asked the students to measure their reaction time (milliseconds) using a reaction time ruler.

   a) Type in the data using the command: time<-c(130,143,150,150,150, 150,155,160,165,170,170,180,190,190,190,200,235,250)

   b) Is there a student that is unusually faster or slower than the other students? Produce a boxplot. Do you see an outlier?

   c) Calculate the mean and the five-number summary. **Interpret** the values of the mean, minimum, maximum and median.

   d) Produce a histogram. Describe and interpret the shape of the distribution.

2. **How long do flies live?** What type of flies? Under what conditions? The datafile *flylife.dat* contains the length in days of the lives of 200 female flesh flies of the species *Sarcophaga crasipalpis* reared in a cage with female flies only. This is the tabulated version of the data:

Days	2	3	5	6	7	8	9	10	11	12	13	14	15
Flies	1	2	2	1	5	1	4	2	5	4	3	7	6

Days	16	17	18	19	20	21	22	23	24	25	26	27	30+
Flies	6	18	11	18	31	21	17	16	7	6	1	3	2

   The last two observations are reported as 30+ because the study lasted one month and those two flies were still alive when the study ended. In this type of situation it is said that the data are censored.

   a) Assuming the data file *flylife.dat* is in the e: drive, read the data file using the command **x<-scan('e:flylife.dat')**

   b) Produce a histogram (insert it into your report). Describe and interpret the shape of the distribution

   c) Which statistic would be more appropriate to calculate in this case? Mean or median? Why?

   d) Calculate and interpret the median and quartiles.

3. **Do bird species with larger females tend to lay larger eggs?** Is there a relationship between the typical mass of the egg and the typical female mass for the 20 species of birds from different families in Table 2.11? Is it linear? If so, how strong is that linear relationship? The typical egg mass is reported in Table 2.11. For the same species, the female mass (grams) are:
99.3,690,61.2,105.5,74.0,29.8,61.5,1387,118.1,1059,
439,299.3,405.4,233.7, 1650,154.3,69,47.6,198,1985.
a) Type in the data (femmass<-c(99.3,690....), eggmass<-c(6.6,51,...)
b) Produce a scatterplot with female mass in the horizontal axis and egg mass in the vertical axis.
  c) Calculate the correlation between egg mass and female mass. Write a short paragraph answering the research question and interpreting the value of r.

LAB
for Chapter 2
R version

4. **Pre-operative hemoglobin and the need for transfusion during or after surgery.** Hemoglobin (Hb) is a protein that contains iron and accounts for approximately 35% of the content of the red blood cells. Its mission is to transport oxygen to the tissues and CO_2 to the lungs. Some patients need a blood transfusion during or after surgery and some do not. Is there a difference in pre-operative hemoglobin between the patients who later need transfusions and those who don't? The data file *HbTransfusion60* contains information for 60 patients who underwent the same type of surgery; 30 needed transfusion and 30 didn't. The first column of the data file contains the values for hemoglobin (Hb), the second column contains an indicator variable: 0=no transfusion, 1=transfusion. The data can be read using the commands: data<-readtable('e:HbTransfusion60.dat')
Hb<-data[,1]
Trans<-data[,2]

a) Prepare side-by-side boxplots using the command **boxplot(Hb~ trans)**
b) Calculate some basic statistics for each group (mean, standard deviation, coefficient of variation, five-number summary).
c) Write a paragraph commenting on any difference you see in pre-operative hemoglobin between the patients who needed transfusions and those who didn't. Which group has higher median hemoglobin? Does one group show more variability than the other? Do all patients in the transfusion group have less hemoglobin than any patient in the non-transfusion group?
d) Explain why we should not calculate the correlation between transfusion and hemoglobin.

Lab on Chapter 2- MINITAB version

Numerical output from MINITAB can be highlighted, copied and pasted into a document. To copy figures, use EDIT>Copy graph.

LAB for Chapter 2 MINITAB version

1. **How fast are these students?** In an Intro Stats class, I asked the students to measure their reaction time (milliseconds) using a reaction time ruler.

   a) Type the data in a column of the worksheet: 130, 143, 150, 150, 150, 150, 155, 160, 165, 170, 170, 180, 190, 190, 190, 200, 235, 250

   b) Is there a student that is unusually faster or slower than the other students? Produce a boxplot (GRAPH>Boxplot). Do you see an outlier?

   c) Calculate the mean and the five-number summary. In the menu, select: STATS>Basic Statistics>Display basic statistics

   Select the column in which you typed the data and click on the button *Statistics*, select the mean and the five statistics that form the five-number summary.

   Interpret the values of the mean, minimum, maximum and median.

   d) Produce a histogram. Describe and interpret the shape of the distribution.

2. **How long do flies live?** What type of flies? Under what conditions? The datafile *flylife* contains the length in days of the lives of 200 female flesh flies of the species *Sarcophaga crasipalpis* reared in a cage with female flies only. This is the tabulated version of the data:

Days	2	3	5	6	7	8	9	10	11	12	13	14	15
Flies	1	2	2	1	5	1	4	2	5	4	3	7	6

Days	16	17	18	19	20	21	22	23	24	25	26	27	30+
Flies	6	18	11	18	31	21	17	16	7	6	1	3	2

   The last two observations are reported as 30+ because the study lasted one month and those two flies were still alive when the study ended. In this type of situation it is said that the data are censored.

   a) Open the data file *flylife*

   b) From the menu, select GRAPH>Histogram to produce a histogram (insert it into your report). Describe and interpret the shape of the distribution

   c) Which statistic would be more appropriate to calculate in this case? Mean or median? Why?

   d) Calculate and interpret the median.

   e) Calculate and interpret the two quartiles.

3. **Do bird species with larger females tend to lay larger eggs?** Is there a relationship between the typical mass of the egg and the typical female mass for the 20 species of birds from different families in Table 2.11? Is it linear? If so, how strong is that linear relationship? The typical egg mass is reported in Table 2.11. For the same species the female mass (grams) are:
99.3,690,61.2,105.5,74.0,29.8,61.5,1387,118.1,1059,
439,299.3,405.4,233.7, 1650,154.3,69,47.6,198,1985.
a) Open the data file *birds20*
b) Produce a scatterplot (GRAPH>Scatterplot) with female mass in the horizontal axis and egg mass in the vertical axis.
   c) Calculate the correlation (STAT> Basic Statistics > Correlation) between egg mass and female mass. Write a short paragraph answering the research question and interpreting the value of r.

4. **Pre-operative hemoglobin and the need for transfusion during or after surgery.** Hemoglobin (Hb) is a protein that contains iron and accounts for approximately 35% of the content of the red blood cells. Its mission is to transport oxygen to the tissues and CO_2 to the lungs. Some patients need a blood transfusion during or after surgery and some do not. Is there a difference in pre-operative hemoglobin between the patients who later need transfusions and those who don't? The data file *HbTransfusion60* contains information for 60 patients who underwent the same type of surgery; 30 needed transfusion and 30 didn't. The first column of the data file contains the values for hemoglobin (Hb), the second column contains an indicator variable: 0=no transfusion, 1=transfusion.
a) Prepare side-by-side boxplots for Hb for the two groups.
b) Calculate some basic statistics for each group (mean, standard deviation, coefficient of variation, five-number summary).
c) Write a paragraph commenting on any difference you see in pre-operative hemoglobin between the patients who needed transfusions and those who didn't. Which group has higher median hemoglobin? Does one group show more variability than the other? Do all patients in the transfusion group have less hemoglobin than any patient in the non-transfusion group?
d) Explain why we should not calculate the correlation between transfusion and hemoglobin.

2.17.3 Exercises for discussion or homework

All these exercises require the use of a computer. They can be worked with either R or MINITAB or some other statistical software.

Exercises for class discussion or homework in Chapter 2

1. **Pulse rate of 210 college students** The data file **pulserate** contains the self-reported pulse rate of 210 students registered in several sections of an introductory statistics course and described in Example 2.5.1. Complete the description by calculating AND interpreting the five-number summary. Produce a boxplot to check if there is any outlier. How many times is the standard deviation contained in the range of the distribution? What percent of the observations are within one standard deviations from the mean? What percent of the observations are within two standard deviations from the mean?

2. **Age of respondents in a survey** The data file **age** contains the self-reported age of 503 respondents to an alcohol and illegal drug survey conducted in 1996 (Example 2.5.2). Calculate AND interpret the five-number summary. Are the two quartiles at the same distance from the median? Why? Calculate the mean of the data. Assume that all those individuals are still alive. What would be their mean age if the survey had been conducted this year?

3. **Is there a relationship between body mass index (BMI) and hemoglobin?** If a person has low values of hemoglobin (the threshold varies between 11 and 13 g/dl depending on gender and age group), the individual is said to have anemia. BMI is an index of body mass and is calculated based on the height and weight of a person. From a data base of patients planning to undergo a certain type of surgery, a random sample of 30 individuals was selected. Their BMI and Hb are listed below (datafile *HbBMI30*). Produce a scatterplot and calculate the correlation coefficient. Interpret the results.

Ind.	1	2	3	4	5	6	7	8	9	10
Hb	12.4	11.0	12.4	14.2	12.7	13.7	11.4	11.8	13.6	13.1
BMI	31.0	38.8	38.1	31.2	23.8	37.5	30.2	40.7	35.9	34.0

Ind.	11	12	13	14	15	16	17	18	19	20
Hb	12.5	13.7	14.3	11.3	9.9	11.4	13.7	14.5	13.8	10.4
BMI	25.3	30.4	24.1	47.8	18.5	27.0	27.3	35.9	28.4	28.3

Ind.	21	22	23	24	25	26	27	28	29	30
Hb	3.7	13.2	14.1	12.8	13.2	13.7	14.1	13.2	12.1	14.2
BMI	25.5	39.1	22.4	26.6	32.9	38.3	42.6	44.2	32.2	24.0

Note: If using R, assuming the file HbBMI30.dat is in drive e:, the data file can be read using the commands:
data<-readtable('e:HbBMI30.dat')
Hb<-data[,1]
BMI<-data[,2]

Exercises for
class discussion
or homework
in Chapter 2

4. **Does distraction make you slower?** The following data were collected by a fifth grader for her science fair project (Price, 2008). She measured the reaction time, using a reaction time ruler, of 28 children (all 10 & 11 years old fifth graders, 14 boys and 14 girls) under two different conditions: quiet time and surrounded by distractions. The order of the treatments (quiet or distraction) was decided at random for each participant. Before the measurements took place, a practice period with the reaction time ruler was allowed for each child. The data are:

				Girls			
ID	1	2	3	4	5	6	7
Quiet	170	220	295	190	130	280	290
Dist.	370	280	205	200	130	230	335
ID	8	9	10	11	12	13	14
Quiet	300	275	253	330	230	245	240
Dist.	300	250	240	310	250	270	330

				Boys			
ID	15	16	17	18	19	20	21
Quiet	180	270	180	230	260	230	250
Dist.	210	260	250	360	300	300	270
ID	22	23	24	25	26	27	28
Quiet	160	220	210	150	250	170	300
Dist.	110	300	260	190	260	220	360

a) Use side-by-side dotplots or boxplots and descriptive statistics to compare the female 5th graders during quiet time with the female college students in question 1 of the Lab. Interpret the graph and calculations. b) Calculate the differences between the reaction time under distraction and during quiet time for the fifth graders (boys and girls). Prepare a plot with the differences. Calculate the mean and median of the differences. Interpret the graph and calculations c) Compare the difference in reaction time (distraction - quiet) for boys and girls. Prepare a plot, calculate statistics and interpret your results.

5. **Do species with smaller eggs tend to lay more eggs?** For the 20 species of birds in Table 2.11, is there any type of relationship between the clutch size and the egg mass? Produce a scatterplot and calculate the correlation to answer this question.

6. **Is there an association between the leg length of a cricket and the distance the cricket jumps?** A group of students measured the length of the hind leg (mm) for a sample of 20 crickets. When the crickets jumped, the students measured the distance (mm) they jumped.
Length of leg: 10,16,17,9,19,18,17,19,18,18,10,13,11,11,11,8,11,12,14,13
Distance jumped: 160,190,180,310,210,350,240,170,360,300,160,140, 180,130,200,120,130,230,350,250
Is there any relationship between the leg length and the distance jumped? Produce a scatterplot and calculate the correlation to answer this question. Assume that the students were not too precise in measuring the distance jumped. What correlation would you prefer to calculate, Pearson's r or Spearman's rank correlation?

7. **Do larger red blood cells tend to have larger nucleus?** In birds (but not in humans) red blood cells, as most cells, have a nucleus. Do larger cells tend to have a larger nucleus? Consider the data set corresponding to different species of birds in the family Emberizidae. In the Cell Size Database in Gregory (2005) the following information is found for typical 'cell short diameter' and 'nucleus short diameter' (both in μm). If working with **R**, use the following commands to input the data. Produce a scatterplot and calculate the correlation.

```
cell<-c(6.2,6.0,6.7,7.4,6.0,6.4 ,5.9,5.9,6.1,6.2,6.5,6.1,6.2,
6.6,7.2,7.5,5.3,6.0,5.9,6.3,7.0,6.9,5.1,6.1)
nucleus<-c(2.2,2.1,2.7,3.2,2.1,2.1,2.3,2.4,2.3,2.2,2.7,2.4
,2.4,2.4,3.0,2.9,2.0,2.3,2.2,2.4,2.3,2.4,2.1,2.1)
plot(cell,nucleus)
n<-length(cell)   ## to count the number of observations
((n-1)/n)*cor(cell,nucleus)
```

Note: The correlation is being multiplied by (n-1)/n, so that the denominator in the correlation is **n** *instead of* **n-1** *because the data correspond to the whole population of species of that family, not a random sample. What do you think? Is there a strong linear relationship? Is it a perfect relationship? Be aware that any conclusion you extract is just for the family Emberizidae since all your data come from that family.*

8. **What is more correlated with the height of a person, foot length or hand span?** The following data correspond to 23 young female college students. What would be your answer to the research question based on this data set? Produce the scatterplots, calculate the correlations, and include a brief description of the distribution of each one of the variables involved.

   height: 63, 66, 67, 61.5, 65, 67, 64, 64.8, 66.5, 64.5, 67, 68, 67, 63, 65, 63, 66.5, 66, 65.5, 63, 65, 68, 62

   foot length: 8, 8, 9.5, 8, 7.1, 8.5, 8.35, 8, 10.1, 8.5, 9.5, 7.5, 8.1, 8, 9.5, 7, 9.50, 9.50, 9.50, 8.5, 8, 9, 7

   hand span: 8, 7.5, 7.5, 6.2, 6.2, 8, 6.40, 7, 8.4, 8.5, 8.25, 7.4, 6.9, 7.5, 8.5, 6, 8.5, 7, 7.25, 7.5, 7, 8, 7.5

9. **Petiole length and leaf blade in sugar maple trees.** The petiole is the stem of the leaf. We wonder if there is a relationship between how heavy the blade part of the leaf is and how long the petiole is. Of course, if there is a relationship, it might vary from species to species. A group of students randomly selected 20 leaves from the sugar maple trees on campus, then went to the lab and measured the length (cm) of the petiole, and the mass (g) of the blade. Is there a relationship between the length of the petiole and the mass of the leaf? How strong is the linear relationship? Produce a scatterplot and calculate the correlation to answer the question. Either type the data or use the data file *sugarmaple*.

Exercises for class discussion or homework in Chapter 2

   petiole length (cm): 6.2, 3.6, 5.3, 5.6, 5.2, 6, 4.2, 6.1, 5.3, 6.9, 4.3, 4.8, 5.5, 5, 5.2, 4.9, 5.5, 4.6, 5.9, 6.4

   leaf blade mass (g): 0.587,0.235,0.480,0.738,0.587,0.793,0.401,0.611,0.617, 0.578,0.373,0.327,0.281,0.528,0.399,0.543,0.413,0.323,0.544,0.577

2.17.4   Small data analysis projects

The data analysis projects require the use of computers. Instructions are provided to work with R but other statistical software could be used. However, the 'stars plot' is not available in most statistical software.

1. **The Phasianidae family.** Hens, wild turkeys, partridges and quails belong all to the Phasianidae family, family # 8 in the database published by Lislevand et al. (2007). From the 72 species in the family, 28 species with no missing data were selected. The data file *hens* contains

the typical mass, and length of wing, tarsus and tail for females, and also the typical clutch size and egg mass.

Use plots and statistics to discuss the association, or lack of, between mass of the female, egg mass and clutch size.

Calculate and analyze the values of the variable $X = \frac{egg\ mass}{female\ mass}$.

Produce plots and calculate statistics for each one of the variables. Write a paragraph on the diversity among species of the same family and the presence of outliers.

Produce scatterplots and calculate correlations to explore the relationship between length of the wing, tarsus, and tail with the mass of the female. Which of those 3 variables seem to be more strongly associated with female mass? Is the relationship linear?

If working with **R**, produce the stars plots and comment about the different types of shapes you see. Select one shape and identify the species that share a similar shape in the stars plot. Sharing the same shape means that the proportions between mass, wing, tarsus and tail are similar.

Write a report or prepare a poster with your findings.

If working with **R**, you can read the data file *hens*, assuming it is in drive e: (in the folder 'DATdatafiles'), using the following commands:

```
## reads the data file
hens<-read.table('e:/DATdatafiles/hens.dat',header=FALSE)
## creates objects with each variable
species<-hens[,1]; mass<-hens[,2]; wing<-hens[,3];
tarsus<-hens[,4]; tail<-hens[,5]; clutch<-hens[,6];
eggmass<-hens[,7]
## create a data frame with some variables for later use.
body<-data.frame(mass,wing,tarsus,tail)
stars(body)
```

2. **The Lybiidae family.** The data file *LybiidaeMales* was prepared with data from the data base by Lislevand et al. (2007). There are about 40 species in that family, only 23 without missing data for the males are included. Use the methods learned in this chapter to answer the following questions:

Describe the distribution of the mass of the males. Calculate and interpret the minimum, median and maximum mass of the 23 species.

The size of which of these parts of the body: tarsus, bill, wing or tail, is more strongly related to the typical mass of the species?

Data analysis projects for Chapter 2

Is there a relationship between the length of the tail and the length of the bill?

Obtain a star plot. Do all the stars look similar? Do you see different groups of species with similar stars?

Write a report or prepare a poster with your findings.

If working with **R** you can read the data file *LybiidaeMales.dat*, assuming it is in drive e: (in the folder 'DATdatafiles'), using the following commands:

```
## reads the data file
birds<-read.table('e:/DATdatafiles/LybiidaeMales.dat',
header=FALSE)
## creates objects with each variable
species<-birds[,1]; mass<-birds[,2]; tarsus<-birds[,3]
bill<-birds[,4]; wing<-birds[,5]; tail<-birds[,6]
## forms a data frame that does not include the species number
body<-data.frame(mass,tarsus,bill,wing,tail)
## now the data frame can be used to produce stars or a matrix
of scatterplots
```

3. **The Falconidae family.** The data file *falcons* contains information about male and female typical mass, length of wing and length of tail. It also contains data for the typical clutch size and typical egg mass for the species of the Falconidae family. Explore the distribution of each variable and the possible associations among variables. Write a report.

4. **Comparing 5 species of trees** A team of students in a freshman biology course collected random samples of leaves from campus. They collected a total of 105 leaves of 5 different species (1=White Oak, 2=Ginko, 3=Sycamore, 4=Sugar Maple, 5=Tulip Poplar), 25 leaves of the first species and 20 from each one of the other four. The datafile *fiveleaves* contains the observations for length of the leaf, maximum width of leaf, length of the petiole (all in centimeters) and mass (g) of the leaf.

Produce plots and calculate statistics that will enable you to compare the leaves of the five species in terms of length, width, petiole and mass.

Produce plots and calculate statistics that will enable you to compare the species in terms of the relationship between length and width of the leaf.

Produce plots and calculate statistics that will enable you to compare the species in terms of the relationship between length of the petiole and mass of the leaf. (This is of interest because the petiole supports the leaf.)

Considering all the species together, which are the two variables that are more strongly correlated?

Write a report summarizing your findings; include the plots and statistics you consider necessary.

If working with **R**, the following commands will be useful to read the data (assuming *fiveleaves.dat* is in drive e: in the folder 'DATdatafiles') and to name each variable:

```
mydata<-read.table('e:/DATdatafiles/fiveleaves.dat',
header=FALSE)
species<-mydata[,1]; length<-mydata[,2]
width<-mydata[,3] ; petiole<-mydata[,4]
mass<-mydata[,5]
```

Data analysis projects for Chapter 2

5. **Exploring the bone mineral density in 8 bones**

Osteoporosis diagnosis is done based on an standardized score (T-score) that is calculated as $T = \frac{BMD - MeanBMD}{StDevBMD}$, where BMD is the bone mineral density in a region of a bone of the individual, and meanBMD and StDevBMD are the mean and standard deviation of the BMD of individuals of the same gender and race in the 20-24 age interval. In this example you will analyze the BMD (not the T-scores) of a sample of 108 white women 50 years or older. The data are in the data file *BMD*. The data file contains information about age and bone mineral density in each one of the following regions for each one of the 108 women: 4 lumbar regions (L1-L4), left femoral neck, left total hip, right femoral neck and right total hip.

Describe the sample in terms of age.

Produce a plot and calculate basic statistics for each one of the 4 lumbar regions. Do you see any pattern in the means of the 4 regions?

Do people have exactly the same BMD in both sides? Calculate the differences between left and right hip, and between left and right femoral neck. Obtain histograms for those differences. Comment.

Discuss the relationship between the BMD of the left and right femoral neck. Do you see any outlier?

Discuss the relationship between the BMD of the left and right hip.

Which is the region that has the strongest relationship with age? How strong is that relationship?

Low bone mineral density is associated with risk of fracture. Overall, which is the region with the lower average BMD?

Write a report summarizing your findings, include plots and calculations.

Data analysis projects for Chapter 2

If working with **R** you can read the data file *BMD.dat*, assuming it is in drive *e* : in the folder *DATdatafiles*, using the following commands:

```
bones<-read.table('e:/DATdatafiles/BMD.dat',
header=FALSE)
age<-bones[,1]
lumbar1<-bones[,2];   lumbar2<-bones[,3]
lumbar3<-bones[,4];   lumbar4<-bones[,5]
LFNeck<-bones[,6];    LTotHip<-bones[,7]
RFNeck<-bones[,8];    RTotHip<-bones[,9]
```

6. **Marijuana** The data file *drugsurvwg* contains information for a few variables and a sample of 503 individuals interviewed by phone around 1996 in an illegal drug and alcohol survey in a state of the Midwest.

   (a) Is this an observational or experimental study? Is this a prospective, retrospective, or cross-sectional study?

   (b) Prepare two-way tables for the following pairs of variables:

       i. **Gender and marijuana.** Calculate the odds ratio comparing the odds of using marijuana for men and women.

ii. **Age group and use of marijuana.** Calculate row percentages. Do you see any pattern for the % of people who have used marijuana with respect to age group?

iii. **Having ever being an smoker and having used marijuana.** Calculate the odds ratio to compare the odds of having used marijuana for smokers vs. non-smokers.

References

Clark-Schubert, N.D. (2009) *Fall Migration Ecology of the Sora (Porzana carolina) at Four Rivers Conservation Area in Missouri.* Master Thesis, University of Arkansas.

Clarke, C.A., Price, D.A., McConnell, R.B., and Sheppard, P.M. (1959) Secretion of blood group antigens and peptic ulcers. *British Medical Journal* **1(5122)**: 603607.

Delk, K.L. (2009) *Effects of Cherries on Arthritis Pain.* Undergraduate honors thesis. East Tennessee State University.

Desper, R., Khan, J. and Schaffer, A.A. (2004) Tumor classification using phylogenetic methods on expression data*Journal of Theoretical Biology* **28(4)**: 477-96.

Gregory, T.R. (2005) Cell size data base *http://www.genomesize.com/cellsize*

Lange, K. (2002) *Mathematical and Statistical Methods for Genetic Analysis* 2nd edition. New York: Springer Verlag.

Lislevand, T., Figuerola,J. and Szkely, T. (2007) Avian body sizes in relation to fecundity, mating system, display behavior, and resource sharing. *Ecology* **88**: 1605.

Kotz, S. and Seier, E. (2008) Visualizing Peak and Tails to Introduce Kurtosis *The American Statistician* **62**: 348-354

Meier, P. (1972) The Biggest Public Health Experiment Ever: The 1954 Field Trial of the Salk Poliomyelitis Vaccine, in *Statistics a guide to the unknown* ed. Judith Tanur et al. pp 2-13. San Francisco: Holden Day

Meldrum, M. (1998) A Calculated Risk: the Salk Polio Vaccine Field Trials of 1954 British Medical Journal **317**: 1233-1236.

Prentice, R.L. et al. for the Women's Health Initiative Investigators (2005) Combined Postmenopausal Hormone Therapy and Cardiovascular Disease: Toward Resolving the Discrepancy between Observational Studies and the Women's Health Initiative Clinical Trial. *American Journal of Epidemiology* **162**: 404-414.

Price, J.N. (2008) Science fair project. Johnson City School District.

Ramsey, F.L. & Schaffer, D.W. (1997) *The Statistical Sleuth.* Pacific Grove: Duxbury

Shacher, G.A. & Staffeld, E.F. (1974) Relation of Gestation Time to Brain Weight for Placental Mammals: Implications for the Theory of Vertebrate Growth *American Naturalist* **108**: 593615.

Spector, W.S. (1956) *Handbook of Biological Data.* Philadelphia: Saunders

Youatt,W.G., Fay,L.D., Howe,D.L. and Harte,H.D. (1961) Hematological data in some small mammals. *Blood* **18**: 758-763.

Chapter 3

Inference by randomization

Populations are generally too large to be practical to study each individual. Treatments cannot be applied, for either practical or ethical reasons, to every single individual in the population. That is why random samples are selected from populations and experiments are conducted with relatively few subjects. The process of making statements about one or more populations, based on samples, or about the superiority of one treatment over another, based on the results of an experiment, is called statistical inference. On the other hand, if the value of a variable is known for each element of the population, it is not necessary to apply methods of statistical inference.

Two important procedures in statistical inference are:

- **ESTIMATION**: *to provide a value or an interval of values for the unknown value of a parameter such as a population mean or proportion.*

- **HYPOTHESIS TESTING**: *to ascertain whether the data constitute evidence (or not) against a null statistical hypothesis written in terms of population parameters.*

If the value of the variable is known for each individual in the population, then there is no need to apply statistical inference

There is more than one approach to statistical inference. In this chapter inference will be made by using randomization methods. Other cases will be studied in chapters 5, 10 and 11. Two important methods involving randomization are the bootstrap to build confidence intervals and the randomization or permutation tests to compare two treatments or populations.

Randomization methods are becoming popular in biological applications. The bootstrap method is also applied to phylogenetic trees.

3.1   The randomization test

There are randomization tests for different types of null hypotheses. We will focus on the randomization test to compare two groups of individuals with regard to a quantitative variable. This test is of special interest because frequently a scientist needs to compare treatment vs. control by performing an experiment. We are assuming that the data come from two independent random samples selected from two populations, or two groups of individuals that have undergone two different treatments. In the case of experiments, we are assuming the subjects were randomly assigned to one of the treatments.

Sometimes the randomization test is applied to two groups of data that do not correspond to random samples, but in that case the results are valid just for the data at hand; no inference for a larger population is drawn. In those cases, the research question to be answered with the randomization test is whether there is a pattern or if the observed differences between the two groups might have happened by chance alone.

Example 3.1 Is phytate bad for chickens?

Chickens and phytate

High values of mucin in the excreta of chickens indicates that the chickens are losing endogenous amino acids which, of course, is not good for them. Diet might alter the levels of mucin in the excreta. Phytate is a form of phosphorous contained in many plants, especially in some grains and seeds. Consider the following research question: Does phytate produce an increase in the average amount of mucin in the excreta of chickens? An experiment to answer that research question was performed by Onyango et al. (2009).

Sixteen chickens were placed in individual cages, and each one was randomly assigned to either the treatment or the control group. They were fed the same diet except that the treatment group had additional phytate added to it.

The response variable measured was the amount of mucin in the excreta. Even if phytate had no effect on the amount of mucin in the excreta, it would be unlikely that the mean amount of mucin would be exactly the same in each one of two groups of eight chickens. Due to individual variability among the chickens and chance alone, a small difference between the means of the two groups would be expected.

The question is: Is the observed difference large enough as to indicate that phytate really increases the average amount of mucin in the excreta of chickens? Or is the observed difference sufficiently small to have happened by chance alone?

The research question 'Does phytate produce an increase in the average amount of mucin in the excreta of chickens?' is not about the sixteen chickens in the experiment, but about the potential effect that phytate could have on hypothetical populations of chickens that could receive either diet.

The null and alternative statistical hypotheses are:

$$H_o : \mu_{Phytate} = \mu_{No-phytate}$$
$$H_a : \mu_{Phytate} > \mu_{No-phytate}$$

One-sided alternative hypothesis

where μ represents the mean amount of mucin in the excreta in the population. The values of the response variable for the sixteen chickens in the experiment are displayed in Table 3.1 and Figure 3.1. The difference between the two sample means is $7.845 - 2.925 = 4.92$.

Table 3.1: Amount of mucin in the excreta of chickens

Treatment	Amount of mucin (g)								$\bar{x}$
Phytate	6.07	7.20	6.61	9.69	9.45	8.95	8.72	6.07	7.845
Control	2.57	2.39	2.51	2.57	1.80	2.37	6.28	2.91	2.925

Source: Onyango et al.(2009)

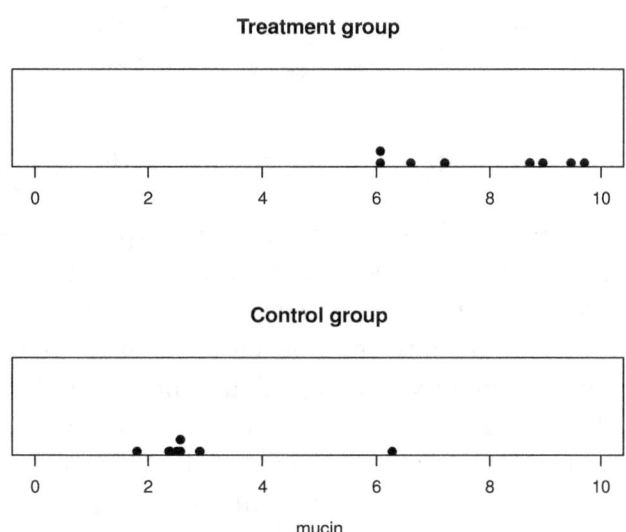

Figure 3.1: Amount of mucin in excreta of chickens for treatment and control groups

3.1.1   Hands-on activity

Randomization tests are performed using a computer because of the numerous calculations involved. However, a hands-on activity can help understand the rationale and algorithm involved in the test. The activity will be illustrated here with the data of case study 3.1, but it could be done with any appropriate data set.

A hands-on activity to understand the randomization test

Get two sets of plastic chips or squares of cardboard that look similar except for the color. Write down the values for the treatment group in one color (one observation per chip) and the values for the control group in the other color, as in Figure 3.2. Each team of students should have a complete set. On the board, draw a horizontal axis, and mark with an X or an arrow the difference between the mean of the two groups (treatment-control) as in Figure 3.4. Under the null hypothesis there is no difference between the

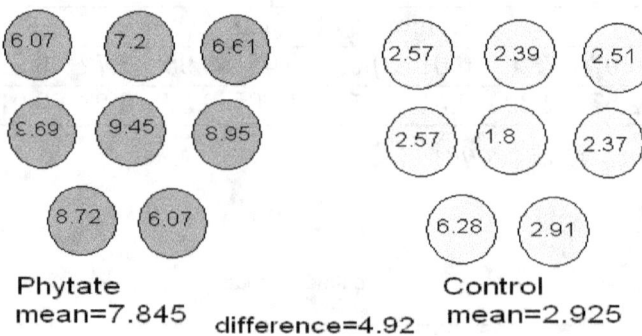

Figure 3.2: Experimental data for treatment (Phytate) and control

effects of the two diets. Thus, under the null hypothesis, each one of the sixteen chickens and their corresponding values for mucin could have come from either of the two groups (treatment or control). We could re-assign the labels 'treatment' or 'control' to the chickens, as long as we have eight of each type. We do this by randomly selecting eight chickens from the sixteen to be labeled 'group 1' ('treatment') and the remaining will form 'group 2.' One of the students in each team selects at random, without looking, 8 chips to form random group 1, the remaining 8 chips will form random group 2. Calculate the means of each random group and calculate the difference between the two means ($\bar{x}_{group1} - \bar{x}_{group2}$), as in Figure 3.3.

Plot the difference between the two random groups on the axis on the board. After all the teams of students have plotted their values, the plot on the board should look similar to Figure 3.4 (that was prepared with the

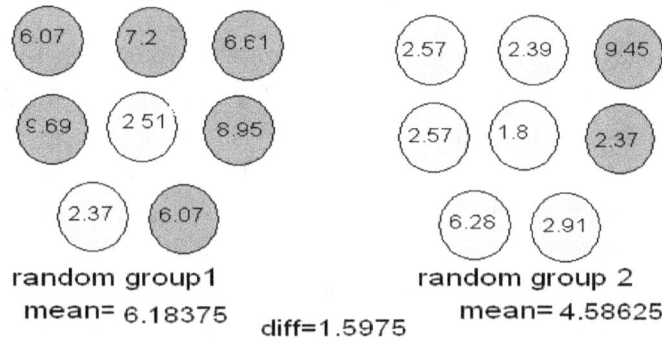

Figure 3.3: Two randomly formed groups with the data

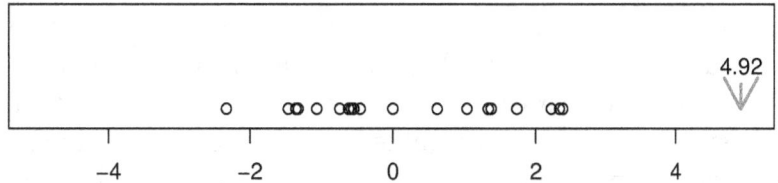

Figure 3.4: Differences between group means for 20 random regroupings of the mucin data

results for 20 teams). Check if the difference between the means between the real treatment and control groups looks like an outlier or an extreme observation in the tails of the distribution, when compared with the differences of the randomly formed groups. Figure 3.4 displays results for 20 random regroupings; of course, the differences you obtain might be different because they are random regroupings. The difference in means between the real treatment and control groups looks like a very extreme value in the context of the empirical distribution of 'difference of means' obtained by randomization. Is a difference of 4.92 likely to happen just by chance if there were no difference in the mean amount of mucin in excreta when chickens are fed phytate? The value 4.92 looks pretty extreme to have happened by chance alone. Figure 3.4 displays the results of only 20 random regroupings; perhaps more regroupings are needed in order to feel more comfortable when answering that question. The help of a computer is needed to perform a large number of regroupings. Before working with a computer, the idea behind the randomization test will be discussed.

The basic idea behind the randomization test

In general, a test for a null hypothesis H_o compares the value obtained for a statistic with the distribution assumed for the statistic if H_o were true. The idea is to check if the observed value looks unusually large (or small) as compared to the values in the distribution. That comparison is summarized with the calculation of the p-value. A randomization test generates the distribution of reference (i.e., the distribution of the statistic when the null hypothesis is true) using randomization.

The null hypothesis in the randomization test is actually stronger than the null hypothesis of equality of means written for Example 3.1. The null hypothesis implicit in the randomization test is that the distribution of the variable is the same in the two populations. If the two distributions are identical, then the two means are equal.

In the case of an observational study to compare two populations, if there were no difference between the two populations, then they would be considered a single population. What would happen if two random samples were to be drawn from that single population? Would the two samples have the same mean? The answer would be probably not, because there would be some difference between the two values just due to chance. In the case of an experiment to compare two treatments, if there were no difference between the effects of the two treatments, the values observed for the response variable could have come from one treatment or the other.

If the two populations were similar or the two treatments had similar effects, then the difference between the two samples would not look too extreme when compared with the difference between groups randomly formed with the data

The randomization test proposes to merge the two samples, assuming that they come from a single population (or that the labels that identify which treatment each subject received are interchangeable) and think of **all the possible ways in which two groups (of the same size as the original groups) could be formed**. The mean for each one of the two possible groups can be calculated (as well as their difference). The distribution of the differences of **all** the possible regroupings is the distribution of reference to evaluate if the difference between the two original samples, or the two real treatments, is sufficiently large to have not likely happened by chance. To apply the randomization test would require the design of a systematic way of counting all the possible regroupings that would produce a difference equal to or larger than the difference observed between the original samples.

An alternative to the systematic consideration of all the possible regroupings is the approximated version of the randomization test that is frequently used. The samples are merged, and two groups (of the same size as the original samples) are formed at random and the difference between the sample means is recorded. The process is repeated multiple times and we end up with an **empirical distribution of the difference of means, obtained by randomization under the assumption that the null hypothesis is true**. Not all the possible regroupings are considered, but a large number of them are. The difference in the value of the statistic for the two original samples (or the subjects in the two treatments) can be compared with the above-mentioned empirical distribution. This allows us to perceive if the observed difference between the samples is likely or unlikely to happen when there is not a real difference between the two populations or treatments. In this way one could judge if the difference in the value of the statistic between the two original groups can be due just to chance or if it is large enough as to indicate a difference between the two populations or treatments.

3.1.2   Applying the randomization test

In Example 3.1, the difference between the means of the treatment and control groups was 4.92. In the hands-on activity, all the differences coming from randomly formed groups were smaller than 4.92. However, since only 20 regroupings are shown in Figure 3.4, it would not be appropriate to extract a conclusion based only on those 20 randomizations. To apply the exact randomization test one needs to think of ALL possible regroupings and count how many of them produce a difference equal to or greater than 4.92.

In how many ways can 16 observations be split into two groups, each with eight observations? Consider that the total number of regroupings of 16 objects into two groups of 8 is simply the number of ways in which 8 objects can be selected from 16 to form the first group because the remaining objects will form group 2. Using mathematical notation, $C_8^{16} = \binom{16}{8} = \frac{16!}{8!(8)!} = 12870$, we can calculate that value using **R** by simply typing **choose(16,8)**. The number of regroupings that produce a difference equal to or higher than the difference between the two true groups, divided by the total number of regroupings, will be the **p-value**, i.e., **the probability of obtaining a result as the one we got or a more extreme one when the null hypothesis is true**. A systematic way of counting how many of those 12,870 regroupings produce a difference equal to or larger than 4.92 is needed. However, a computer could be used to generate a large number of random regroupings even when an 'approximated' p-value will be calculated instead of the exact p-value.

Testing for equal means using randomization

Keep in mind the steps followed in the hands-on activity in Section 3.1.1:

1. Calculate the difference in means between the two original groups (4.92 in the example).

2. Merge the two groups.

3. Select at random the same number of chips that were in the first group (8 in the example).

4. The remaining observations will constitute group 2.

The algorithm for the randomization test

5. Calculate the mean of each group and the difference between the means.

6. Compare the difference in means between the two random groups with the difference in means between the two original groups (4.92).

7. Repeat steps 3-6, keeping track of how many times the difference between the two random groups is equal to or larger than the difference between the two original groups. (If the alternative hypothesis was two-sided, we would also need to keep track of the number of extreme values at the other side of the distribution; in the example these would be the differences smaller than -4.92)

8. Calculate the approximated p-value as the number of regroupings identified in step 7, divided by the total number of regroupings.

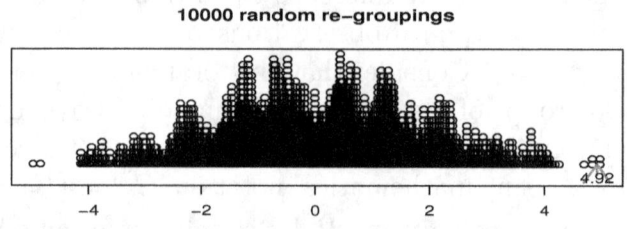

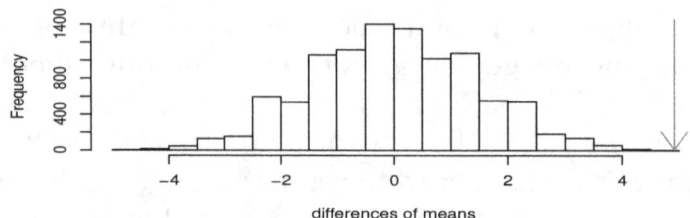

Figure 3.5: Empirical distribution obtained by randomization

The list of steps above can be considered an **algorithm**, or list of detailed instructions to perform a task. Once the algorithm has been written, it can be translated into a language the computer can understand (such as **R**). The set of commands is called a **program**. The program at the end of this section mimics the hands-on activity with chips, only that the regrouping is done 10,000 times instead of a few times. Comments have been added to each line, after the symbol #, to explain the function of each command. The last 6 lines of the program prepare the statistical graphs in Figure 3.5 for the distribution of the difference of means for 10,000 random regroupings. **Algorithms** The same procedure could be applied with medians instead of means if the **and** hypotheses were written in terms of medians instead of means. **programs**

Two representations of the empirical distribution are included in Figure 3.5: the dotplot and the histogram. An arrow marks the difference between the two original groups (*treatment* and *control*). The approximated p-value is the proportion of points in the dotplot that are to the right of the arrow. In the case of Figure 3.5, the approximated p-value is $2/10000 = 0.0002$ because there are only two dots at or beyond the value 4.92. Working with the dotplot it is easier to count the number of values in the empirical distribution that are equal to or larger than 4.92. The histogram presents a smoother version of the same empirical distribution. In the histogram, the approximated p-value would be the area to the right of 4.92. The reason we look only at one side (values equal to or greater than 4.92) of the empirical distribution is because the alternative hypothesis is one-sided and of the type $'>'$. **The smaller**

The approximated p-value is very small indeed. Thus, the null hypothesis **the p-value,** of equal means is rejected because it is very unlikely that a difference of 4.92, **the stronger** or a larger one, could happen by chance alone when the null hypothesis **the evidence** is true. The conclusion is that phytate does increase the mean amount of **against** H_o mucin in the excreta of chickens, which in turn means that chickens lose more endogenous amino acids when they ingest additional phytate.

When is a p-value considered to be small?

An important question is: 'How small does the p-value need to be in order **Significance** to reject the null hypothesis?' The smaller the p-value, the stronger the **level** evidence given by the data against the null statistical hypothesis. With a $\alpha =$ value such as the one obtained for the example (0.0002) there is little doubt. $\mathbf{P}(\text{rejecting } H_o)$ However, sometimes the p-value is not so small, and a threshold is needed **when it is true** as a reference to decide if it can be considered small enough to reject the null hypothesis. The threshold is set at the maximum value to be tolerated for the probability of making a very special type of wrong decision: rejecting a null hypothesis when it is true. That wrong decision is called a **Type I error** and α is the name usually given to its probability. The value for α,

If the consequences of Type I error are serious we want to work with an small α

To compare two groups of 'before/after' observations, apply the randomization test on the differences *after* − *before*

also called **significance level**, is set as soon as the statistical hypotheses are written and even before collecting the data. A very common value for α is 0.05 but another value could be chosen. Remember, α is the probability of an error; thus, the more serious the consequences of that particular error are, the smaller the value to be assigned to α. However, to be able to work with a very small value of α, a large sample size is needed because otherwise the probability of another type of error (**Type II error**) can become very large. More will be discussed in Chapters 5 and 11 about α and the errors that can be made while testing hypotheses.

(*Note: Another name given sometimes to the p-value is 'achieved significance level' (ASL). In SPSS and some other statistical software, the calculated p-value is displayed under the label 'sig' for significance*)

Revisiting Example 2.9.1 Cherry juice vs. placebo

Two treatments, *cherry juice* and *placebo* (Delk, 2009), are being compared. In each group there are two observations for each individual (*before* and *after*). In a case like this, the differences *after*−*before* are calculated for each individual in each group and the randomization test is applied to those differences. The new variable is the difference *after*−*before*. In Example 2.9.1, the average change for the subjects in the cherry juice group was 1.94 and in the placebo group it was only 0.1. Thus, the difference between the two groups was 1.84. It is this difference of 1.84 that will be compared to the differences between the randomly formed groups. In this particular example, looking at Figure 2.24, one realizes that no other regrouping will yield a difference of means greater than the one between treatment and control. Thus, the null hypothesis of no difference is rejected.

A program to apply the approximated randomization test

To obtain an empirical distribution by randomization and the corresponding approximated p-value, you simply need to copy and paste the following set of commands into **R** and press $< Enter >$. Each time the program is executed, the output might differ a little because the regroupings are produced randomly and we only compute a fraction of them. Remember that an approximated p-value, not an exact p-value, is being calculated.

R: randomization test for means

Using statistical software:

The following commands will generate 10000 random regroupings and produce a dotplot and a histogram, as in Figure 3.5. It also calculates the approx-

imated p-value. It can be easily adapted to work with medians instead of means. To use the program for a different example, replace the data in A and B.

```
## RANDOMIZATION TEST to compare groups A and B
## when  the alternative is Ha:u(A)>u(B)
A<-c(6.07,7.20,6.61,9.69, 9.45,8.95,8.72,6.07)
B<-c(2.57,2.39,2.51,2.57,1.80,2.37,6.28,2.91)
nrep=10000     # number of regroupings
## NO MORE INPUT IS NECESSARY
n1<-length(A)            ## number of observations in A
n2<-length(B)  ## number of observations in B
n<-n1+n2                     ## total number of observations
meanA<-mean(A)        ;       meanB<-mean(B)
truedif<-meanA-meanB      ##  difference of the two means
## performs the randomization test
alldata<-append(A,B) ## merges the two groups
chips<-1:n                   ## creates n chips
difmeans<-double(nrep)       ## creates storage space
for (i in 1:nrep){           ## commands will be executed 10000 times
chipsgroup1<-sample(chips,n1,replace=FALSE)  ## selects group 1
chipsgroup2<-chips[-chipsgroup1]          ## rest goes to group 2
group1<-(alldata[chipsgroup1])            ## values in  group 1
group2<-(alldata[chipsgroup2])            ## values in group 2
difmeans[i]<-mean(group1)-mean(group2)       ## difference in means
 }
## assigns value 1 to the reg-roupings in which the
##  difference of means is equal to or greater than the difference
## the A and B groups
numgreater<-difmeans>= truedif  ## for two-sided alternative hypothesis
## use abs(difmeans)>= abs(truedif) in previous line
## calculates the proportion or approximated p-value
pvalue=mean(numgreater) ; pvalue
par(mfcol=c(2,1)) ## 2 plots in figure
stripchart(difmeans, method='stack',pch=1,at=0,main=" ",ylim=c(-0.3,2))
arrows(truedif,-0.2,truedif,0,col="red",lwd=2)
text(truedif,-0.25,c(truedif))
hist(difmeans, main="  ",xlab="differences of means")
arrows(truedif,1000,truedif,10,col="red")
text(truedif,1100,c(truedif))
```

3.1.3   Randomization test for paired data

Assume that a matched-pairs design with n pairs of individuals was used in an experiment. In each pair, each individual has received either treatment A or treatment B. We will use the same data set that R.A. Fisher used in 1935 to introduce the randomization test in his classical book on the design of experiments (Fisher, 1966).

Example 3.2 Darwin's *Zea mays* **plants**

Darwin (1876) knew that inbreeding is not good for humans and wondered whether something similar happens with plants. He produced 15 corn (*Zea Mays*) plants by cross-fertilization and 15 plants by self-fertilization and measured how tall the plants were after a given number of weeks. Recently there has been some debate concerning the issue of the design used by Darwin and if it really was a matched-pairs design (Jacquez & Jacquez, 2002). The data set (heights in inches) is in Table 3.2. The mean of the differences is 2.616667 inches.

Table 3.2: Darwin's data on *Zea Mays* plants

Pair	1	2	3	4	5	6	7	8
Cross	23.5	12	21	22	19.125	21.500	22.125	20.375
Self	17.375	20.375	20	20	18.375	18.625	18.625	15.250
Difference	6.125	-8.375	1	2	0.750	2.875	3.500	5.125

Pair	9	10	11	12	13	14	15
Cross	18.25	21.625	23.25	21	22.125	23.0	12
Self	15.25	18.000	16.25	18	12.750	15.5	18
Difference	1.75	3.625	7.00	3	9.375	7.5	-6

The null and alternative statistical hypotheses in the paired data case are about the mean difference. Assuming that the hypothesis of the scientist is that plants coming from cross-fertilization grow taller, the statistical hypotheses are:

$$H_o : \mu_d = 0$$
$$H_a : \mu_d > 0$$

where d=Cross-Self.

The basic idea of the randomization test for paired data is now explained using Darwin's data. If there were no difference between the effects of treatment A and B, cross- and self-fertilization in this case, then the labels within each pair could be exchanged. For example, under the null hypothesis of no difference, the values for the first pair, 23.5 and 17.375, could have come from 'self' and 'cross' respectively, instead of from 'cross' and 'self.' In one case (the true data), the difference between the elements of the first pair is 1.75, in the other case it would be -1.75. The randomization test for paired observations considers that, if $H_o : \mu_d = 0$ is true, each difference could be either positive or negative. In the example, there are 15 pairs of data; if there are two possibilities for the sign of each difference, there is a total of $2^{15} = 32768$ possible outcomes and thus 32,768 possible values for the mean of the differences $\bar{d}$. It is like tossing a coin 15 times in order to decide if you will exchange the labels ('cross' and 'self') or not. If 'heads' comes up, the labels remain the same and the difference keeps its original sign. If 'tails' comes up, we exchange the labels 'cross' and 'self' and the difference changes sign.

In the figure above you see the results of tossing a coin 15 times: tails were obtained in tosses # 1, 2, 8 and 9. For those four differences the sign (either + or -) is reversed, the other 11 differences keep their original sign. Thus the differences, after this random exchange of labels, are: -6.125, +8.375, 1, 2, 0.750, 2.875, 3.5, -5.125, -1.75, 3.625, 7, 3, 9.375, 7.5, -6, and the mean difference is $\bar{d}=2$.

We have not examined the 32,768 possibilities, but Figure 3.6 displays the mean difference for 10,000 random exchanges of labels, equivalent to having repeated the experiment of tossing the coin 15 times to decide which labels are exchanged 10,000 times. Only in 268 of those 10,000 times, the mean difference was equal to or larger than the true mean difference from Table 3.2. Thus, the approximated p-value is 0.0268. The simulation that produced Figure 3.6 was run using the program in **R** at the end of this section. If we run the program again, the result may be a little different. Fisher (1966) thought about the 32,768 possible outcomes and calculated that in only 863 of them the mean difference was equal to or larger than the mean difference for the original data. Using that information and a one-sided alternative hypothesis, the exact p-value is 863/32768=0.02633667.

Note: Fisher used randomization as a way of validating the results obtained with the t-test, to be learned in Chapter 11. He did not work with

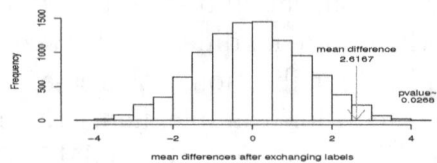

Figure 3.6: Finding the approximated p-value with Darwin's data

an alternative hypothesis, and thus considered both tails of the randomization distribution, concluding that in 5.267% of the possibilities available, the mean difference was equal to or greater than the one observed.

Using statistical software:
The commands below were used to produce Figure 3.6.

R:
matched-pairs randomization test

```
## RANDOMIZATION for paired data, to test ud=0 vs. ud>0, for d=A-B
A<-c(23.5,12,21,22,19.125,21.5,22.125,20.375,18.25,21.625,23.25,
21,22.125,23,12)
B<-c(17.375,20.375,20,20,18.375,18.625,18.625,15.25,16.5,18,16.25,
18,12.75,15.5,18)
nrand<-10000  ## number of randomizations
d<-A-B   ## calculates differences
n<-length(d)  ## how many observations
truemdif<-mean(d)        ##  mean of the true differences
randmeansd<-double(10000)        ## creates storage space
for (i in 1:10000){      ## commands to be executed 10000 times
flips<-rbinom(n,1,0.5)   ## randomly decides where flips happen
nflips<-2*flips-1        ## 1:original sign  -1:flip
newdif<-nflips*d   ## differences after flipping
newmeand<-mean(newdif)   ## mean of the new differences
randmeansd[i]<-newmeand  }   ## stores the mean differences
numgreater<-randmeansd>= truemdif  ## comparing with true dif
pvalue<-mean(numgreater)  ## calculates pvalue as proportion
pvalue  ##  shows pvalue on the screen
hist(randmeansd,main=" ",xlab="mean differences exchanging labels")
arrows(truemdif,800,truemdif,10,col="red")  ## puts arrow in plot
rtruemdif=round(truemdif,4)   ## rounding true mean difference
text(truemdif,900,c(rtruemdif))     ## writes difference in plot
text(max(randmeansd),400,'pvalue~')  ## writes label for p-value
text(max(randmeansd),300,c(pvalue))  ## writes p-value in plot
```

3.2 The Bootstrap

Bootstrapping is a term used to describe getting yourself out of a difficult situation. It originates from a story in which the protagonist claims to have survived by pulling himself up, from a hole or the mud, by his bootstraps. The statistical method called Bootstrap was designed by Bradley Efron in 1979. It can be applied when we do not have anything else to rely on but our own data. Based on the data, an empirical bootstrap distribution for the statistic is generated by resampling from the original sample. The empirical bootstrap distribution is used to make inferences.

The bootstrap method uses an empirical distribution for the statistic based just on the data at hand

3.2.1 Resampling

Assume that the value of a population parameter needs to be estimated. For example, the mean of a population needs to be estimated from the values observed for the individuals in the sample. Assume that a single value (**point estimate**) is not wanted as sole 'guess' for the value of the parameter, but a whole interval of possible values (**confidence interval**) is desired. To estimate with confidence, or estimate using a confidence interval, means to make an statement like this: *'We think, with 95% confidence, that the value of the population mean is between _____ and _____.'* The problem is: How do we come up with such an interval? In this chapter the problem will be addressed using the bootstrap method. A different method will be studied in Chapter 11.

Bootstrap resampling: taking random samples with replacement, and of the same size, from the sample

To calculate a confidence interval, we must first come up with a distribution. An empirical distribution can be generated through resampling; but what is resampling? Assume that a random sample of n individuals is selected from a population. This would be referred as the 'original' sample. For each individual in the sample, the value of a variable X is recorded yielding a set of n values. From that original sample, a random sample (also of size n) is obtained by sampling with replacement. This means that an individual is randomly selected from the sample, the value of the variable X is recorded, and the individual returned to the original sample before selecting the next individual. The new sample is called a **bootstrap sample** obtained by **resampling**, i.e., sampling from the original sample. The process is repeated over and again, usually 1,000 or 2,000 times. For each bootstrap sample, the statistic of interest, for example the sample mean, is calculated. Preparing a dotplot or a histogram with the values of the statistic for the bootstrap samples allows us to observe how the value of the statistic varies from bootstrap sample to bootstrap sample. This empirical distribution of the values of the statistic obtained by resampling is later used to calculate confidence intervals for the parameter of interest.

Example 3.3 How long on average are the leaves of this rhododendron?

Length of rhododendron leaves

Assume that we would like to estimate the mean length of the leaves of the shrub in Figure 3.7. The population of interest is formed by the leaves of that particular plant but we do not have the patience or the time to measure every single leaf on site, nor should we pick all the leaves. A random sample of 10 leaves is selected and their length is measured in centimeters. The values are: 9.7, 10.4, 8.9, 10.7, 9.85, 10.3, 9, 10, 10.45, and 8.65. The ten observations in the sample and their mean ($\bar{x}= 9.795$) are displayed in Figure 3.8a; the standard deviation is $s= 0.7201273$. The sample mean is a **statistic**

Figure 3.7: Leaves of a shrub for which the mean length will be estimated

that is the usual **estimator** of the population mean. The population mean is the mean length of ALL the leaves in the shrub. That **population mean** is the **parameter**, whose exact value is not known and will be estimated. The numerical value of the sample mean, in this case 9.795 centimeters, can be used as a **point estimate** of the population mean. A sample mean rarely is exactly equal to the mean of the population. The mean of different samples drawn from the same population are expected to be a little different, depending on which individuals (leaves in this case) are in the sample. It is preferable to be more cautious and give a **confidence interval**, instead of a single value as our estimate of the population mean. In this chapter, bootstrapping will be used to produce such an interval.

3.2.2 Hands-on activity

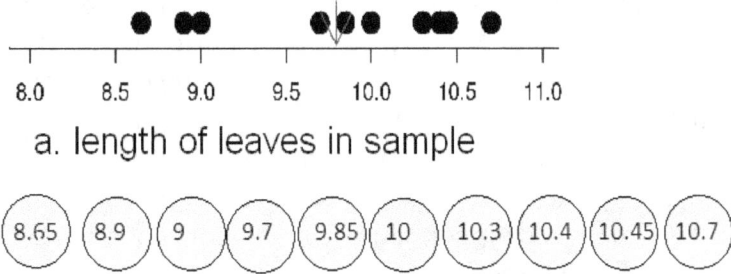

a. length of leaves in sample

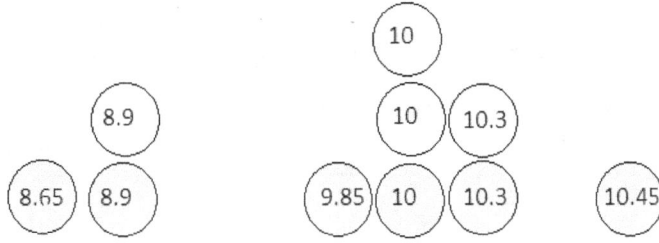

b. plastic chips with sample values

c. One bootstrap sample

Figure 3.8: Sample and bootstrap sample

First the lengths of the leaves in the sample are copied onto plastic chips as shown in Figure 3.8b. Working in pairs, each team has a set of 10 plastic chips. One of the students holds the chips and the other student selects 10 chips, one by one. After each selection, the value on the chip is written down and the chip is returned. This is called 're-sampling' or taking samples **A hands-on** from the sample. Because the sample size is equal to the size of the original **activity for** sample, the sampling has to be done 'with replacement.' Each team will **bootstrapping** end up with a **bootstrap sample**. In Figure 3.8c, a bootstrap sample is displayed: some values appear more than once and some values do not show up at all. Students write down the values in the bootstrap sample and calculate the mean.

The total number of samples that could be drawn with replacement from a sample of size 10, is $10^{10} = 10000000000$ or 1e+10 because there are 10 possible values available in each one of the 10 selections. Usually we do not generate that many bootstrap samples (typically 1,000 or 2,000). However, for illustration purposes we will first select only 100 samples. The means of

100 bootstrap samples are shown in Figure 3.9. The values in Figure 3.9 are called **bootstrap replicates of the sample mean** and form **an empirical distribution obtained by resampling from the original sample**. The bootstrap replicates of the sample mean have been sorted. If we want to focus on the central 90% of the empirical distribution we need to cut out the lower 5% and the upper 5%. Since there are only 100 dots in the dotplot, the upper 5 and lower 5 dots can be identified. Marks have been put between the values in the 5th and 6th positions (the midpoint is 9.425), and between the 95th and 96th positions (the midpoint is 10.0425) because those marks will cut off the lower and upper tails. This is how confidence intervals will be built in the following section.

Order the bootstrap replicates of the statistic and find the quantiles according to the desired confidence

```
 [1]   9.240   9.380   9.385   9.385   9.390 | 9.460   9.475   9.490   9.510   9.515
[11]   9.530   9.555   9.560   9.570   9.570   9.580   9.600   9.610   9.615   9.620
[21]   9.625   9.625   9.625   9.635   9.635   9.635   9.640   9.645   9.655   9.660
[31]   9.665   9.670   9.675   9.680   9.680   9.685   9.685   9.695   9.695   9.705
[41]   9.705   9.710   9.720   9.725   9.730   9.735   9.755   9.755   9.760   9.760
[51]   9.765   9.765   9.775   9.790   9.795   9.800   9.805   9.805   9.810   9.810
[61]   9.815   9.820   9.820   9.825   9.825   9.830   9.835   9.835   9.835   9.835
[71]   9.840   9.840   9.855   9.855   9.865   9.865   9.880   9.880   9.880   9.880
[81]   9.885   9.920   9.965   9.970   9.970   9.970   9.975   9.975   9.985   9.995
[91]  10.000  10.000  10.005  10.020  10.050 10.055  10.095  10.100  10.240  10.265
```

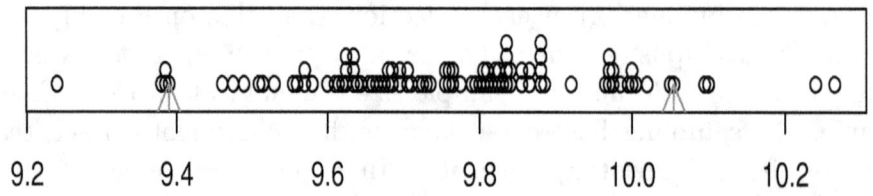

Figure 3.9: Means of 100 bootstrap samples

Of course, the empirical distribution would be more illuminating if it had more values, that is, if more bootstrap samples were selected. Imagine that there are 2000 teams of students working in this activity and each

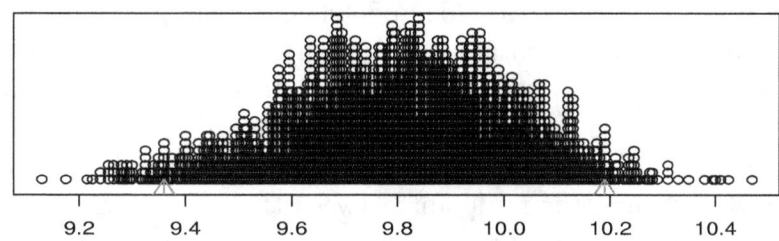

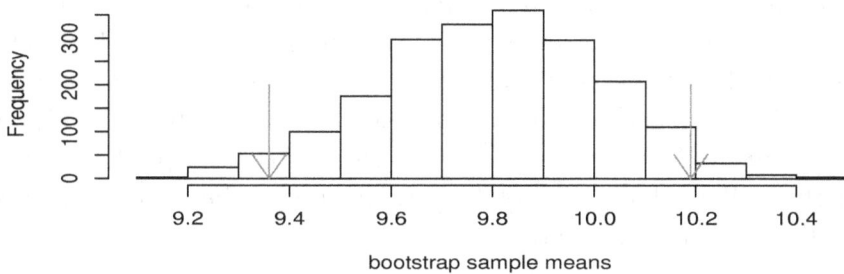

Figure 3.10: 2000 bootstrap replicates of the sample mean

team draws a bootstrap sample and calculates its mean. The computer can be used to simulate those 2,000 results. Figure 3.10 displays the dotplot and the histogram of 2,000 bootstrap replicates of the sample mean for the rhododendron example. The mean and the standard deviation of the original sample are $\bar{x}=$ 9.795 and $s=$ 0.7201273, respectively. The mean of the 2,000 bootstrap sample means in Figure 3.10 is 9.800558, a value very close to the mean of the original sample. The standard deviation of the 2,000 bootstrap sample means is 0.2140413, a value smaller than the standard deviation of the sample. Of course, the value could vary slightly if another set of 2,000 bootstrap samples were to be generated. There is less variability among the sample means than among the individual observations in the sample. This can be also seen by comparing the horizontal scale in Figures 3.8a and 3.9. Even more, $0.7201273/\sqrt{10}=$ 0.2277242, a value quite close to the standard deviation of the means of the bootstrap samples and where 10 is the sample size. The standard deviation of the means of the bootstrap samples (or bootstrap replicates of the sample mean) is sometimes called **the bootstrap standard error of the sample mean.**

3.2.3   Bootstrap confidence intervals

Figure 3.10 displays the set of 2000 bootstrap replicates of the sample mean or empirical distribution obtained by resampling. This figure describes how the mean varies from bootstrap sample to bootstrap sample. The values of the 2000 means of the bootstrap samples in Figure 3.10 can be sorted. Once those means are ordered, the 2.5 and 97.5 percentiles can be identified. Since there are 2000 values, the midpoint between the values that occupy the 50th and 51th positions excludes the lower 2.5% of the bootstrap replicates. At the higher end, the midpoint between the values that occupy the 1949th and 1950th positions excludes the higher 2.5% of the bootstrap replicates. These are the 2.5 and 97.5 percentiles, also called the 0.025 and 0.975 quantiles. Quantiles are easily calculated with software. In the case of the bootstrap replicates of the sample mean in Figure 3.10, the 0.025 and 0.975 quantiles are 9.36 and 10.19 respectively. Those values are marked with arrows in Figure 3.10. That means that 2.5% of the bootstrap replicates of the sample mean have a value 9.36 or lower and, at the other extreme, 2.5% have a value 10.19 or higher. Thus, 95% of bootstrap replicates of the sample mean are between 9.36 and 10.19 centimeters.

The interval $(9.36, 10.19)$ is used as a 95% confidence interval for the population mean. We say that we are 95% confident that the mean length of all the leaves in that specific shrub is somewhere between 9.36 and 10.19 centimeters. Remember that the leaves were randomly selected from a single shrub; thus, the statement is about the mean length of the leaves of that particular shrub. In general, **inference is valid only for the population from which the sample was randomly selected.**

The interval $(9.36, 10.19)$, calculated for the example, is called a **bootstrap percentile confidence interval.** An alternative simple method is described at the end of Section 11.1. There are other more complicated methods to build confidence intervals based on the bootstrap samples; the interested reader is referred to Manly (2007). When the distribution of the variable in the population is extremely skewed, some of the more complicated methods produce more reliable results.

Using statistical software:
The following commands were used to produce Figure 3.10 and calculate the confidence interval, as well as the mean and standard deviations of the bootstrap replicates of the sample mean. The data, the number of bootstrap replicates (nboot) and the confidence can be changed in the first three lines of commands. Then, copy and paste the commands into R.

Confidence interval: a range of values in which, with some confidence, the value of the parameter is thought to be located

Conclusions from inference are valid only for the population from which the random sample was selected

R: bootstrap confidence intervals

```
## PERCENTILE BOOTSTRAP CONFIDENCE INTERVAL FOR THE MEAN
## INPUT :
x<-c(9.70,10.40,8.90,10.70,9.85,10.30,9.00 ,10.00,10.45,8.65)
nboot=2000    ## number of bootstrap replicates
confidence=0.95
## NO MORE INPUT IS NEEDED
n=length(x)
sampm<-double(nboot)              ## prepares storage space
for (i in 1:nboot){               ## repeat 2000 times
subsam<-sample(x,n,replace=TRUE)    ## bootstrap sample
sampm[i]<-mean(subsam)               ## stores the sample mean
 }
## calculates the endpoints of the confidence interval
low<-quantile(sampm,(1-confidence)/2)
hi<-quantile(sampm,confidence+(1-confidence)/2)
par(mfcol=c(2,1))  ## to do 2 plots in the same figure
stripchart(sampm,method="stack", at=0.1,cex=1, pch=1,main=" ")
## marks the endpoints of the interval
arrows(low, -0.5,low,0.1,col="red")
arrows(hi, -0.5,hi,0.1,col="red")
hist(sampm, xlab= "bootstrap replicates ", main=" ")
## marks the endpoints of the interval
arrows(low, 200,low,0,col="red")
arrows(hi, 200,hi,0,col="red")
## prints the interval
low; hi
## prints mean and sd for  sample and bootstrap replicates
mean(x); sd(x)
mean(sampm); sd(sampm)
```

3.2.4 Confidence and width of the interval

In the previous section, 90% and 95% confidence intervals for the mean were calculated. However, a different percentage could be used for the confidence interval. The desired confidence determines the size of the tails (5% each for 90% confidence, and 2.5% each for 95% confidence) that are cut off at each extreme from the distribution of the bootstrap replicates in order to find the endpoints of the interval. The effect of the confidence on the width of the interval will be shown next using the onion cell example.

Example 3.4 Length of cells in an onion root

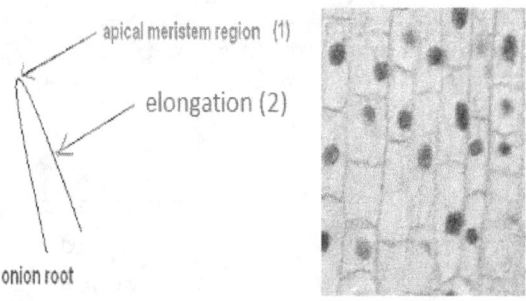

Figure 3.11: Cells in the elongation region of an onion root

Figure 3.11 shows the location of the elongation region and some of its cells. Assume that somebody wants to estimate the mean length (millimeters) of the cells in that region. A random sample of 40 cells from several onion roots were selected and their lengths were measured. The lengths (mm) are:

0.032	0.043	0.023	0.056	0.040	0.076	0.026	0.034
0.033	0.045	0.046	0.021	0.056	0.060	0.037	0.028
0.045	0.074	0.044	0.055	0.035	0.044	0.027	0.058
0.033	0.047	0.035	0.054	0.075	0.045	0.052	0.062
0.026	0.039	0.067	0.071	0.020	0.042	0.022	0.046

Length of onion cells

The stem-and-leaf plot below and the first histogram in Figure 3.13 display the 40 observations in the sample.

```
The decimal point is 2 digit(s) to the left of the |

   2 | 01236678
   3 | 23345579
   4 | 02344555667
   5 | 245668
   6 | 027
   7 | 1456
```

The mean and the standard deviation are 0.04435 and 0.01558525, respectively. The five-number summary (minimum, lower quartile, median, upper quartile and maximum) of the lengths of the 40 cells is:

$$0.02000, \quad 0.03300, \quad 0.04400, \quad 0.05525, \quad 0.07600$$

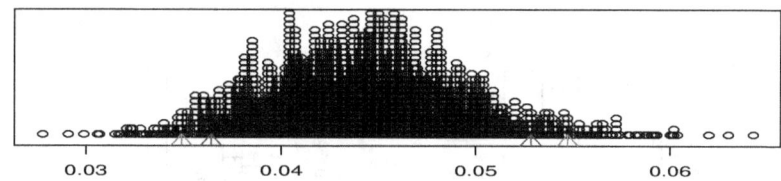

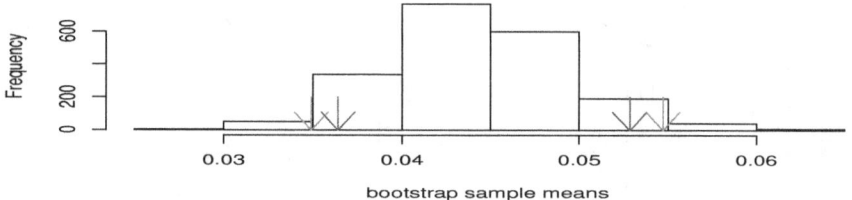

Figure 3.12: 95% and 90% confidence intervals

For the onion cells example, 2000 bootstrap samples were generated to obtain the empirical distribution of bootstrap replicates of the sample mean in Figure 3.12. To calculate the 95% confidence interval, a tail with 2.5% of the total number of bootstrap replicates is cut off at each extreme. To calculate the 90% confidence interval, a tail with 5% of the total number of bootstrap replicates is cut off at each extreme. Consequently, the interval with lower confidence will be shorter because larger tails would be cut off. Greater confidence requires wider intervals. For the onion example, the 90% and 95% confidence intervals for the mean are:

> **The greater the confidence, the wider the confidence interval**

Confidence	Interval
90%	(0.0364 , 0.0528)
95%	(0.0349 , 0.0547)

3.2.5 Bootstrapping other statistics

The bootstrap method can be used with statistics other than the mean (e.g. the standard deviation, median, etc.) In order to apply it to the median and other quantiles it is better to have larger samples than the one used in the rhododendron example. The sample median and quartiles are, after all, a single observation or the mean of two observations; with small samples, the empirical distribution could look quite non-smooth and may have gaps.

> **If working with medians, avoid small samples**

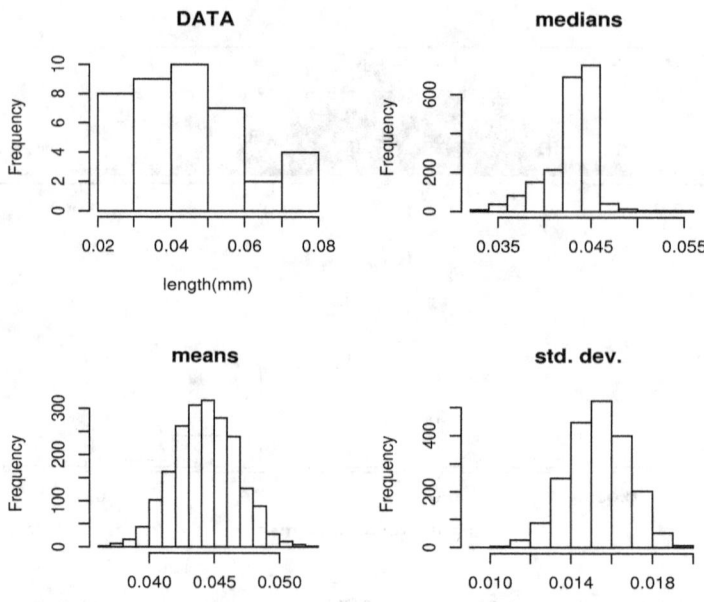

Figure 3.13: Sample and empirical distributions of the bootstrap replicates of the sample mean, median and standard deviation

To work with other statistics using the program in **R** in Section 3.2.3, just replace the command 'mean' by the desired statistic in the line 'stores the sample mean.'

Figure 3.13 displays the empirical distribution for the bootstrap replicates for the median and the standard deviation. Those empirical distributions were used to calculate the 95% confidence intervals below:

Parameter	95% confidence interval
Median	(0.0370 , 0.0465)
Standard deviation	(0.0124 , 0.0181)

3.2.6   Confidence intervals and hypotheses

Confidence intervals can be used, under certain conditions, to answer hypothesis testing questions in the context of studies for a single population. Also, if in a formal test of hypothesis the null hypothesis is rejected, it is recommended to estimate a confidence interval afterwards in order to discuss the issue of practical significance.

Assume that before drawing the sample of 40 onion root cells, described in the previous section, somebody made the statement: 'the average length of cells in the elongation part of the onion root is 0.045 millimeters.' After having drawn the sample, does the previous statement sound reasonable? Approaching the problem as a case of hypothesis testing, the null and alternative statistical hypotheses can be written as:

Using a confidence interval to answer a test of hypothesis question

$$Ho : \mu = 0.045$$
$$Ha : \mu \neq 0.045$$

The 95% confidence interval for the population mean μ calculated in the previous section is (0.0349 , 0.0547). The value 0.045 is in that interval. Thus, the data do not constitute evidence against the null hypothesis and we do not reject it (working with $\alpha = 0.05$). Notice that both the alternative hypothesis and the confidence interval are two-sided. Notice also that the 95% confidence matches the $\alpha = 0.05$. Those two agreements are necessary in order to use a confidence interval to answer a test of hypothesis question.

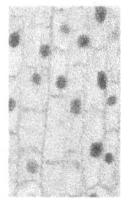

Revisiting Example 3.3 Rhododendron leaves

Assume that before the 10 leaves were collected and measured, somebody had looked at the shrub in Figure 3.7 and said: 'the leaves of that shrub are on average 12 centimeters long.' His statement can be translated into two statistical hypotheses:

$$Ho : \mu = 12$$
$$Ha : \mu \neq 12$$

Based on the random sample of 10 leaves, the (95%) confidence interval calculated by using the bootstrap method was (9.33, 10.195). Does the value 12 look reasonable? No, because the value 12 falls outside of the 95% confidence interval. The null hypothesis $Ho : \mu = 12$ should be rejected because the data constitute evidence against the null hypothesis (given $\alpha = 0.05$).

For paired data, bootstrap the differences

Revisiting Example 2.7.1 Cherry juice and arthritis

An experiment with cherry juice was described in Section 2.7. The variable measured is pressure needed to produce pain. Table 2.5 contains the data for the baseline values and the values at the end of the four-week experiment, as well as the differences for each individual. In Chapter 2, we saw this example at the descriptive level, now we are able to make inference. Assume that the research question is: 'Does cherry juice have any effect on how much

pressure patients can endure?' This is a paired-data case in which the null
and alternative hypothesis are:

$$Ho : \mu_d = 0$$
$$Ha : \mu_d \neq 0$$

where μ_d is the mean difference *after—before* of pressure needed to feel pain
in a population of potential users of the cherry juice to treat arthritis pain in
the knee. The alternative hypothesis considers that the effect of cherry juice
could be positive or negative.

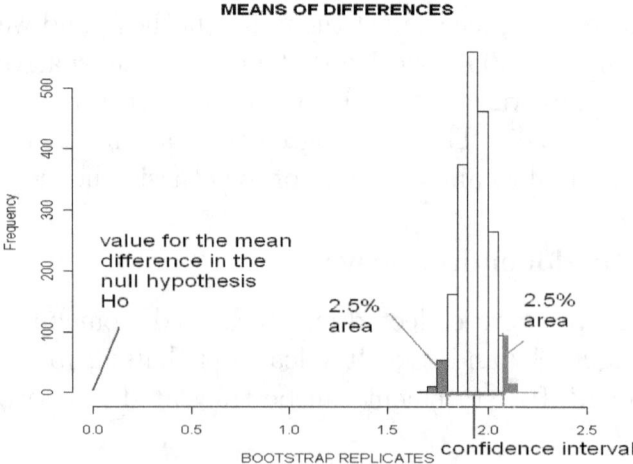

Figure 3.14: Testing a hypothesis using a bootstrap confidence interval

The differences *after—before* shown in Table 2.5 for the participants in
the experiment are: 2,2,2.4,1.8,1.9,1.8,2,1.5,1.9,2.1. Those differences are the
values to be entered in the bootstrap program in order to calculate a confidence interval for the mean difference in the population. The 95% confidence
interval based on 2000 bootstrap samples is (1.8,2.08). The empirical distribution of bootstrap replicates is in Figure 3.14. The confidence interval does
NOT contain the value 0. Actually, the confidence interval is relatively far
away from the value for the parameter μ_d specified by the null hypothesis.
Thus, the null hypothesis of mean difference equal to 0 can be rejected. The
conclusion is that, on average, there is a change in the pressure on the knee
that patients with arthritis can endure when they drink cherry juice daily.
In this case, a confidence interval was used to test a hypothesis.

If a formal hypothesis test is done and the null hypothesis is rejected, it is convenient to calculate a confidence interval afterwards in order to gauge how far from the null hypothesis reality seems to be. When a null hypothesis is rejected, it is usually said that **the results are statistically significant**. The issue of **practical significance** is a different one and not decided by the statistician but from the point of view of the specialist in the field of application. When a null hypothesis is rejected, confidence intervals are used to answer questions such as 'OK, they are not equal, but how different are they?'

> **If the null hypothesis is rejected ('that is not the value'), calculate a confidence interval ('the value is likely to be in this interval')**

In the cherry juice example, we conclude that after 4 weeks of drinking a daily dose of cherry juice there is a change in the average pressure in the knee a patient can endure, and that we are 95% confident that the additional pressure they can tolerate is on average between 1.8 and 2.08. A nurse or a doctor could claim that it is only a small improvement and that tolerating 1.8 units of additional pressure on the knee does not warrant a daily dose of cherry juice. If that is the case, the results would be 'statistically significant' but not of 'practical significance.' However, if the health specialists think that such a reduction of pain is beneficial, then the results are both of 'statistical significance' and 'practical significance.'

> **Statistical significance and practical significance**

3.2.7 Bootstrapping the correlation coefficient

When dealing with a single quantitative variable, the interest is generally focused on building confidence intervals for the mean, median or standard deviation of the population. When studying two quantitative variables, it might be of interest to calculate a confidence interval for the correlation coefficient that quantifies the strength of the linear relationship between the two variables.

> **How strong is the linear relationship between two variables?**

Revisiting Example 2.12.1 Altitude and RBC

The correlation coefficient ($r = 0.8028$), between altitude and number (millions) of red blood cells, was calculated for a sample of 17 individuals using expression (2.8). What if a different set of 17 people had constituted the sample? Would the value of r have been different? What can be said about the correlation between altitude of residence and number of RBC for the whole population from which the sample was taken? Can a confidence interval be calculated to estimate the correlation in the population?

Remember that the observations used to calculate the **correlation coefficient** come in pairs, each value of x is attached to a value of y; the altitude

> **When data are paired, resample the pairs**

for one individual cannot be mixed with the number of red blood cells of another in this context. The resampling process needs to be applied to the individuals. A random sample of size 17 is selected with replacement from the 17 individuals in the original sample. For each one of the bootstrap samples the value of r is calculated. The process is repeated 1,000 or 2,000 times. The values of r are sorted and the 2.5 and 97.5 percentiles are calculated. Those percentiles are the values that separate the upper and lower 2.5% tails of the distribution of bootstrap replicates as in Figure 3.15. The 95% confidence interval for the correlation in altitude and RBC example is (0.587, 0.934). Thus, instead of making a risky statement saying that the correlation between altitude and millions of red blood cells for all individuals in the population is 0.8028 (as it was for the sample), it is said that 'we are 95% confident that the correlation (in the population) between altitude and number of red blood cells is a value in the interval (0.587 ,0.934).' The confidence interval does not contain the value 0. Thus, we conclude that the correlation between altitude and RBC is significantly different from 0; there is an association between the altitude at which humans reside permanently and the number of red blood cells they have. However, the confidence interval looks quite wide, this is in part because the sample size is small ($n = 17$).

Larger samples produce narrower confidence intervals

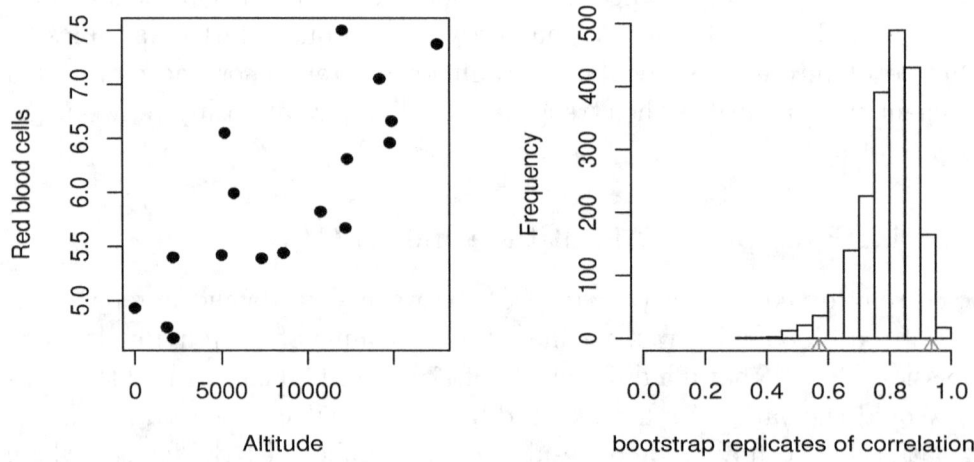

Figure 3.15: Confidence interval for r : altitude & RBC

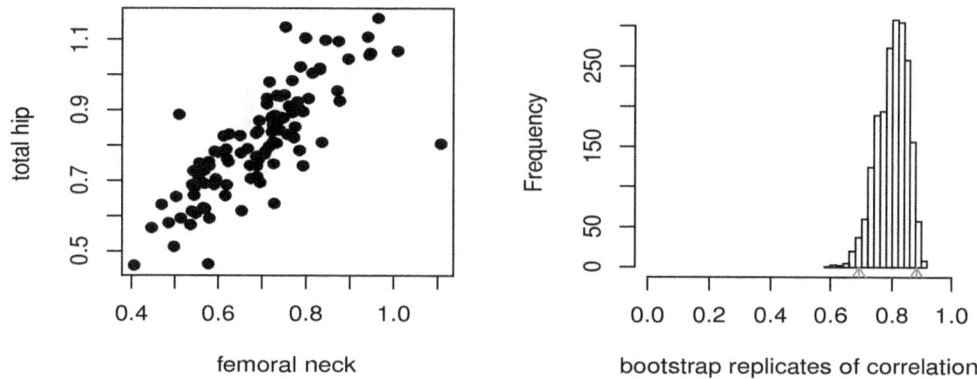

Figure 3.16: Confidence interval for r: BMD in hip & femoral neck

Example 3.5 Bone mineral density in left and right hip

The same program in **R**, prepared for the altitude and RBC example, was used to calculate the bootstrap percentile confidence interval for the correlation between the bone mineral density (BMD) in the right hip and the right femoral neck (*data file BMD*) for 108 women. Compare the width of the confidence intervals in Figures 3.15 and 3.16. The value of r for the 108 women is 0.7976317, a value not too different from the r=0.8028 for the RBC and altitude example. However, the 95% confidence interval (0.684,0.886) is narrower, in part because the sample size is much larger ($n = 109$).

Using statistical software:
The following commands produce the scatterplot and the bootstrap percentile
confidence interval for r as in Figure 3.14. The data, the confidence and the
number of bootstrap replicates can be changed in the first lines of commands.

R: bootstrapping the correlation coefficient

```
## BOOTSTRAPPING THE CORRELATION COEFFICIENT
## INPUT
x<-c(0,1840,2200,2200,5000,5200,5750,7400,8650,10740,12000,
12200,12300,14200,14800,14900,17500)
y<-c(4.93,4.75,5.40,4.65,5.42,6.55,5.99,5.39,5.44,5.82,7.50,5.67,
6.31,7.05,6.46,6.66,7.37)
nboot=2000  ## number of bootstrap replicates
confidence=0.95
```

```
## NO MORE INPUT IS NEEDED
n<-length(x)  # counts number of observations
who<-seq(1,n,by=1) #creates a list labels 1:n
rboot<-double(nboot) # creates storage space
## now we select random samples from the labels
## and identify the corresponding values of x and y
for (i in 1:nboot){
subsam<-sample(who,n,replace=TRUE)
xwho<-(x[subsam])
ywho<-(y[subsam])
rboot[i]<-cor(xwho,ywho)
 }
low<-quantile(rboot,(1-confidence)/2) ## lower end of interval
high<-quantile(rboot,confidence+(1-confidence)/2)  ## upper end
par(mfcol=c(1,2)) ## 2 plots in figure
plot(x,y,pch=19) ## scatterplot
hist(rboot, main= ' ',xlab= 'bootstrap replicates of r',xlim=c(0,1))
arrows(low,-100,low ,0,col="red")
arrows(high,-100, high ,0,col="red")
low;high   ## prints the interval
```

3.3   Fisher's exact test for 2×2 tables *

The randomization test was used in Section 3.1 to compare two groups with respect to a quantitative variable. In this section, two groups will be compared in terms of a categorical variable using Fisher's exact test.

3.3.1   The lady tasting tea story

Can the lady tell if milk or tea was poured first?

R.A. Fisher introduced this case in his book on design of experiments (Fisher, 1966), first published in 1935. It is now a classic in the statistical literature: 'the lady testing tea' problem. According to some sources (Fisher-Box, 1978), the lady was Dr. Muriel Bristol, a biologist specializing in algae and a colleague of Fisher at the Rothamsted Experimental Station. Other sources refer to her just as a lady in a social gathering in Cambridge (Salsbury, 2006). She claimed that she could tell if tea or milk was poured first in a cup of tea with milk; some doubted her statement. Somebody proposed to conduct an experiment. There seems to be no exact record of the experiment as conducted that day. However, the case inspired thought and discussion on experimental design and perhaps was the origin of definition of the exact test.

Chapter II of Fisher (1966) starts with the claim of the lady and the following experiment. A total of eight cups of tea with milk are prepared, in four cups milk is poured first, in the other four cups tea is poured first ('replication' principle). All the other variables such as proportion of tea and milk, appearance of the cups, temperature, type of tea and milk are to be maintained constant ('control of other variables' principle). Assume also that the cups are presented in a random order ('randomization' principle). After tasting each cup of tea, the lady will state which was poured first, milk or tea.

Applying the principles of replication, randomization and control of other variables

In Fisher (1966) the actual outcome of the experiment is not reported. However, an enlightening discussion on experimental design is included and a few hypothetical results are explored. Agresti (1996) discusses a possible result for the experiment (3 correct identifications in each case) that is represented in Figure 3.17: she correctly classifies 3 of the 4 'milk first' cups and 3 of the 4 'tea first' cups. In Figure 3.17 some cups have been shaded to indicate that tea was poured first, but in an experiment they should be indistinguishable by sight alone. First, the result of 3 correct identifications will be discussed. Later the result found in another source will be discussed.

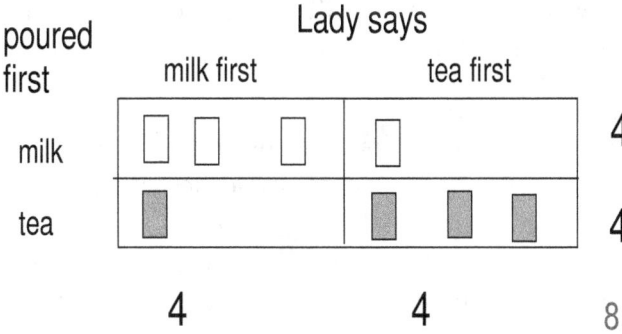

Figure 3.17: The lady identifies correctly 3 cups in each group

The question is: **What is the probability of three correct guesses, in each group, just by chance assuming she does not really have a discriminatory ability?** Because of the way the experiment was designed, the person knew there were 4 cups of 'milk first' and four of 'tea first.' The marginal frequencies are fixed, the row and column totals are all equal to 4. There is a total of 8 cups involved, and 4 are in the first row (milk was poured first); from those 4, 3 have been assigned by the lady as being in the 'milk first' category and one in the 'tea first' category. Consider the many ways in which this can happen, if done randomly. Mathematical notation introduced in previous sections will again be used here. The symbol ! represents the factorial, 4! means $4 \times 3 \times 2 \times 1$. The notation C_4^8 represents the number of ways in which 4 objects can be selected from a total of 8 objects, and it is calculated as $\frac{8!}{4!4!} = 70$.

1. Focus on the totals at the bottom of the table. In how many ways can we select 4 cups, out of a total of 8, to be in the first column? $C_4^8 = \frac{8!}{4!4!} = 70$. Using **R**, it is enough to write *choose(8,4)* to get the answer (70).

2. From the 4 cups in the first row (the ones in which milk was poured first), in how many ways can 3 be selected to appear in the cell of that row? $C_3^4 = \frac{4!}{3!1!} = 4$.

3. In the second row there are 4 cups, those in which tea was poured first. In how many ways can one cup be selected to appear in the first cell? $C_1^4 = \frac{4!}{1!3!} = 4$ (any one of the 4 cups could be misclassified by the lady).

4. The 4 ways in 2) can be combined with the 4 ways in 3), thus there are 16 ways in which the result depicted in Figure 3.17 can occur by chance alone without the lady really being able to distinguish if tea or milk was poured first. Thus, the probability of the 3 correct guesses in each group is 16/70.

In this story, the research question is whether the lady has the ability she claims, the null and alternative statistical hypotheses are:

Ho: the lady does not have the ability to discriminate
Ha: the lady can tell if tea or milk has been poured first (i.e., she does have the ability to discriminate)

The definition of **p-value** is **the probability of getting the result obtained or a more extreme one when the null hypothesis is true**. The probability of the result in Figure 3.17 happening by chance alone is $16/70 = 0.2285714$. A more extreme result is depicted in Figure 3.18, where all the cups were correctly identified. The probability of that more extreme result is $1/70 = 0.01428571$ because there is only one way in which all the cups are correctly identified. Thus, the p-value is $0.2285714 + 0.01428571 = 0.2428571$, not small by any standard. The null hypothesis is not rejected.

Regardless of her relatively good performance (3 correct identifications), if this had been the result of the experiment, she would have not accumulated enough evidence to disprove those who doubted her claimed discriminatory ability.

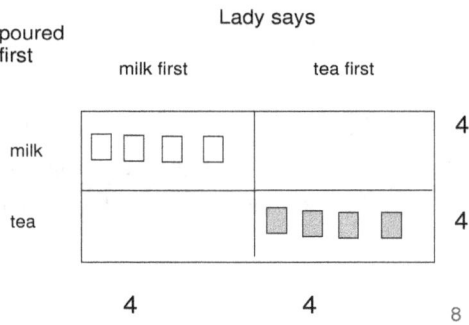

Figure 3.18: The lady correctly identifies all the cups

Salsburg (2001) relates a conversation with H.F. Smith, who said that he was present when tea was served. According to Salsburg (2001), Prof. Smith mentioned that the lady WAS able to identify correctly ALL the cups she was presented. Assume that the experiment was designed with 8 cups as described before AND that Prof. Smith's recollection is true (the situation depicted in Figure 3.18). What is the probability of this happening just by chance? That question has already been answered: $1/70 = 0.01428571$. If this were the case, the p-value would be small and the null hypothesis would be rejected; the lady would have provided sufficient evidence against the skepticism of others.

3.3.2   Writing the statistical hypotheses *

In the previous subsection we have described the null and alternative statistical hypotheses as:

Ho: the lady does not have the discriminatory ability
Ha: the lady can tell if tea or milk has been poured first

A more formal notation can be used to write the statistical hypotheses in terms of the odds ratio (Agresti, 1996). The symbol θ is used to represent the odds ratio in the population. In this specific story, consider the odds of the lady classifying a cup as 'milk first.' An odds ratio of 1 represents the situation where the odds of classifying a cup as 'milk first' is the same for the two groups of cups according to the way they were prepared ('milk first' or 'tea first'), that is, the lady cannot discriminate. Thus, the null and alternative hypotheses can be written as:

Are the odds the same for both groups?

$$Ho : \theta = 1$$
$$Ha : \theta > 1$$

The alternative hypothesis reflects the idea that if the lady can tell if tea or milk is poured first, the odds of assigning a cup to the category 'milk first' are higher for a cup in which milk was actually poured first than for a cup in which tea was poured first.

Using software: Even some smartphones have Fisher's test implemented. There are several free web-based software applications available that can be found by doing a search for 'Fisher's exact test.' Usually a 2x2 table appears on the screen and the data are typed into the table. The name 'exact' comes from the fact that the p-value is calculated using a known probability distribution considering all the possible permutations, not an approximated or simulated one. Most statistical software perform Fisher's exact test.

MINITAB & R: Fisher's exact test for 2x2 tables

In MINITAB, the test can be performed for raw data (not already tabulated data) using STAT > Tables > Cross Tabulation and clicking in Options. It is relatively easy to perform the test in **R**. *First, the values of the cells in the table are entered column by column, indicating the number of rows for the table as in the command* **matrix** *below. By typing* **fisher.test**, *the test is performed.*

```
teacups<-matrix(c(3,1,1,3),nr=2)
fisher.test(teacups,alternative='greater')
```

The output reports $p - value = 0.2429$, the same as the p-value previously calculated, except that rounded to 4 decimals. To look at the data, type **teacups**. The output is:

```
> teacups
     [,1] [,2]
[1,]    3    1
[2,]    1    3
```

For the outcome in Figure 3.18 the output is : $p - value = 0.01429$.

```
teacups<-matrix(c(4,0,0,4),nr=2)
fisher.test(teacups,alternative='greater')
```

3.3.3 The effect of the sample size

The results of hypothetical experiments with 8, 12 and 16 cups in each group
are displayed in Table 3.2 and Figure 3.19 for the lady tasting tea story. In all
those hypothetical experiments the lady is assumed to correctly identify 3/4
of them ('milk first' or 'tea first'), the only difference is the sample size. As the
sample size increases, the probability of correctly identifying 3/4 of the cups
or more in each group, by pure chance, decreases. The same proportion of
correct identifications, but with larger samples, constitutes stronger evidence
against the null hypothesis.

Table 3.3: Results on hypothetical experiments in the lady tasting tea story

Number of cups per group	Correctly identified in each group	p-value
4	3 out of 4	0.2429
8	6 out of 8	0.06597
12	9 out of 12	0.01956
16	12 out of 16	0.006057

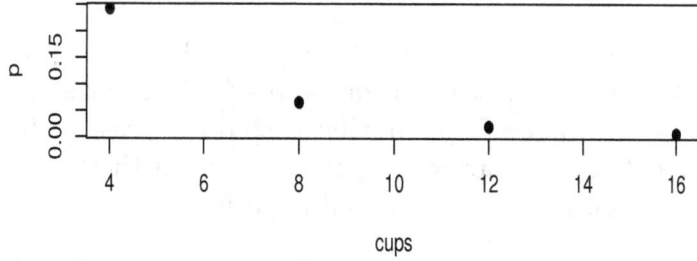

Figure 3.19: Probability of correctly guessing 3/4 or more of the cups, without having the discriminatory ability

3.4   Exercises

3.4.1   Review questions

1. The purpose of the randomization test in Section 3.1 is:
A) To build a confidence interval.
B) To randomly assign the subjects to either treatment or control groups.
C) To check if the subjects were randomly selected.
D) To compare two populations or treatments with regard to a quantitative variable.

2. Consider the following null and alternative hypotheses:

$$H_o : \mu_{elongation} = \mu_{apical}$$
$$H_a : \mu_{elongation} > \mu_{apical}$$

where μ represents the mean length of cells. A random sample of 10 cells is selected at random from each region (elongation and apical) of the roots of an onion. The observations (in millimeters) are:
elongation: 0.107, 0.052, 0.127, 0.043, 0.033, 0.058, 0.045, 0.057, 0.052, 0.038
apical: 0.033, 0.021, 0.031, 0.033, 0.032, 0.020, 0.025, 0.010, 0.021, 0.026
The mean of the 10 cells from the elongation region is 0.0612, and the

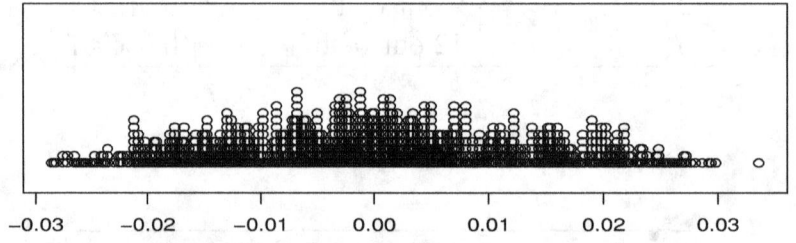

mean of the 10 cells from the apical region is 0.0252. A program in **R** was used to perform the randomization test; calculations for only 1,000 regroupings were done. Explain what each dot represents in the figure given. Calculate the approximated p-value from the figure. What is your decision with regard to the null hypothesis?

3. Assume that two populations A and B are being compared. The null and alternative hypotheses are:

$$H_o : \mu_A = \mu_B$$
$$H_a : \mu_A \neq \mu_B$$

Review questions for Chapter 3

where μ represent the population mean.

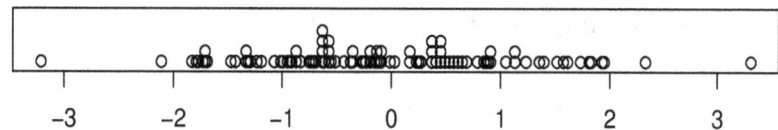

A random sample of size 10 was drawn from each population, the observations are:
Sample from A: 6, 7.20, 4, 9.4, 9, 8.6, 8.5, 6, 4, 5
Sample from B: 2.6, 2.3, 4, 3, 1.8, 3, 6.5, 3, 8.6, 7
The sample means are 6.77 and 4.18, respectively. The bootstrap program in **R** was used to perform the randomization test; only 100 regroupings of the data were done. Calculate the approximated p-value from the figure above (remember that the alternative hypothesis is two-sided). What is your decision with regard to the null hypothesis?

4. Describe in your own words how the randomization test for the equality of means is performed.

5. We are comparing two forests (A and B) in terms of the diameter at breast height (1.37 meters above ground) of the trees. The null and alternative statistical hypotheses are:

$$H_o : \mu_A = \mu_B \qquad H_o : \mu_A > \mu_B$$

We want to use $\alpha = 0.05$. A random sample of 30 trees was selected from each forest and their diameters measured. The sample means are 20 inches and 16 inches respectively. A randomization test was conducted using a computer; the difference $\bar{x}_1 - \bar{x}_2$ was equal to or larger than 4 in only 35 of the 10,000 random regroupings. The approximated p-value and your conclusion about the null hypothesis are:
A) 4/35, do not reject H_o
B) 0.0035, do not reject H_o
C) 0.0035, reject H_o
D) 4/10000, reject H_o

Review questions for Chapter 3

6. An experiment is conducted to compare a treatment with a control. Each one of 40 mice is randomly assigned either to a calorie-restricted diet or to the regular lab mouse diet. The response variable is 'length

of life' of the mice. The null hypothesis is that diet does not make a difference in terms of mean lifetime; the alternative is that, on the restricted diet, mice live longer. We want to work with $\alpha=0.05$. For the twenty mice eating the calorie restricted diet, the average length of life was 43 months. For the twenty mice in the control group, the average length of life was 27 months. A randomization test was conducted using a computer, and only in 28 of the 10000 regroupings was the difference of the means of the randomly formed groups equal to or larger than 16 months. Calculate the approximated p-value and state your conclusion about the null hypothesis.

A) 16/28, do not reject the null hypothesis.

B) 0.0028, reject the null hypothesis.

C) 0.0028, do not reject the null hypothesis.

D) 16/10000, reject the null hypothesis.

7. What is a bootstrap sample?

A) It is a sample drawn with replacement from the original sample and of the same size of the original sample.

B) It is just a regular random sample from a population.

C) In the bootstrap method we draw many different samples of the same size from the population.

D) It is a random sample drawn without replacement from the original sample and half its size.

Review questions for Chapter 3

8. What is the purpose of the statistical method called 'bootstrapping'?

9. The reaction time (milliseconds) of a random sample of 18 female college students, measured with a reaction time ruler, were: 130, 143, 150, 150, 150, 150, 155, 160, 165, 170, 170, 180, 190, 190, 190, 200, 235, 250.

A program in **R** was used to do bootstrapping and only 100 bootstrap samples were selected. Use the dotplot and the list of the bootstrap replicates of the sample mean to calculate a 90% confidence interval for the mean reaction time in the population from where the sample was selected. Explain what each dot in the dotplot represents.

```
 [1] 157.78 160.06 161.56 161.56 161.89 162.56 163.00 163.56 164.39 164.61
[11] 164.72 164.89 165.17 165.28 165.83 166.00 166.83 167.11 167.50 168.22
[21] 168.33 168.39 168.61 168.61 169.06 169.28 169.61 170.06 170.06 170.17
[31] 170.28 170.33 171.28 171.56 171.72 172.11 172.22 172.28 172.39 172.67
[41] 172.78 172.94 172.94 173.06 173.89 174.06 174.17 174.17 174.33 174.33
[51] 174.61 175.00 175.06 175.06 175.44 175.72 176.00 176.39 176.39 176.56
[61] 176.56 176.61 176.67 177.00 177.39 177.67 177.78 177.78 177.94 178.06
[71] 178.33 178.33 178.61 178.67 178.89 179.17 179.33 180.17 180.44 180.83
[81] 181.00 182.50 183.06 183.06 183.33 184.17 184.72 186.00 186.11 186.28
[91] 186.28 186.39 186.67 187.67 187.94 188.22 188.94 189.78 190.83 193.50
```

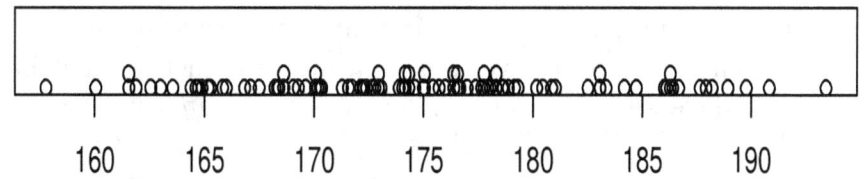

10. A random sample of 20 leaves from a sugar maple tree was selected and their mass (in grams) recorded. Boostrapping was applied in order to build a 95% confidence interval for the mean; the confidence interval was (0.43, 0.57). Which of these possible interpretations is more adequate?

    A) We are 95% confident that all the leaves in the tree weigh between 0.43 and 0.57 grams.

    B) We are 95% confident that all the leaves in the sample weigh between 0.43 and 0.57 grams.

    C) 95% of the leaves in the tree weigh between 0.43 and 0.57 grams.

    D) We are 95% confident that the mean weight of all the leaves in the tree is between 0.43 and 0.57 grams.

    E) We are 95% confident that the mean weight of the 20 leaves in the sample is between 0.43 and 0.57 grams.

11. Somebody asserts that the mean length of the petiole of the leaves of a sugar maple tree is 7 centimeters. We write the null statistical hypothesis $H_o : \mu = 7$. The alternative statistical hypothesis is two-sided. The test will be conducted using $\alpha = 0.05$. We take a random sample of 20 sugar maple leaves from that tree, measure the length of the petiole, and using the bootstrap method calculate a 95% confidence interval for the mean length of the petiole: (4.95,5.7). What is your decision with respect to the null hypothesis?

    A) Reject the null hypothesis.

    B) We cannot answer the question based on the confidence interval.

    C) I don't know, it is a borderline situation.

    D) Do not reject Ho.

12. We want to know if participating in an exercise and diet program that combines walking and low sugar diet changes, on average, the cholesterol level of people who have high cholesterol level. We want to work with α=0.05. The null and alternative hypotheses are:

    $$H_o : \mu_d = 0 \qquad H_a : \mu_d \neq 0$$

Review questions for Chapter 3

A random sample of 25 patients with high cholesterol level was selected at random and put in a program that combines walking and low sugar diet. The cholesterol level was measured for each patient before and after the treatment, and the differences *before-after* were calculated for each patient. Thus, the values we have for each patient are already the differences *before-after*. Using a computer, 2,000 samples (with replacement) of size 25 were selected from the sample, the mean difference was calculated for each bootstrap sample and based on those values, the following 95% confidence interval was calculated : (15,32). What is your decision about the null hypothesis?

A) Reject it because the value 0 is not in the confidence interval.
B) Do not reject it because the value 0 is not in the confidence interval.
C) Do not reject it because the confidence interval includes 0.
D) Do not reject it because 32-15 is not 0.

3.4.2   Exercises for discussion or homework

The programs in **R** included in this chapter to perform the randomization test and to apply the bootstrap method (both for means and correlations) are available from our website:

`http://faculty.etsu.edu/seier/IntroStatBioBook.htm`

You can copy and paste any program into a word processor, make the appropriate changes (data, number of replicates, confidence), and copy and paste it into **R**. If you want to save the programs, do it as **.txt** files. Using the extension **.doc** might change the aspect of the quote marks in the program because of the several fonts available, something that can cause problems in **R**.

1. **Estimating mean reaction time.** The reaction time (milliseconds) of a random sample of 18 female college students measured with a reaction time ruler were: 130, 143, 150, 150, 150, 150, 155, 160, 165, 170, 170, 180, 190, 190, 190, 200, 235, 250. Use the program in **R** to calculate a 95% bootstrap percentile confidence interval for the mean reaction time in the population from which the sample was drawn.

2. **Reaction time in boys and girls in the 5th grade.** Read Exercise 4 in Chapter 2. Assume that this is a random sample of 5th graders from a school district. Use the randomization test to answer this question: Is there a significant difference (assuming $\alpha = 0.05$) in the mean reaction time of boys and girls under quiet conditions?

3. **Does distraction affect boys and girls in the same way?** Read Exercise 4 in Chapter 2. Does distraction affect the reaction time of boys and girls of this age in a similar way? For each one of the 28 5th graders, calculate the difference in reaction time: *distraction − quiet*. Make a list of those differences for the boys and a list of those differences for the girls, and use dotplots to compare the two groups. Use the randomization test to compare the two groups of differences, and interpret the results.

4. **Sexual dimorphism?** Consider a species of birds. The research question is whether there is sexual dimorphism (i.e., if the males and females appear different) with respect to the length of the wing. Specifically, we want to find out whether, for the population of adult birds of this species, the mean length of the wing is larger in males than in females. Thus, the two statistical hypotheses are: $H_o : \mu_{males} = \mu_{females}$ and $H_a : \mu_{males} > \mu_{females}$. A random sample of 10 male birds is selected and a random sample of 10 females is selected. The average length of the wing for the 10 males is 12.3 centimeters, and the average length of the wing for the 10 females is 10.8 centimeters. The 20 measurements are merged and randomly split into two groups of 10 measurements each one. The mean length for each group is calculated, as well as the difference between those two means. The procedure is repeated 10,000 times with the help of a computer. Out of the 10,000 times, it happens that only in 20 regroupings the difference between the means of the two groups is equal to or larger than 1.5 centimeters. What is your decision with respect to the null hypothesis? Is there sexual dimorphism with respect to the wing in this species?

5. **Is the mean length of cells different in the two regions?** There are two regions in the onion root, the 'apical meristem' region and the 'elongation' region. Are the cells in region 1 and 2 of the same length on average, or do cells in the apical region tend to be smaller on average? The null and alternative hypothesis are:

$$H_o : \mu_{elongation} = \mu_{apical}$$
$$H_a : \mu_{elongation} > \mu_{apical}$$

where μ represents the mean length of cells. The following are the lengths (mm) for two random samples, one from each region, of 10 cells.

apical: 0.033, 0.021, 0.031, 0.033, 0.032, 0.02, 0.025, 0.01, 0.021, 0.026

elong.: 0.107, 0.052, 0.127, 0.043, 0.033, 0.058, 0.045, 0.057, 0.052, 0.038
Use the **R** program to perform a randomization test with this data set.
State your conclusion about the null hypothesis.

6. **Does staking the potato plant help?** A student with a passion for
   gardening carefully planned some experiments and observational studies with potato plants and recorded the data. One of his hypotheses
   was that if he staked the potato plants (as you would with tomato
   plants), he could increase the mean yield of the plant. Write the null
   and alternative statistical hypotheses. The data set below corresponds
   to 8 potato plants grown without fertilizer and in exactly similar conditions except for the staking; 4 of them were staked and 4 grew without
   stakes. The observations correspond to the yield in pounds after 131
   days of planting.
   No stakes (control group): 0.40 1.55 1.10 1.05
   Staked (treatment group): 3.15 2.85 1.75 2.50
   Source: Data produced by V. Burke
   The mean yields (in pounds) per treatment are: 1.025 (no stakes) and
   2.563 (staked), and the difference between them is 2.563-1.025=1.538.
   Sketch a plot of the data by hand. There are 8 observations in two
   groups of 4. How many 'regroupings' of the observations could you
   make? In how many of those regroupings was the difference between
   the two means larger than 1.538? What is your conclusion? You do
   not need to use a computer to solve this exercise, the data set is small
   enough to manage by hand.

7. **Is there a linear relationship between body mass index (BMI)**
   and hemoglobin (Hb)? Read Exercise 3 in Chapter 2. Build a 95%
   bootstrap percentile confidence interval for the correlation coefficient.
   Does the interval include the value 0? Interpret the results.

Exercises for
discussion
or homework
in Chapter 3

8. **Is there a linear relationship between the length of the leg of**
   a cricket and the distance the cricket jumps? Read Exercise 6
   in Chapter 2. Calculate a 90% bootstap percentile confidence interval
   for the correlation coefficient. Interpret your results. Would a 95%
   confidence interval be wider or narrower?

9. **Are height and foot length correlated among young female**
   students? Read Exercise 8 in Chapter 2 (Section 2.16.4). Assume
   this is a random sample from female students in a college. Calculate a

95% confidence interval for the correlation coefficient and interpret the result.

10. **How strong is the relationship between the length of the petiole and the mass of the leaf blade in sugar maple trees?** Read Exercise 9 in Chapter 2. Calculate a 95% bootstrap percentile confidence interval for the correlation coefficient and interpret it. Would a 98% confidence interval be wider or narrower? Would a 90% confidence interval be wider or narrower?

11. **Leaf area and yield of potato plants** A student with a passion for gardening suspected that there is a relationship between the total leaf area (cm^2) of a potato plant and the yield in pounds of the plant. He carefully estimated the total sum of the areas of the leaves in a plant and weighed (pounds) the potatoes produced by the plant at harvest. He did this for 25 potato plants. The data set is on Table 3.4 (data file: *leafandyield*). Produce a scatterplot and calculate the correlation. Calculate a 95% confidence interval for the correlation between leaf area and potato yield in the population of potato plants of this variety, cultivated in this particular region of the country under the same conditions of soil and fertilization.

Table 3.4: Leaf area and yield in potato plants

Plant No.	Total lef area (cm^2)	Yield(lb)	Plant No.	Total leaf area (cm^2)	Yield (lb)
1	2658.5	0.75	14	8503.0	3.30
2	11250.6	4.20	15	10078.0	3.90
3	7142.1	3.20	16	3966.8	1.40
4	5810.4	2.60	17	8765.1	3.70
5	3442.8	1.50	18	11313.8	4.10
6	12665.1	4.25	19	7793.7	3.40
7	9177.5	3.80	20	8092.0	2.80
8	4411.6	1.90	21	10686.3	3.50
9	5618.6	2.00	22	12318.2	4.50
10	10107.3	4.00	23	9675.5	4.10
11	9899.2	3.90	24	9346.2	3.60
12	4991.5	2.10	25	10251.4	3.50
13	6263.2	2.75			

Source: Data produced by V. Burke

12. Read question 4 in the lab for Chapter 2. Perform a randomization test to compare the transfusion group with the non-transfusion group with regard to mean hemoglobin.

13. A study with *Sarcophaga crasipalpis* (flesh flies) consisted in placing 20 male flies in a cage and 20 female flies in another cage. All the flies had just emerged from the pupal case. After two weeks, only 2 males and 16 females were alive. Prepare a two-way table to display this information. The rows are: 'male' & 'female.' The columns are 'alive' & 'dead.' Apply Fisher's exact test to test the null hypothesis that the odds of dying within 2 weeks are the same for males and females.

3.4.3   Small data analysis projects

All these projects require you to define a population or populations, and variable(s) of interest, conduct an observational study or experiment, apply one of the methods studied in this chapter, and write a short report. A poster could be used instead of a typed report to tell the story.

1. Read Example 3.1 (chickens and phytate). Think of two treatments or populations you would like to compare in terms of the mean value of a quantitative variable. Write the research question and the statistical hypotheses. Conduct a small observational study or experiment. Describe your data with summary statistics and statistical graphs. Use the program in **R** provided in the book to perform the randomization test and arrive at a conclusion. Write a report telling the whole story.

2. Read Example 3.3 (estimating the mean length of rhododendron leaves). Come up with your own small project. Define a population and a quantitative variable of interest. Draw a random sample from the population and measure the variable for the individuals in the sample. Plot the data and calculate the mean. Use the **R** program provided in the book in order to build a bootstrap percentile confidence interval for the value of the mean. Interpret the confidence interval you obtained. Write a short report describing the population, variables, data, and the confidence interval.

3. Read the 'altitude and RBC' and 'bone mineral density in femoral neck and hip' examples in Section 3.4. Come up with your own project. Think of a population and two quantitative variables between which you are interested in determining if there is an association. Select a

random sample from the population and measure both variables for each individual. Produce a scatterplot and calculate the correlation for the data in the sample. Use the program in **R** to calculate a 95% confidence interval for the correlation in the population. Interpret the confidence interval. Write a short report describing the population, variables, data and your findings.

4. Read the 'lady tasting tea' story in Section 3.5. Think of two groups (humans or any other species - animal or plants) you would like to compare (males and female, older and younger, two different species) in terms of a categorical variable (of the 'yes'/'no' type). Either select a random sample from each group for an observational study or design and perform a small experiment. Prepare a 2 × 2 table with your data and use a bar chart to illustrate the data. Write the null and alternative hypotheses. Use Fisher's exact test to arrive at a conclusion. Write a report stating the research question and statistical hypotheses, and describe your study, data, and conclusions.

5. Conduct a small experiment with tomato seeds. Think of two different conditions under which to plant the seeds. Write the null and alternative hypotheses on the odds of germination. Plant 20 seeds under each one of the two conditions. Wait about two weeks and check how many seeds of each group have germinated. Prepare a two-way (2 × 2) table with your data. Use Fisher's exact test to arrive at a conclusion. Write a report stating the research question and statistical hypotheses, and describe your study, data, and conclusions.

Data analysis projects for Chapter 3

References

Agresti, A. (1996) *An Introduction to Categorical Data Analysis.* New York: Wiley.

Delk, K. (2009) *Effects of Cherries on Arthritis Pain.* Undergraduate honors thesis. East Tennessee State University.

Darwin, C. R. (1876). *The effects of cross and self fertilization in the vegetable kingdom.* London: John Murray. *http://darwin-online.org.uk*

Efron, B. (1979) Bootstrap methods: Another look at the Jacknife. *Annals of Statistics* **7**: 1-26

Fisher, R.A. (1966) *The Design of Experiments.* Eight edition. London: Oliver and Boyd.

Fisher-Box, J. (1978) *R.A. Fisher: The Life of a Scientist.* New York: Wiley.

Holmes, S. (2003) Bootstrapping Phylogenetic Trees. *Statistical Science* **18**: 241-255.

Jacquez, J.A. , Jacquez, G.M. (2002) Fisher's randomization test and Darwin's data- A footnote to the history of Statistics. *Mathematical Biosciences* **180**: 23-28.

Manly, B. (2007) *Randomization, Bootstrap and Monte Carlo Methods in Biology, Third Edition.* Boca Raton: Chapman & Hall/CRC

Onyango, E., Asem, E. and Adeola, O. (2009) Phytic acid increases mucin and endogenous amino acid losses from the gastrointestinal tract of chickens. *British Journal of Nutrition* **101**: 836-842.

Salsburg, D. (2002) *The Lady Tasting Tea-How Statistics Revolutionized Science in the Twentieth Century.* New York: Holt.

Chapter 4

The binomial distribution

This chapter is about the basic concepts of probability and a very useful probability model: the binomial distribution. The focus is on situations or experiments that involve randomness and have two possible outcomes: success and failure. The binomial distribution makes it easy to calculate the probability of getting a given number of successes when the experiment is repeated several times.

4.1 Introduction

Why study probability?

Probability is a branch of mathematics that deals with randomness. The motivation for the development of probability was the analysis of games of chance. Probability can be studied at different levels and has a wide range of applications in the life sciences, particularly in genetics. Frequently the question: 'What is the probability of...?' , is used in the sense of 'How likely is this to happen?' In order to answer this question it is very important that the context of the problem is clearly identified and that:

The origin of probability is in the analysis of games of chance

1. All the possible outcomes are considered.

2. The chances of each outcome are represented by a number between 0 and 1, and that the sum of those numbers is 1.

For example, when a seed is planted, the two possible outcome are: the seed germinates or the seed does not germinate. Assume there is information that approximately 85% percent of seeds of this type have germinated in the past under similar conditions. Since there are only two possible outcomes, their probabilities will be called p and $1 - p$:

$$p = \text{P(seed will germinate)} = 0.85$$
$$1 - p = \text{P(seed will not germinate)} = 0.15$$

What if several seeds are planted? If ten seeds were to be planted instead of one, a natural question is: **What is the probability that all ten of them will germinate?** We need to calculate the probability that the first will germinate, AND the second will germinate, AND the third will germinate, and so on. A key assumption is that the seeds behave independently one from the other. The probability that all of them will germinate is $0.85 \times 0.85 \times 0.85 \times ... = (0.85)^{10} = 0.1968744$.

Somebody could ask: **What is the probability that 9 germinate and 1 does not germinate?** Well, consider that the first nine will germinate and the last one will not; the probability of that happening is $(0.85)^9 0.15 = 0.03474254$. But wait, something is missing here, who says that it has to be precisely the last one the one that will not germinate; it could be the first or the second, or any other one. All the possible cases in which one seed does not germinate have to be considered, there are 10 such possibilities and all those cases have the same probability $(0.85)^9 0.15$. Thus, the probability that only 9 will germinate is $10(0.85)^9 0.15 = 0.3474254$.

Probabilities can be calculated from scratch as in the seed exercise. However, time can be saved by solving probability problems using probability 'models' that describe typical situations rather than solving each one separately. Some examples similar to the germination of the seed (with a different value for p) are: a child inheriting a certain condition, a predator capturing a prey or not, a rat in a maze taking the left or right path, a woman getting infected or not with the papilloma virus, etc. Instead of figuring out how to calculate the probability for each problem, it would be useful to have a probability model that can be used in all those examples.

Probability is useful in hypotheses testing
Probability is also useful in making decisions about statistical hypotheses. A null statistical hypothesis is tested by doing an observational study or an experiment to get reliable data. The question used to test H_o is: 'What is the probability of getting observations as the ones we got, or more extreme ones, just by chance, when H_o is true?' The answer to this question is called 'p-value' and helps deciding if the data constitute evidence against the null statistical hypothesis.

When did the study of probability begin? *

Probability is a branch of mathematics that deals with randomness. Games of chance have been played for a long time. Dice made of fired pottery dating from around 3000 BCE have been found, and there is evidence that bones called 'astragalus' were used before the introduction of dice. The first known book on the study of games of chance is 'Liber de Ludo Aleae' (Book on the Games of Chance) written by Gerolamo Cardano circa 1550 and published after the author's death in 1663. Cardano was a physician and a mathematician who lived in the Italian peninsula. That book contains the outline of a first theory of probability. Blaise Pascal, a French mathematician (1623-1662), also became interested in the study of probability due to the questions posed to him by a famous gambler (A. Gombaud, the Chevalier de Mere). Pascal maintained correspondence on this topic with Pierre de Fermat, another famous mathematician of that time. In 1657, Christiaan Huygens published in Holland 'De Ratiociniis in Ludo Aleae' (The value of all chances in games of fortune). In 1718 Abraham De Moivre published his book 'Doctrine of Chances or, a Method of Calculating the Probability of Events in Play.' The book 'Ars Conjectandi' (The Art of Conjecturing) by Jacob Bernoulli was published in Basel, Switzerland, in 1713. Later, Daniel Bernoulli (1700-1782) made further contributions to the study of probability. Probability theory has continued to develop; more than ten journals on probability alone (and several more on probability and statistics) are currently published.

It is a common occurrence in science, mathematics, and technology that something developed for one purpose is later found to be useful in other fields. Currently the breadth of the applications of probability goes well beyond the games of chance and plays a major role in the life sciences. It should be noted that probability is the main mathematical tool of statistics.

Situations involving randomness

Consider the situation that occurs when a coin is tossed. Assume the coin is fair, meaning that there is no bias as to which side comes up, but there is an equal chance for heads or tails. The outcome will be either heads (H) or tails (T), but it is not known exactly which one will come up the next time the coin is tossed. There are other situations in which randomness is present: having a child who could have brown eyes or not, planting a bulb and not knowing if it will bloom during the spring, observing a bee leaving the hive and not knowing if it will go to a given food source or not. In many of these cases the two possible outcomes are **not** equally likely.

There are other situations where there may be more than two possible outcomes, for example: 'eye color of a baby to be born,' or 'eye color of a person we will pick at random from a population': the possible outcomes

Randomness: there are several possible outcomes and we don't know which one will happen

would be brown, blue, hazel, green and 'other'; selecting a person from a population and asking for his/her type of blood, which could be O, A, B or AB (in the USA, approximately 44%, 42%, 10% and 4% of the population respectively).

It might not even be possible to list one by one all the possible outcomes. The height of a planted maize plant is to be measured after 3 months. The height will be some number between 0 and 250 centimeters, that is, it might be any value in the interval [0,250]. A cup of fresh water is to be taken from a stream and the number of a given type of invertebrates counted. The number of organisms could be 0,1,2,... We define the variable X:*number of invertebrates in a cup of water*. The minimum possible value for X is 0, but it does not have an upper bound ($X \geq 0$).

The terms **random experiment** and **random phenomenon** are often used in relation to situations involving randomness, like the ones described in this section. Three important definitions are:

Random experiment

Sample space

Event

- **Random Experiment** or random phenomenom: is an experiment or situation where all the possible outcomes are known, but we do not know which outcome will actually happen.

- **Sample space**: S, the set of all the possible **outcomes** from a random experiment. The outcomes, or **elementary outcomes**, are also called **simple events**.

- **Event**: any subset of the sample space. An event can have one, several or none of the outcomes or simple events.

Some examples:

1. When tossing a fair coin, $S=\{H, T\}$. An event A could be formed by a single outcome such as $A=\{H\}$.

2. When rolling a die, $S=\{1,2,3,4,5,6\}$. Throughout the book we are assuming that dice are cubes and that they are 'fair' in the sense that all the six faces are equally likely to come up on top when the die is rolled. Events A and B could be defined as A: 'to get an even number', $A = \{2, 4, 6\}$, and B: 'to get a six' , $B=\{6\}$.

Complement of an event: the outcomes that are NOT in the event

3. For eye color, the sample space is $S = \{brown, blue, green, hazel, other\}$

   If event A is defined as $A = \{brown\}$, the complement of A is formed by all the other colors $A' = \{blue, green, hazel, other\}$.

4. DNA is a long sequence of nucleotides that can be C,A,T or G. If a location is selected at random and the nucleotide is determined, the sample space is $S = \{C, A, G, T\}$. An event D is defined so that $D = \{C, G\}$.

5. Planting a maize plant and observing how tall it will be after 3 months. It will be some number between 0 and 250 centimeters, $S = [0, 250]$. An event T could be 'tall plant,' i.e., T: the plant measures more than 160 centimeters, $T = (160, 250]$.

6. The local temperature (Celsius) tomorrow at noon. If it is summer, it will be some value in the interval 20-40 C^o, but the exact value is not known. The sample space is $S = [20, 40]$. An event H could be defined as H: the temperature is at least 30 C^o, $H = [30, 40]$.

4.2 Probability

There are several definitions of probability of an event, with some authors calling them 'interpretations of probability.'

4.2.1 Classic definition

Probability is the proportion of outcomes in the sample space that are in the event:

number of favorable outcomes/ number of possible outcomes

The classic definition of probability can be used only under two conditions

In order to be able to apply this definition, two conditions are needed:

- There is a finite number of outcomes.

- All the outcomes are equally likely to happen.

This definition is very simple and useful in cases such as rolling a regular die with 6 sides. There are 6 possible outcomes, and if the die is fair, all the possible outcomes are equally likely. The probability of getting a 4 is 1/6, which is the same for each one of the other numbers on the die. This definition cannot be applied in cases such as the eye color. There are 5 possible outcomes (brown, blue, green, hazel and other) but since the colors are not equally frequent, the probability of brown eyes **is NOT** 1/5.

4.2.2   Probability as the limit of the relative frequency

If a fair coin is tossed just a few times, the results might not be exactly half heads and half tails. But if the coin is tossed a very large number of times, both the number of heads and the number of tails are roughly half the total number of tosses. Figure 4.1 displays the results of tossing a (virtual) fair coin 10,000 times using a computer and recording the relative frequency of heads after each toss. The results of the first few and last tosses appear in the Table 4.1.

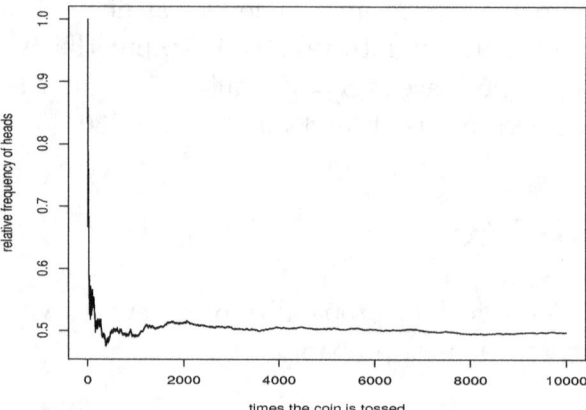

Figure 4.1: Cumulative relative frequency of heads after each one of 10000 tosses of a fair coin

Table 4.1: Tossing a coin 10,000 times

toss number	outcome	x	cumulative counts	cumulative relative frequency
1	H	1	1	1/1=1
2	H	1	2	2/2=1
3	T	0	2	2/3=0.6666667
4	H	1	3	3/4=0.75
...	...	...	...	...
9998	H	1	4975	4975/9998= 0.4975995
9999	H	1	4976	4976/9999= 0.4976498
10000	T	0	4976	4976/10000=0.4976000

Notice that at the beginning there can be a lot of variation in the cumulative relative frequency. When the number of tosses (n) becomes very large, then the proportion of heads becomes more and more stable. As n becomes larger and larger, the relative frequency becomes closer to the value 0.5 associated with the 'fair coin' situation. The value 0.5 is said to be the 'limit' **Idea of limit** of the relative frequency $heads/n$ when n approaches $infinity$, meaning as n becomes larger and larger.

Probability is defined as the limit of the relative frequency when n approaches 'infinity' (∞). This definition makes sense in a lot of applications, but there are situations in which the assumption that something can be repeated a very large number of times does not sound reasonable.

Probability as the limit of a relative frequency

4.2.3 Axiomatic definition of probability

Since the two preceding definitions (proportion of favorable outcomes and limit of the relative frequency) cannot always be applied, a more universal definition is needed. In mathematics, an axiom is a proposition considered self-evident or 'assumed to be true.' Generally, an axiom is later used in 'proving' other statements. First some definitions and notation are needed.

Definitions and notation

In Figure 4.2 the circles represent two events. The first Venn diagram represents disjoint or mutually exclusive events, meaning **events that do not have any outcome in common**. The second Venn diagram represents two events that are not disjoint, meaning they share some outcomes.

Mutually exclusive or 'disjoint' events do not have any outcome in common

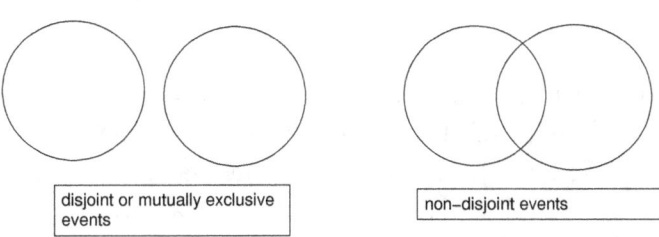

Figure 4.2: Venn diagrams representing disjoint and non-disjoint events

Figure 4.3 represents three operations that can be done with events. In the first Venn, diagram both events (A and B) have been completely shaded; this represents the union of the two events. In the second Venn diagram in

Figure 4.3, only the region that belongs to both events has been shaded. This represents the 'intersection' of the events. The third diagram in Figure 4.3 displays an event (A) and its complement.

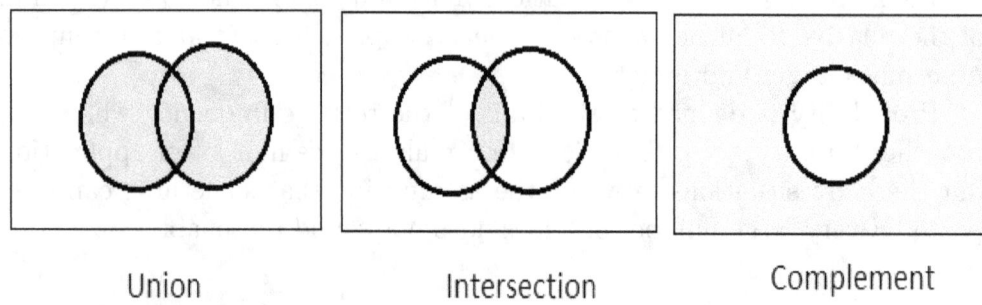

Union   Intersection   Complement

Figure 4.3: Union ($A \cup B$), intersection ($A \cap B$), and complement (A') of events

The main three operations with events are:

The union of two events $A \cup B$ includes all the outcomes that are in either A or B

- The 'union' of two events A and B, $A \cup B$, is formed by all the outcomes in A or B. It is enough that an outcome is in one of the events in order to belong to their union.

- The intersection of two events A and B, $A \cap B$, is the part of the sample space that is common to both events. An outcome has to belong to both events in order to belong to their intersection.

The intersection $A \cap B$ includes only the outcomes that are in both A and B

- The complement of an event is represented by either A^C or A' or $\bar{A}$, and it is formed by the outcomes from the sample space that are not in the event.

In the example of rolling a 6-sided fair die, $S = \{1, 2, 3, 4, 5, 6\}$. If $A = \{6\}$, then the complement of A will be formed by all the outcomes that are not in A, $A' = \{1, 2, 3, 4, 5\}$. If event B is defined as 'getting a face with an even number,' $B = \{2, 4, 6\}$. The union of A and B will be formed by all the outcomes in A and all the outcomes in B, $A \cup B = \{2, 4, 6\}$. The intersection of A and B will be formed by the outcomes that both have in common (Figure 4.3b); in the example, $A \cap B = \{6\}$. In this example, **A is contained in B**, $A \subset B$, because all the elements in A are also in B. If the event C is defined as 'getting a face with an odd number,' $C = \{1, 3, 5\}$. A number between 1 and 6 cannot be even and odd at the same time, thus, B and C are disjoint or mutually exclusive.

Impossible events

On occasion, somebody might ask about the probability of an event that has no outcome in the sample space. For example, if a die (with sides 1, 2, 3, 4 ,5, 6) is rolled and somebody asks: 'What is the probability of getting a 7 when the die is rolled just once?' The number 7 is not in the sample space, thus **Events** $P(7) = 0$. The symbol $\emptyset$ (called 'empty set') is used to represent impossible **with** events, $P(\emptyset) = 0$. We also use the notation $\emptyset$ to represent the intersection of **probability** two mutually exclusive or disjoint events, $A \cap B = \emptyset$. **zero**

The axioms

The probability of an event A is a number $P(A)$ such that: **Probability is a**
number between

1. $P(A) \geq 0$ (the probability of an event is non-negative) **0 and 1**
   assigned to
2. $P(S) = 1$ (the probability of the whole sample space is 1) **an event,**
   following
3. If A and B are mutually exclusive events: $P(A \cup B) = P(A) + P(B)$ **certain rules**
   (the probability of the union of disjoint events is the sum of their probabilities)

Axiom 1 says that the probability of an event can never be negative, meaning that the minimum value for a probability is zero. Axiom 2 says that the probability of the whole sample space is 1. Axiom 3 says that when **A probability** two events cannot happen at the same time, the probability that either of **can never** them happens is the sum of the probabilities of each one. There is a more **be negative** sophisticated version (studied in more advanced probability courses) of axiom 3 that considers more than two mutually exclusive events.

The probability of an event is a number, but that number is usually **If two events** not assigned arbitrarily. It can be based on specific numerical data and **are** criteria as the classic or the relative frequency definitions. It can also be a **mutually** more 'subjective' or 'personal' probability based on previous experience and **exclusive,** several sources of information. We start by assigning probabilities to the **the probability** individual outcomes or elementary events, and from there the probability of **of their union** other events is calculated. For example, Snee (1974) reports that from 592 **is the sum** students in a statistics course, 220 had brown eyes, 215 had blue eyes, 93 had **of their** hazel eyes and 84 had green eyes. If we were to select at random a student **probabilities** from that population, the probabilities corresponding to each eye color would be 220/592, 215/592, 93/592, and 64/592. From there, we can calculate the probabilities for some other events. Assume that each student has a single eye color, in other words to have brown eyes and hazel eyes are mutually exclusive events, the probability of having brown AND hazel eyes is 0. From Axiom 3, the probability that a randomly selected student from that population has 'brown or hazel eyes' is $(220 + 93)/592 = 0.5287162$. Based on the same information and applying the definition of complement of an event, the probability of <u>not</u> having blue eyes is $P(Blue') = 1 - 215/592 = 0.6368243$.

The classic definition of probability and the definition of 'probability as the limit of relative frequencies' are used, when applicable, in order to calculate the probability of the outcomes. However, sometimes a person needs to decide the probability of an outcome based on his/her experience, combined with several sources of information. The experiences of different people might be different; that is the reason behind the term 'personal' or 'subjective' probability. If two people with a long experience in agriculture are asked the question: 'What is the probability that a Pima cotton plant will grow healthy in the eastern part of Tennessee?,' the value they assign to the probability might be a little different because their experiences and information might be a little different. DeGroot & Schervish (2002) write: 'the final assignment of numerical probabilities is the responsibility of the scientist himself/herself.' Of course, the values assigned to the probabilities have to obey the axioms.

4.2.4   Independent events and the multiplication rule

Independent events: the probability that BOTH occur is the product of their probabilities

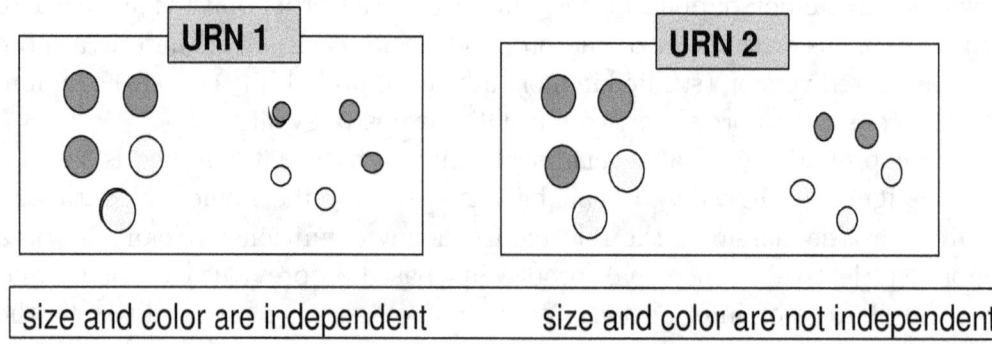

Figure 4.4: Independence and non-independence between size and color

Figure 4.4 represents two urns where the dark color balls are red and the light color balls are yellow. In the first urn, color and size are independent; in the second one, color and size are not independent. Why? In the first urn 'color and size' are independent because the proportion of reds among both the large and the small balls is the same. In the first urn, $P(red\,and\,large) = 3/10$ and $P(red) \times P(large) = (6/10)(1/2) = 3/10$. However, that is not true in the second urn where there are fewer red balls among the small balls than among the large balls. In the second urn, $P(red\,and\,large) = 3/10$ and $P(red) \times P(large) = (5/10)(1/2) = 5/20$. In general, we say that **two events A and B are independent if $P(A \cap B) = P(A) \times P(B)$.**

4.2.5 Some consequences of the axioms

There are several properties of probability that are consequences of the 3 axioms; four of them are listed in the box below. The first two refer to the situations depicted in Figure 4.5. The fourth consequence refers to the situation depicted in Figure 4.6. The reasons why these consequences are true are explained later.

Probability of the complement is equal to 1−P(event)

1. $P(A^C) = 1 - P(A)$ or $P(A') = 1 - P(A)$. (The probability of the complement of an event is 1 minus the probability of the event).

   If the events are not mutually exclusive, the probability that either one occurs is equal to the sum of their probabilities minus the probability that both occur

2. If $A \subset B$ then $P(A) \leq P(B)$. (If one event is contained in another, its probability cannot be larger than the probability of the bigger event).

3. $P(A) \leq 1$. (The probability of any event is at most 1).

4. $P(A \cup B) = P(A) + P(B) - P(A \cap B)$. (In general, the probability of the union of two events is the sum of their probabilities minus the probability of their intersection)

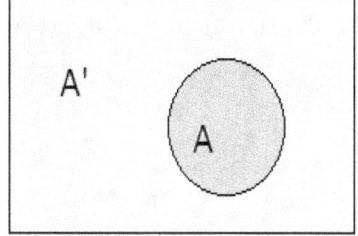 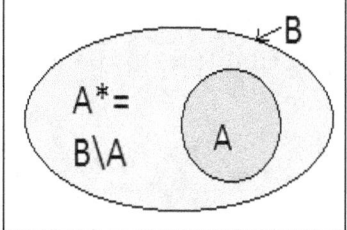

Figure 4.5: a. Complement of an event. b. Event A contained in event B

Why are those consequences true? *

1. The probability of the complement of an event is 1 minus the probability of the event: $P(A^C) = 1 - P(A)$ or $P(A') = 1 - P(A)$. Why? Well, look at Figure 4.5a. An event and its complement together form the whole sample space, i.e., $A \cup A' = S$. Thus, the probability of their union is 1 because that is the probability of the sample space according to axiom 2. Also, A and A' are disjoint because they do not have any element in common, so the probability of the union is the sum of the probabilities. Therefore, $P(A) + P(A') = 1$, and we conclude that $P(A') = 1 - P(A)$. This is a simple but very useful property because sometimes it is easier to calculate the probability of the complement and subtract it from 1 than to try to calculate the probability of the event directly.

2. If one event is contained in another, its probability cannot be larger than the probability of the bigger event: If $A \subset B$ then $P(A) \leq P(B)$. Look at the second diagram in Figure 4.5 and think of B as formed by two parts: A and the part of B that is not in A (represented with A^* or $B \setminus A$. A^* is not the complement of A, but the complement of A with respect to B. $B = A \cup A^*$, and A and A^* are disjoint; thus, $P(B)$ can be written as the sum of the probabilities of A and A^*, $P(B) = P(A) + P(A^*)$. Since probabilities are non-negative numbers, $P(A^*) \geq 0$; thus, $P(A) \leq P(B)$.

3. The probability of any event is at most 1, $P(A) \leq 1$. This follows from the previous consequence. All events are contained in the sample space S, and the probability of the sample space is 1, then the probability of any other event cannot be greater than 1.

4. $P(A \cup B) = P(A) + P(B) - P(A \cap B)$. Look at Figure 4.6. The proof simply consists of expressing $A \cup B$ in terms of disjoint events so that their probabilities can be added.

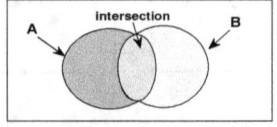

Figure 4.6: Intersection of two events

4.3 Binomial distribution

4.3.1 Bernoulli experiments

Think of an experiment that has only two possible outcomes, or in the case of several outcomes, these can be grouped in two groups called 'success' and 'failure.' For example, tossing a coin has two possible outcomes, heads (H) and tails (T). 'Success' can be defined as heads and 'failure' as tails. If a die is rolled, the possible outcomes are 1, 2, 3, 4, 5, 6. However, if we are playing a game in which one needs to get a 6 in order to win, 'success' can be defined as 6, and all the other outcomes will be grouped under 'failure.'

A Bernoulli experiment has two possible outcomes: success or failure

A Bernoulli variable

A Bernoulli variable X is defined such that $X = 0$ when 'failure' happens, and $X = 1$ when 'success' happens. The name Bernoulli comes from the famous family of mathematicians already mentioned in Section 4.1. In the coin example, if the outcome is tails then $X = 0$, if the outcome is heads then $X = 1$.

Repeating Bernoulli experiments

Consider the following two examples of a Bernoulli experiment that is repeated n times:

Repeating a Bernoulli experiment n times

1. Tossing a coin is a random experiment where success can be defined as 'heads' and 'tails' will be considered failure. Tossing a coin 3 times is a sequence of random experiments. One might care for the outcomes in the specific order in which they happen such as H, H, T, or one might care just for the number of success (2) in the $n = 3$ replicates. Here we focus on the number of successes in the n independent replicates of a success/failure type of experiment. Counting the number of successes is equivalent to adding the values that the Bernoulli variable takes, because each '1' represents a success and each '0' a failure.

2. A random sample of size n will be selected from a very large human or animal population. Each individual in the sample will be classified as having or not having a given trait. The selection of each individual can be considered a random experiment with two possible outcomes, having or not having the trait. To select n individuals is to repeat the experiment n times. The quantity of interest is the proportion of individuals in the sample who have the trait.

Hands-on activity

Take a coin (preferably a 'quarter') and toss it 3 times. Report the results here:

toss #	outcome
1	
2	
3	

How many heads did you get in total?

How many heads did the other students in the class get?

Figure 4.7 displays the results for 10 students in a statistics class. Produce a similar figure for this class. Are there values that happen more frequently than others?

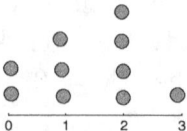

Figure 4.7: Number of heads in 3 tosses of a fair coin for each one of 10 students

Listing all the possible outcomes

Assume a fair coin is tossed 3 times. The eight possible outcomes are:

$$\text{HHH, HHT, HTH, THH, TTH, THT, HTT, TTT.}$$

How many heads are in each outcome? Obviously:

$$3, 2, 2, 2, 1, 1, 1, 0.$$

How many outcomes have 0 heads? Just one: TTT.

How many outcomes have 1 head? There are 3: TTH, THT and HTT.

How many outcomes have 2 heads? There are 3: THH, HTH, HHT.

How many outcomes have 3 heads? Just one: HHH.

The answers to those questions are summarized in Table 4.2. Looking at Table 4.2, the results of the hands-on activity are easier to understand. Some values for the 'number of heads' were more frequent than others because there are more outcomes that lead to certain values.

Table 4.2: Possible number of heads when a coin is tossed 3 times

Outcomes	Number of heads	Number of outcomes
TTT	0	1
TTH, THT, HTT	1	3
THH, HTH, HHT	2	3
HHH	3	1

Looking for a pattern

- If we toss the coin only two times the sample space is

$$S = \{HH, HT, TH, TT\}.$$

  The possible number of heads are 0,1,2, and the number of outcomes that produce each one of those values are 1,2,1 respectively. When we toss the coin 3 times, the number of outcomes that produce 0,1,2,3 heads are 1,3,3,1 respectively. Is there a pattern there?

- If there is a pattern, maybe we can use it when we toss the coin more times. For example: What would happen if instead of tossing the coin 3 times we toss it 7 times? We know the possible number of heads are 0,1,2,3,4,5,6,7 but we do not know how many outcomes produce each one of those values. Is there a shortcut to find out how many outcomes contain exactly 2 heads when we toss the coin 7 times?

The answers to those questions will be explored in the next section.

4.3.2 Combinations and Pascal's triangle

Start with 3 ones arranged as a triangle. Call the top row, row #0, and the second one, row #1.

$$\begin{array}{ccc} & 1 & \\ 1 & & 1 \end{array}$$

Now add more rows using the following rules:

- Start and end each row with a 1.

- Get the other elements of the row by adding the two adjacent numbers in the previous row.

Using those rules, the first 4 rows will look like this:

$$
\begin{array}{ccccccc}
 & & & 1 & & & \\
 & & 1 & & 1 & & \\
 & 1 & & 2 & & 1 & \\
1 & & 3 & & 3 & & 1
\end{array}
$$

Pascal's triangle

Continuing until row 7, the numbers in that row will be: $1, 7, 21, 35, 35, 21, 7, 1$. Those numbers can be used to answer the question at the end of the previous section. This array is called the 'Pascal Triangle,' and it is a helpful tool to calculate in how many different ways x successes can be located in n trials or repetitions of a random experiment of the success/failure type.

Brief history of Pascal's triangle *

The array of numbers was named after Blaise Pascal (1623-1662), but it was also independently developed by other mathematicians. There are references to the same triangle in China around 1050, and it was known among Arab mathematicians at about the same time (Burton, 2003). Al-Tusi, a Persian mathematician, published it in the 1200's. It was published in Germany around 1527. Tartaglia (1556) published the triangle through the eighth row, and Cardano published it to the 17th row giving credit to somebody else. However, Pascal did a very careful study of its properties.

Factorials

Combinations

Combinations

The numbers in the nth row of Pascal's triangle can be represented with the following notation:

$$C_x^n = \binom{n}{x} = \frac{n!}{x!(n-x)!} \quad for \ \ x = 0, 1, 2, ...n \tag{4.1}$$

where $n!$ is called the factorial of n and is defined as

$$n! = n \times (n-1) \times (n-2) \times ...2 \times 1. \tag{4.2}$$

The symbol $\binom{n}{x}$ is known as the 'number of combinations' or 'the number of ways in which x objects can be chosen from n different objects.' The ideas related to combinatorics seem to have appeared around the 9th century in India. The location of x successes in n trials is similar to the problem of selecting x from n objects or the problem of allocating x balls in n boxes (assuming one box cannot contain more than one ball). *Using* **R**, *the number of combinations can be calculated typing 'choose(n,x).'*

Assume that there is a row of n little boxes and there are only x balls. Each ball is to be placed in a different box, but there are more boxes than balls. In how many different ways can you choose the x boxes that will contain the balls? As an example consider Figure 4.8 where $n = 3$ and $x = 2$. In which boxes can the balls go? The three options are displayed in Figure 4.9, $C_2^3 = \binom{3}{2} = \frac{3!}{2!(3-2)!} = 3$. The Pascal triangle can help to save time when calculating the number of ways that a particular number of successes can happen in n replicates of a Bernoulli experiment.

Figure 4.8: Two balls to place in 3 boxes

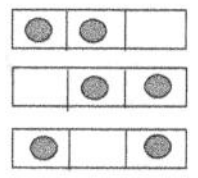

Figure 4.9: Possible locations for the 2 balls

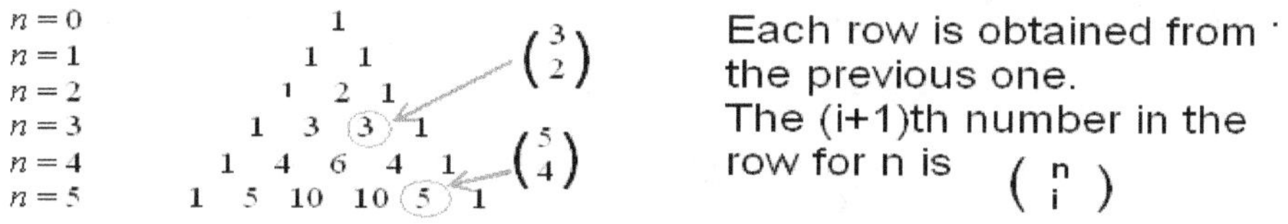

$n=0$					1					
$n=1$				1		1				
$n=2$			1		2		1			
$n=3$		1		3		③		1		
$n=4$	1		4		6		4		1	
$n=5$	1		5	10		10	⑤		1	

$\binom{3}{2}$

$\binom{5}{4}$

Each row is obtained from the previous one.
The (i+1)th number in the row for n is $\binom{n}{i}$

Figure 4.10: Pascal's triangle

4.3.3 Probability distributions

The random experiment of tossing a fair coin 3 times was analyzed in the previous two sections. A number representing the number of heads was assigned to each outcome in the sample space. In probability language, a **random variable** was defined. The random variable is

X: *number of heads in 3 tosses of a fair coin.*

$$\text{S} = \{TTT, TTH, THT, HTT, HHT, HTH, THH, HHH\}$$
$$X: \quad 0 \quad\quad 1 \quad\quad 1 \quad\quad 1 \quad\quad 2 \quad\quad 2 \quad\quad 2 \quad\quad 3$$

A random variable is a rule that assigns a number to each outcome

Discrete random variables only take integer values

A **random variable** is a variable such that the value it takes depends on the outcome of a random experiment. When the random variable only takes integer values, as in the case X: *number of heads in three tosses of a coin*, it is called a **discrete** random variable. The behavior of a discrete random variable can be described by its **distribution**, i.e., listing its values, x, and their corresponding probabilities, $p(x)$. Looking at Table 4.2 and knowing that the coin is fair and that the tosses of the coin are independent, the probability of any one of those 8 outcomes is $0.5^3 = 0.125$. The probabilities for each possible value of the variable X are displayed in Table 4.3 and Figure 4.11. Expression (4.3) can be used to calculate the probabilities in Table 4.3. X is a discrete random variable, and $p(X)$ is its **mass function** or **probability function**. The term 0.5^3 is the probability of each outcome and $\binom{3}{x}$ represents the number of ways in which x heads can happen in 3 tosses of a coin.

$$P(X = x) = \binom{3}{x} 0.5^3 \qquad for \qquad x = 0, 1, 2, 3 \qquad (4.3)$$

The probability function or mass function describes the distribution of a discrete random variable

Notice that all the values of the probabilities in Table 4.3 are non-negative and that they add up to 1; this is a property of all mass or probability functions.

Table 4.3: Probability distribution of X:*number of heads in 3 tosses of a coin*

x	0	1	2	3
p(x)	0.125	0.375	0.375	0.125

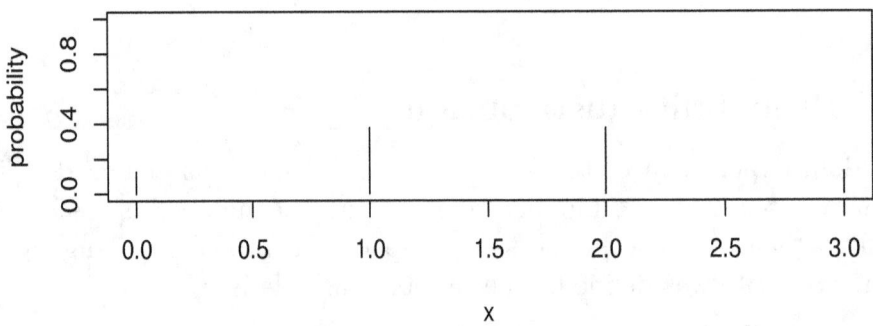

Figure 4.11: Probability or mass function

4.3.4 The binomial distribution

A **discrete probability distribution** lists the values that a **discrete random variable** can take and the probability of each value, or provides a mathematical expression to calculate those probabilities. The mathematical expression to calculate the probabilities is called **mass function** or **probability function**. Probability distributions are used as models to describe the behavior of random variables. The binomial distribution is a probability model suitable to apply in situations like the case of the three independent tosses of a coin. In order to use the binomial model to calculate probabilities, the following conditions must be fulfilled:

- There is a basic random experiment (called a 'Bernoulli' experiment) with two possible outcomes (or group of outcomes): SUCCESS and FAILURE. It is our choice which outcome to label as 'success' as long as we are consistent. 'Success' does not necessarily mean what colloquially is called success. In the Bernoulli experiment context, 'success' could be something like having a disease. In a Bernoulli experiment, the probability of 'success' is called p and the 'probability of failure' is called q or $1 - p$. An example of a Bernoulli experiment is to plant a seed and to define success as 'the seed germinates.' If $p = 0.87$ then the probability of failure is $q = 0.13$.

- The probability of success p is known.

- The experiment is repeated a known number of times, n, the replicates are independent from each other, and the probability of success p is the same in all the replicates.

- The variable of interest is X: *number of successes* in n trials. The objective is to calculate the probability of a given number of successes in n trials.

Necessary conditions to apply the binomial probability model

Table 4.2 lists the possible outcomes when a coin is tossed three times, Table 4.4 includes the more general case of three replicates of a Bernoulli experiment with outcomes in terms of success (S) and failure (F). Table 4.3 displays the probability distribution for the variable X:*number of heads in three tosses of a fair coin*. Table 4.5 displays the probability distribution for the more general case in which $n = 3$, but the probability of success is p.

Notice that in Table 4.4 the outcomes with the same number of successes x have each the same probability. That probability can be calculated once and then multiplied by the number of ways in which the x successes can be

Table 4.4: Possible outcomes in three replicates of a Bernoulli experiment

For the coin example			General case n=3		
Outcome	of heads	probability	Outcome	successes	probability
HHH	3	$\left(\frac{1}{2}\right)^3$	SSS	3	p^3
THH	2	$\left(\frac{1}{2}\right)^2\left(\frac{1}{2}\right)$	FSS	2	$p^2(1-p)$
HTH	2	$\left(\frac{1}{2}\right)^2\left(\frac{1}{2}\right)$	SFS	2	$p^2(1-p)$
HHT	2	$\left(\frac{1}{2}\right)^2\left(\frac{1}{2}\right)$	SFS	2	$p^2(1-p)$
TTH	1	$\left(\frac{1}{2}\right)\left(\frac{1}{2}\right)^2$	FFS	2	$p(1-p)^2$
THT	1	$\left(\frac{1}{2}\right)\left(\frac{1}{2}\right)^2$	FSF	2	$p(1-p)^2$
HTT	1	$\left(\frac{1}{2}\right)\left(\frac{1}{2}\right)^2$	SFF	2	$p(1-p)^2$
TTT	0	$\left(\frac{1}{2}\right)^3$	FFF	2	$(1-p)^3$

Table 4.5: Probability distribution of X: number of successes in 3 replicates of a Bernoulli experiment

x	0	1	2	3
$p(x)$	$(1-p)^3$	$3p(1-p)^2$	$3p^2(1-p)$	p^3

placed in the n trials. Expression (4.1) becomes handy to count the number of ways in which the x successes can be located in the n replicates of the experiment. The quantities $\binom{n}{x}$ defined in expression (4.1) are also known as the **binomial coefficients**. Expression (4.3) described the probabilities when $n = 3$ and $p = 0.5$; for the general case, the mass function of the binomial distribution is:

Binomial probability model

$$P(X = x) = \binom{n}{x}p^x(1-p)^{n-x} \quad for \quad x = 0, 1, 2, 3, ...n \quad (4.4)$$

Example 4.1

Consider that a coin is being tossed, not 3 times but 7 times. What is the probability of getting 2 heads?

$$P(X = 2) = \binom{7}{2}0.5^2(1 - 0.5)^5 = 0.1640625$$

Example 4.2

It is known that approximately 37% of the U.S. population has blood type 0+. What is the probability that if 7 individuals are randomly selected, exactly 2 of them have blood type 0+? In this case,

$$P(X = 2) = \binom{7}{2}(0.37)^2(0.63)^5 = 0.2853156$$

4.3.5 Calculating binomial probabilities

Having to calculate $P(X = x) = \binom{n}{x}p^x(1-p)^{n-x}$, as in Examples 4.1 and 4.2, can seem a daunting task. There are some ways in which those calculations can be simplified.

Using Pascal's triangle *

Pascal's triangle can be used to calculate $\binom{n}{x}$. For example, using Pascal's triangle it is known that 2 successes in 7 trials of a Bernoulli experiment can happen in 21 different ways, $\binom{7}{2} = 21$. That knowledge can be used to calculate $\binom{7}{2}p^2(1-p)^5 = 21p^2(1-p)^5$. The symbol p is replaced by its value in the given problem. For Example 4.1, p is replaced by $p = 0.5$. Hence

$$P(x = 2) = 21(0.5)^2(0.5)^5 = 0.1640625$$

For Example 4.2, $p = 0.37$. Thus,

$$P(x = 2) = 21(0.37)^2(0.63)^5 = 0.2853156$$

Alternative ways of simplifying the calculation of binomial probabilities:

Pascal's triangle

binomial tables

Binomial probability tables

Binomial probability tables can be found in the web or in the appendix of practically any statistics book. Locate the row corresponding to the value of n and x (7 and 2 in the example, respectively) in the left column of the table. Locate the column corresponding to the value of p. In the intersection of the selected row and the column is the value of the probability. Most binomial tables include columns for $p = 0.1, 0.2, 0.25, 0.3, 0.4, 0.5$. It is unlikely to find a table with a column for $p = 0.37$ in order to solve Exercise 4.2. However, using software you can create your own binomial table for any value of n and p. Table 4.6 and Figure 4.12 display the probabilities for $n = 7$ and both values of p (0.5 and 0.37) necessary to solve Examples 4.1 and 4.2. In Figure 4.12, observe how the shape of the distribution changes with the value of p.

168 CHAPTER 4. THE BINOMIAL DISTRIBUTION

Table 4.6: Binomial table for $n = 7$ and $p = 0.5, 0.37$

x	$p = 0.5$	$p = 0.37$
0	0.0078125	0.0393898064
1	0.0546875	0.1619358707
2	0.1640625	0.2853155817
3	0.2734375	0.2792771567
4	0.2734375	0.1640199174
5	0.1640625	0.0577974947
6	0.0546875	0.0113148535
7	0.0078125	0.0009493188

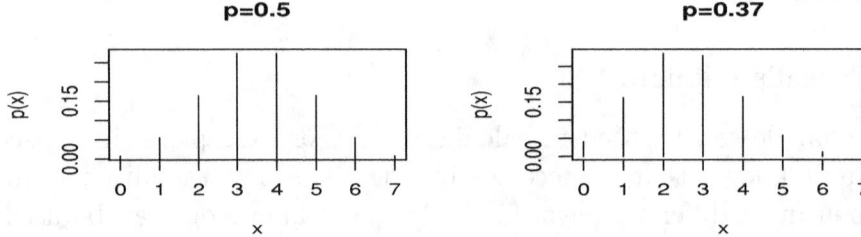

Figure 4.12: Two binomial distributions, n=7, p=0.5 and 0.37

Using statistical software:

In R, the command used to calculate the probability of x successes in n replicates of a Bernoulli experiment, when the probability of success is p,

R & MINITAB: *is **dbinom(x,n,p)**. For example **dbinom(2,7,0.5)** will produce the proba-*
Producing *bility of obtaining 2 success in 7 trials when the probability of success in each*
your own *trial is 0.5, as in the case of tossing a fair coin seven times.*
binomial
tables   *To create a binomial table for $n = 7$ and $p = 0.5$, and all the possible val-*
ues of x with their corresponding probabilities, use the following commands:

```
x<-c(0,1,2,3,4,5,6,7)    # stores values from 0 to 7
px<-dbinom(x,7,0.5)      # calculates  probabilities
table<-cbind(x,px)       # puts the columns x and p(x) together
table                    # prints the table
plot(x,px,'h')           # plots the mass function
```

The command dbinom$(x, 7, 0.37)$ will produce the binomial probabilities for $n = 7$ and $p = 0.37$, as in the last column of Table 4.6

     In MINITAB, type the possible values of x (from 0 to 7) in one column and from the menu select CALC > Probability Distributions > Binomial.

In the window that opens: select the option 'Probability,' in 'Number of trials' type 7, in 'Event probability' type 0.5 for the coin example (or 0.37 for the blood type example). In 'Input column' type the name of the column where the values of x are, for example C1, and in 'Output column' type the name of the column where the probabilities will be stored, for example C2.

4.4 Exercises

4.4.1 Review questions

1. What is a random experiment? What is the sample space? What is an event? Create an example to illustrate these concepts.

2. When are two events, A and B, said to be independent? When are they said to be mutually exclusive?

3. Assuming that A and B are events, what does each one of the following expressions mean: $P(A \cup B)$, $P(A \cap B)$, $P(A)$?

4. Draw a Venn diagram considering two events A and B . Assume that $P(A) = 0.2$, $P(B) = 0.3$, $P(A \cap B) = 0.1$. Answer the following questions:

   (a) Are A and B mutually exclusive ?

   (b) Are A and B independent?

   (c) Calculate $P(B')$, the probability that B does not happen.

   (d) Calculate $P(A \cap B')$, the probability that A happens but not B.

   (e) Calculate $P(A \cup B)$, the probability that A or B happen (at least one or both of them happen).

   (f) Calculate $P(A' \cap B')$, the probability that neither A nor B happens.

5. Review the necessary conditions for the binomial probability model to be appropriate to apply to a given situation. Explain why the binomial distribution could be used in the following case to calculate the required probabilities: Consider DNA to be a sequence of nucleotides; each one of them can be either C,A,G or T. Assume that a sequence of 20 nucleotides is randomly generated using a roulette wheel like the one in Figure 4.13. Assume that the roulette wheel is calibrated so

that the probability of each one of the four letters is the same. We define the variable X: number of As in the sequence of 20 nucleotides. We are interested in calculating the probability of exactly 12 A's in the sequence.

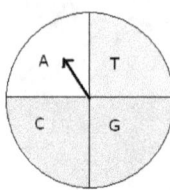

Figure 4.13: Roulette to generate an artificial DNA sequence

6. In the following cases say why the binomial probability should NOT be used to calculate the required probabilities (in each case indicate which condition fails):

Review questions for Chapter 4

   (a) A coin is tossed until heads is obtained. What is the probability of having to toss the coin 3 times until the first head shows up.

   (b) We have a box with 10 balls that are almost identical except for the color; 4 of them are red. We select 3 balls, one by one, without replacement. Calculate the probability of getting 2 red balls.

7. When a certain type of bulbs are planted in November, the probability that they bloom in the spring is only 0.3 . Assume each bulb blooms independently of the others. Use a binomial table (which value of n?, which value of p?) to answer the following questions:

   (a) If 20 bulbs of that type are planted, what is the probability that only 7 bloom?

   (b) If 20 bulbs of that type are planted, what is the probability that 7 or less bloom?

4.4.2 Exercises for discussion or homework

1. The surface of peas can be wrinkled or smooth, and their color can be yellow or green. Define the events:
   A: pea is green   B: pea is wrinkled
   Describe in words the meaning of :
   a) A' b)$A \cap B$ c) $A' \cap B$
   d) $(A \cap B)'$ e) $A \cup B$

2. Assume that in a given human population the probability of having trait A is 0.15 , the probability of having trait B is 0.25 and the probability of having both is 0.1. Answer the following questions:

   (a) Are A and B independent? Why?

   (b) Are A and B disjoint or mutually exclusive? Why?

   (c) What does A' represent? Calculate $P(A')$.

   (d) What does $A \cap B$ represent? Calculate $P(A \cap B)$.

   (e) What does $A \cup B$ represent? Calculate $P(A \cup B)$.

   (f) A person from that population will be selected at random. Calculate the probability that the person has trait A but not trait B. (*Hint: Draw a Venn diagram.*)

   (g) A person from that population will be selected at random. Calculate the probability that the person has trait B but not trait A.

   (h) What percent of the population has neither A nor B?

3. Consider the following DNA sequence that is 20 nucleotides long: AGT-GCCAGCAAACACCTGCT. Now assume that a random sequence of 20 letters is generated with the roulette wheel of Figure 4.13. The simulated sequence will be checked, position by position, against the sequence given above. What is the probability of getting exactly 10 coincidences just by chance? What is the probability of getting 10 coincidences or more by chance alone? (*Hint: If you are generating the sequence with the roulette wheel, the probability of a coincidence in a given position is 1/4. For example, the first letter of the true sequence is A, the probability of getting an A with the roulette is 1/4).* **Exercises for discussion or homework in Chapter 4**

4. Assume that you are raising *Drosophila melanogaster* in a lab and have 15 larvae in a stock bottle and the ratio of male:female for fruit flies is known to be 1:1. (i.e., probability of male is 0.5).

   (a) What is the probability that exactly 7 will be males?

   (b) What is the probability that 7 or fewer will be males?

   (c) If there are 25 larvae instead of 15, what is the probability that only 10 will be males?

   (d) If there are 25 larvae instead of 15, what is the probability that 10 or fewer will be males?

**Exercises
for discussion
or homework
in Chapter 4**

5. For a given type of seeds, the probability of germinating is 0.85. Thirty seeds are planted. Assume each one will germinate independently of the others. Use statistical software to produce a binomial table appropriate to answer the following questions: What is the probability that exactly 20 of them will germinate? What is the probability that 20 or less will germinate?

6. A biologist does an experiment with bees. During 4 days of training, food is placed at a given location. On the fifth day ('testing day') no food is placed at that location. Based on previous experiments, the probability of a bee returning on the fifth day to the location is assumed to be 0.35. If there are 20 bees, and each bee behaves independently of the others. What is the probability that exactly 7 will go to the food location on the testing day?

7. Assume that when a female mallard is offered two pieces of bread, one plain and one dyed green, the probability that it picks the green bread first is 0.5. Assume also that each mallard behaves independently of the others. In a study, 10 female mallards will be offered two pieces of bread, one of each color. What is the probability that 7 out of them will approach the green bread first? What is the probability that 7 or more of them will approach the green bread first?

8. In the USA, approximately 37% of the people have blood type O+. What is the probability that, in a random sample of 10 individuals, exactly 3 are O+? What is the probability that only 1 is O+? What is the probability that at least 1 is O+?

References

Burton, D.M.(2003) *The History of Mathematics.* Boston: Mc Graw Hill.

De Groot, M. and Schervish M.J. (2002) *Probability and Statistics*, Third Edition. New York: Addison Wesley.

Snee, R. D. (1974) Graphical display of two-way contingency tables. *The American Statistician* **28**: 912.

Chapter 5

Inference with the binomial

In general, a test for a null hypothesis H_o compares the value obtained for a statistic with the distribution assumed for the statistic if H_o were true. The idea is to check if the observed value looks unusually large (or small) as compared to the values in the distribution. That comparison is summarized with the calculation of the p-value. The binomial distribution, introduced in the previous chapter, is the natural distribution to use to calculate the p-value when the variable under study is of the 'yes/no' type. The test is called 'the exact test' for proportions.

Random categorical variables that are of the 'yes/no' type are quite common, such as 'being infected with the HIV virus,' 'being a current smoker,' 'developing a serious side effect when using a medicine.' The quantity of interest (or parameter) is the proportion of individuals in the population of reference that fall into the 'yes' category. In some cases the proportion is expressed as a percent. In public health studies, the proportion of individuals with a disease is sometimes called the 'prevalence' of the disease in the population or subpopulation of interest. The concept of prevalence is also applied to the use of something, such as tobacco, alcohol, or illegal drugs.

5.1   Hands-on activity

We perform this activity with a 'two-color' plastic chip (one side red and one side yellow), but it can be done with a coin if such chips are not available, just replace 'red' with 'heads' and 'yellow' with 'tails.' The manufacturer did not include any information about the probability of each side when the chip is flipped. We write the null and alternative statistical hypotheses in terms of p (the probability of getting the red side when flipping the chip):

$$H_o: \qquad p = 0.5$$
$$H_a: \qquad p \neq 0.5$$

Hands-on activity

Each student is given a chip and has to test the null hypothesis about his/her own chip. Assume that one of the students in the classroom flips a chip 10 times and obtains 8 reds and 2 yellows. Is that result enough evidence against the null hypothesis H_o that says that red and yellow are equally likely? The binomial distribution with $n = 10$ and $p = 0.5$ is a handy way to answer that question.

What is the probability of getting 8 or more reds when the probability of red is 0.5?

Use the binomial table below and calculate:

$$P(X=8)+P(X=9)+P(X=10)=0.0546875$$

However, it should be noticed that the alternative hypothesis H_a is two-sided. To get 2 or fewer reds out of 10 would be as 'extreme' as getting 8 reds or more. We need to calculate

$$P(X=0)+P(X=1)+P(X=2)=0.0546875$$

in order to calculate the p-value. Thus, the probability of getting the result the student got (8 reds) or a more extreme one, when the null hypothesis is true is $0.0546875 \times 2 = 0.109375 \sim 0.11$. That is NOT a very small probability, so the outcome of 8 reds the student got is not considered enough evidence against the null hypothesis for that particular chip.

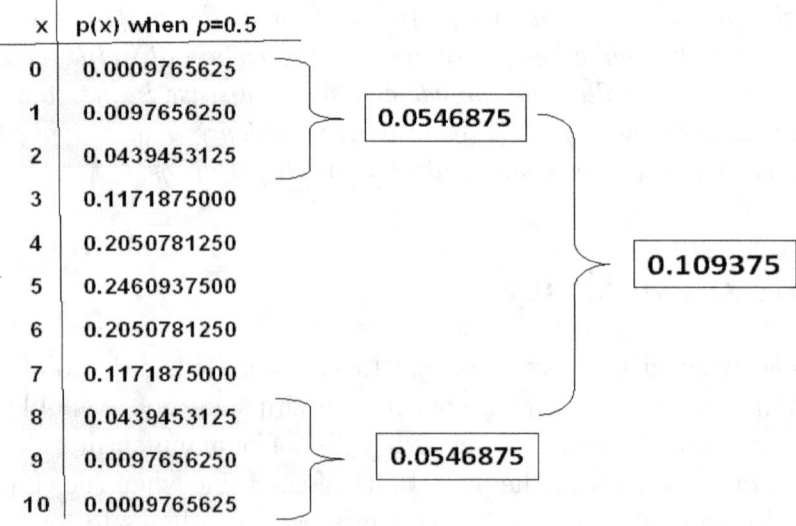

x	p(x) when p=0.5
0	0.0009765625
1	0.0097656250
2	0.0439453125
3	0.1171875000
4	0.2050781250
5	0.2460937500
6	0.2050781250
7	0.1171875000
8	0.0439453125
9	0.0097656250
10	0.0009765625

Now, toss your own chip or coin 10 times, calculate the p-value and make a decision (reject or not reject) about H_o for your own chip or coin.

5.2 Writing and testing hypotheses- a ducks story

Example 5.1 Female mallards and green bread

In a certain species of ducks (mallards), males have green heads and females are a plain color. The green coloring of the male heads might be attractive to the females. The question is: Are female mallards attracted to the color green in food? A student is given the assignment, in her Ethology class, to conduct a small study to address this question. The hypothesis testing process for this example is explained next as in Seier & Robe (2002).

Translating a research question into statistical hypotheses

Writing the statistical hypotheses

In relation to the research question there are two possible contrasting ideas:
Idea 1: Female mallards have no preference between plain and green bread.
Idea 2: Female mallards prefer green bread.

When a female mallard is confronted with two pieces of bread, one plain and one green, the probability of picking the green one first will be called p. The two contrasting ideas can be written in terms of p.

Statistical hypotheses about a population proportion

Idea 1 or Ho:	$p = 0.5$
Idea 2 or Ha:	$p > 0.5$

These confronting ideas are called **statistical hypotheses**. The first one states that the female mallards like the green and the plain bread equally. This statement is called the **null hypothesis**, it represents an idea of no preference or no difference. The symbol H_o is used to represent the null hypothesis. The second idea says that the ducks prefer the green bread; it states something different than the null hypothesis, it is called the **alternative hypothesis**. The symbol used for the alternative hypothesis is H_a or H_1. It is important to remember that the statistical hypotheses refer to the whole population, not just to the individuals in the sample. Decisions about the hypotheses are expressed in terms of H_o, which is either rejected or not.

Gathering data to make a decision

The student will go to a lake near campus where mallards are quite abundant. Ten female ducks will be randomly selected. Each one will be offered two pieces of bread: one plain and one green. The student writes down which piece of bread each duck approaches first. The information is summarized by reporting how many female mallards in the sample approached the green bread first. The variable to observe is 'number of successes in n trials,' thus the binomial distribution will be the distribution of reference with respect to which the p-value is calculated.

Arriving at a conclusion

Using the binomial distribution to test $H_o : p = p_o$

If 9 out of the 10 female mallards go for the green bread first, the proportion of female mallards in the sample that prefer the green bread is $\hat{p} = 0.9$. Remember that hypotheses are about populations, in this case the population of female mallards; hypotheses are not about the sample. What is the probability that 9 out of the 10 mallards in the sample go for the green bread first when the null hypothesis, $p = 0.5$ (no real preference between plain and green bread in the population), is true? This question can be answered using the binomial distribution, provided each mallard is assumed to behave independently of the others. However, it is not enough to calculate $P(x = 9)$ when $p = 0.5$. Why? Nine out of 10 seems to indicate that female mallards tend to prefer green bread to plain. If more than 9 had picked the green bread first, it would be a situation even farther from what is expected under the null hypothesis. A number higher than 9 would have given stronger evidence against the null hypothesis. The probability that 9 (the value the student supposedly observed) **or more** female mallards pick the green bread first needs to be calculated.

We want to know not only what the chances are of getting the result obtained, but also what are the chances of getting a result that is even more extreme, that is, farther from what the null hypothesis indicates. What is the probability, assuming that in general female mallards have no preference between green and plain bread, that 9 *or more* female mallards in a sample of 10 would pick the green bread first just by chance? In other words, if X represents the number of female mallards that pick the green bread first, what is $P(X \geq 9)$?

The binomial distribution, learned in the previous chapter, now comes in handy. Table 5.1 is a binomial table for $n = 10$ and $p = 0.5$ (see also Figure 5.1). **Notice that the value of p in the table is the value of**

p specified in the null hypothesis. Why? Because we need to calculate the **p-value** which is the probability of getting 9 or more ducks in a sample of 10 who favor green just by chance **when the null hypothesis is true,** i.e., when female mallards have no preference between green and plain color in bread.

Table 5.1: Binomial distribution for n=10 and p=0.5

x	$p(x)$
0	0.0009765625
1	0.0097656250
2	0.0439453125
3	0.1171875000
4	0.2050781250
5	0.2460937500
6	0.2050781250
7	0.1171875000
8	0.0439453125
9	0.0097656250
10	0.0009765625

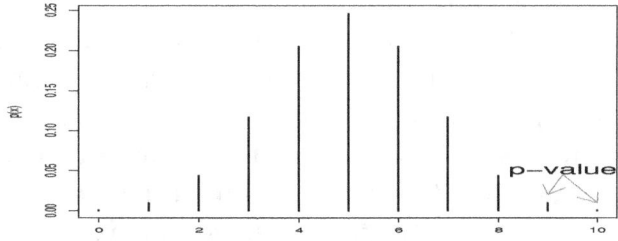

Figure 5.1: Binomial distribution n=10 and p=0.5

Adding the probabilities in the last two rows of table 5.1, $P(X \geq 9) = 0.0097656250 + 0.0009765625 = 0.0107421875 \approx 0.01$. That probability is very small indeed. Thus, if 9 or more female mallards go for the green bread first, it is hard to think that this happens just by chance and that in general female mallards have no preference between green and plain bread. Therefore, the null hypothesis that represents no preference should be rejected. **The probability of getting the result obtained or a more extreme one when the null hypothesis is true is called the 'p-value.'** In the example, the p-value is 0.01074219 .

Now assume that only 7 out of the 10 female mallards had gone for the green bread first. The p-value is equal to $P(X \geq 7) = 0.1171875000 + 0.0439453125 + 0.0097656250 + 0.0009765625 = 0.171875 \approx 0.17$. That probability definitely does NOT look small, which means it is not so rare to get 7 or more female mallards going for the green bread first just by chance, assuming that in general female mallards have no preference between green and plain bread. In this case the data do not constitute enough evidence against the null hypothesis; thus, it is not rejected.

The two previous situations are pretty clear: 0.01 is small and 0.17 is large. However, sometimes a borderline situation happens. Assume that only 8 out of the 10 female mallards had gone for the green bread first. Now the p-value is calculated as $P(X \geq 8) = P(X = 8) + P(X = 9) + P(X = 10) = 0.0546875$. Is this value small or large? Well, now it is not so easy to arrive at a conclusion. The common practice is to compare the p-value to a prefixed value. Assume that, before collecting the data, the student arbitrarily decided to work with $\alpha = 0.05$ (significance level). Since 0.0546875 is larger than 0.05, the null hypothesis is not formally rejected. However, it is kind of a borderline situation. The p-value is so close to the given significance level that it suggests that there might be some preference for the green color, except that the evidence is not strong enough as to formally reject the null hypothesis of no preference.

The obvious questions are: How small does a p-value need to be to reject the null hypothesis? Why was the value 0.05 given as a threshold? A criterion with regard to p-values is required, and that criterion needs to be set as soon as the hypotheses are written, even before data are collected. **A scientist should never look at the data first and then set a criterion. Statistics is used to arrive at an honest conclusion about hypotheses, not to forcefully 'prove' pre-conceived ideas.** In order to set a threshold, we need to learn about the two types of error in the next section. First we will see how to perform the test using software.

Using statistical software:

Now that the rationale of the test for one proportion using the binomial distribution is understood, the test can be performed directly using statistical software without needing to produce a binomial table. The computer will

calculate the p-value. For the mallards example, it is enough to type:

binom.test(9,10,p=0.5,alternative=c("greater"))

The output is:

number of successes = 9, number of trials = 10, p-value = 0.01074.

In MINITAB, use: STAT > Basic statistics > 1 proportion, to enter the number of events (9), trials (10), and hypothesized proportion (0.5). In 'options' indicate the nature of the alternative hypothesis (greater than). For a picture of the MINITAB screen see page 194.

5.3 The two types of error

To test a null hypothesis H_o is to decide, based on data, if we reject it or not reject it. Obviously it is desirable to make the correct decision, but sometimes the wrong decision is made. Two different possible mistakes in the mallards example would be:

- H_o **is rejected but it is the wrong decision because H_o is true.** In other words, female mallards actually have no preference between green and plain bread but the test concludes that they prefer the green bread. This is called the **Type I error**.

- H_o **is not rejected but it is the wrong decision because H_o is false.** In other words, the test concludes that there is not enough evidence to dismiss the hypothesis of no preference, when actually female mallards do tend to prefer the green bread. This is called the **Type II error**.

Two different ways of being wrong: Type I and Type II errors

The DECISION	The TRUTH	
	Ho is true	Ho is false
Do not reject Ho	correct! ⌣	**Type II error**
Reject Ho	**Type I error**	correct! ⌣

It is desirable to keep the chances of making mistakes very small. In the approach used here and in most introductory statistics textbooks, the final decision is expressed in terms of H_o ('reject' or 'not reject' H_o). The null hypothesis reflects a 'no difference,' 'status quo' or neutrality situation. When H_o is rejected, a statement is being made saying that something is better, or worse, or different, depending on the nature of the alternative hypothesis. If two medicines are being compared in a pharmaceutical study, a 'Type I error' would be to think that one medicine is better when actually both have similar effectiveness. Type I error is usually considered a serious error and it is desirable to have some control over the probability of its occurrence.

α or 'significance level' is the probability of Type I error

The value of α is set by the researcher

The probability of making a Type I error is called α or **significance level** and it is set at the beginning of a study, before looking at the data or

**α is NOT
calculated
from
the data**

even collecting them. Traditionally, a very common value for α is 0.05 but in studies (such as medical research) where the consequences of Type I error are very serious, a smaller α such as 0.01 might be preferable. The researcher is free to set his/her own α. **If the practical consequences of Type I error are very serious, a small value should be chosen for α.**

What would happen if the value of α were made very, very small? In order to reduce the risk of rejecting H_o when it is true, H_o would never be rejected. In that case, the risk of not rejecting it when it is false will increase. Not rejecting H_o when it is false is called **Type II error** and its probability is called β. For a given sample size, if α is made smaller, β will become larger. A way of having reasonable values for the probabilities of both types of error is to have a lot of evidence (data) so that a reasonable decision can be made.

**β
is the
probability
of Type II
error**

5.4   What is considered 'small' in the p-value world?

**H_o is rejected
when the
p-value
is smaller
than α**

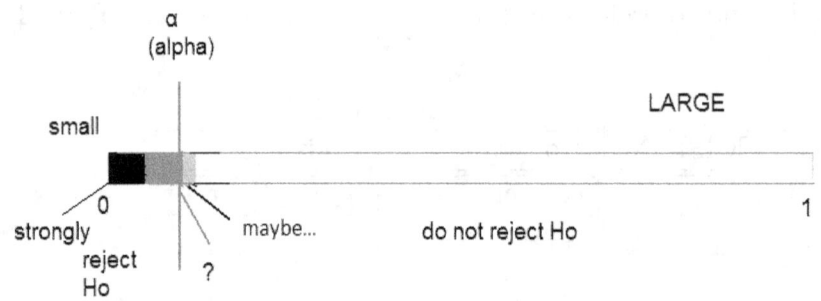

Figure 5.2: Interpretation of p-value

**The smaller the
p-value,
the more
enthusiastically
H_o
is rejected**

Remember that the p-value is the probability of getting the results obtained from the sample or more extreme results, when the null hypothesis is true. Probabilities only take values in the interval [0,1]. Values very close to 0 (such as 0.0001 or 0.002) are of course small and leave little doubt about rejecting the null hypothesis. Large p-values in the upper half of the [0, 1] interval, such as 0.6 or 0.7, leave no doubt about not rejecting H_o. But when is a p-value considered small enough as to reject the null hypothesis? The common practice is to compare the p-value with the value of α. The pre-set value of α (probability of Type I error) divides the interval $[0,1]$ in two regions for the p-value as in Figure 5.2: small (reject H_o) and large (do not reject H_o). If the p-value is too close to α, at either side of it, one might become a little insecure and ask for more data. Also when the p-value is barely larger than α, one wonders if with a larger sample size one might have rejected the null hypothesis. It is not only a matter of the p-value being below the pre-fixed value of α, it is a matter of how close to zero the p-value is that makes the researcher comfortable with the idea that the data constitute evidence against the null hypothesis. To be aware of the power of the type of test that he/she is applying (a concept explained in Section 5.7) might help the scientist to have a clearer picture of the situation.

5.5 More examples of the test for a proportion

There are two types of variables, categorical and quantitative. Categorical variables refer to characteristics that are not expressed naturally in numbers, such as gender, being HIV positive or not, smoking status (*former*, *current* or *never an smoker*), blood type (O,A,B,AB), whether one suffers side effects when taking a medicine. In the case of categorical variables, the parameter that describes the population is a proportion. For example, the proportion of people in the population who are HIV positive, the proportion of people who have blood type O+, the proportion of pine trees in a forest that are affected by a disease, the proportion of the adult population of a state who are current smokers, the proportion of fruit fly eggs that will hatch as male flies, the proportion of patients who suffer side effects when using a medicine. Statistical hypotheses H_o and H_a can be written about the value of the proportion in the population. For example, 'the proportion of current smokers in Tennessee is at most 0.23 or is over 0.23,' 'the proportion of males among fruit flies is 0.5 or it is less than 0.5,' 'the proportion of patients that suffer side effects when using a given medicine is 0.01 (1%) or it is above 0.01.' Data is collected through a survey (also called 'observational study') or by doing an experiment. The number of individuals in the sample who exhibit the trait is recorded, such as the number of current smokers, the number of male flies, or the number of subjects suffering side effects. The decision about the null hypothesis is made in a similar way as in the case of the female mallards story. Applications of hypothesis testing go beyond biology to include opinion polls, public health, and social surveys. When **Beyond** the sample size is very large, there is an alternative method to test those **Biology** hypotheses, which will be studied in Chapter 11. When the sample size is small, the method learned through the mallards story is particularly useful. Of course, the method of testing hypotheses about a population proportion using the binomial distribution can be applied to large samples as well. Binomial tables can be easily produced with statistical software. Moreover, most statistical software have the option of testing hypotheses for proportions using the binomial distribution; it is called the **exact test for proportions**. More examples of the binomial test will be seen next. Additional examples are found in the exercises at the end of the chapter.

Example 5.2 Male:female ratio in fruit flies

When working with fruit flies in a genetics laboratory, it is generally assumed that the probability of male is $p = 0.5$ (or male:female ratio is 1:1). A student wants to test that hypothesis. She writes the null and alternative statistical hypotheses:

$$H_o : p = 0.5$$
$$H_a : p \neq 0.5.$$

In this case the alternative hypotheses is two-sided, meaning that the question is whether the probability of male is 0.5 (H_o) or different (either larger or smaller) from 0.5 (H_a). She randomly selects 25 larvae. Assume that only 10 are males when the flies emerge. Should the hypothesis of 1:1 ratio be rejected? Are the data (10 males out of 25 flies) enough evidence against the null hypothesis (assuming $\alpha = 0.05$)?

The binomial table for $n = 25$ and $p = 0.5$ is shown in Table 5.2. In that table, probabilities are written in scientific notation, for example $1.43e - 02$ represents 0.0143. The probabilities are also displayed in Figure 5.3. What is the probability of 10 or fewer males out of 25 flies when the probability of being male is $p = 0.5$? That probability can be obtained from Table 5.2 by adding the probabilities corresponding to $x = 0, 1, ...10$. However, remember that the alternative hypothesis is two-sided. Thus, not only the values $x \leq 10$ need to be considered, but also the values that are as or more extreme than the observed value of 10, but at the other side of the table, i.e., the probabilities for $x = 15, 16, 17, ..., 25$. Adding those probabilities, and rounding to 4 decimal places, the p-value is 0.4244. A p-value of 0.4244 is large (and larger) than $\alpha = 0.05$, and the null hypotheses $H_o : p = 0.5$ cannot be rejected. However, this does NOT 'prove that $p = 0.5$,' it is simply concluded that 10 males out of 25 flies does not constitute enough evidence against $H_o : p = 0.5$ as to reject it. In science, scientific theories

Testing if the male:female ratio is 1:1 among fruit flies

A two-sided alternative hypothesis

Table 5.2: Binomial distribution, n=25 and p=0.5

x	p(x)
0	2.980232e-08
1	7.450581e-07
2	8.940697e-06
3	6.854534e-05
4	3.769994e-04
5	1.583397e-03
6	5.277991e-03
7	1.432598e-02
8	3.223345e-02
9	6.088540e-02
10	9.741664e-02
11	1.328409e-01
12	1.549810e-01
13	1.549810e-01
14	1.328409e-01
15	9.741664e-02
16	6.088540e-02
17	3.223345e-02
18	1.432598e-02
19	5.277991e-03
20	1.583397e-03
21	3.769994e-04
22	6.854534e-05
23	8.940697e-06
24	7.450581e-07
25	2.980232e-08

are not 'proved,' they can only be 'disproved.' In the approach to statistical inference adopted here and in most introductory textbooks, null hypotheses are not accepted, they are either 'rejected' or 'not rejected.'

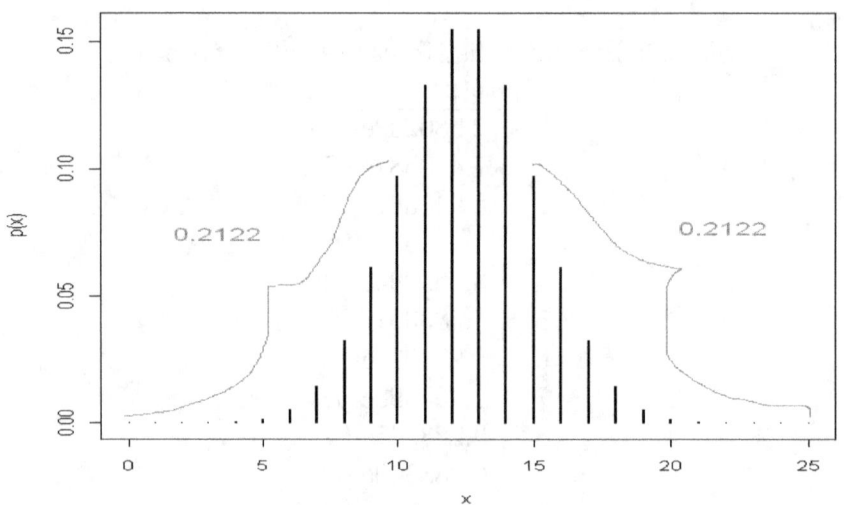

Figure 5.3: Binomial $n=25$ and $p=0.5$ and p-value for Example 5.2

Using statistical software:
The command **binom.test(10,25,p=0.5,alternative=c("two.sided"))** *performs the test with R, for Example 5.2, without having to work with the binomial table.*

R:
exact test
for a
proportion

```
The output is:              Exact binomial test
data:   10 and 25
number of successes = 10, number of trials = 25, p-value = 0.4244
alternative hypothesis: probability of success is not equal to 0.5
```

Example 5.3 The germination rate of seeds

germination rate

A one-sided alternative hypothesis

Assume that the probability that a certain type of seed will germinate is 0.85. A sack of seeds has been found in the basement of a house. The forgotten seeds maybe are too old to germinate with a probability of 0.85. The null and alternative hypotheses are:

$$H_o : p = 0.85$$
$$H_a : p < 0.85$$

Table 5.3: Binomial distribution, n=10 and p=0.85

x	p(x)
0	5.766504e-09
1	3.267686e-07
2	8.332598e-06
3	1.259148e-04
4	1.248655e-03
5	8.490856e-03
6	4.009571e-02
7	1.298337e-01
8	2.758967e-01
9	3.474254e-01
10	1.968744e-01

Not wanting to go through all the work and expense of planting seeds that will not give a good yield, 10 seeds are randomly selected from the sack and planted. Only 3 seeds germinate. The binomial probabilities for $n = 10$ and $p = 0.5$ are shown in Table 5.3.

In this case, because the alternative hypothesis is one-sided ($H_a : p < 0.85$), the p-value is the probability of getting 3 successes (the value in the sample) or a more extreme value (less than 3) when $p = 0.85$. Adding the probabilities in the first 4 rows of the table, $P(0) + P(1) + P(2) + P(3)$, gives a p-value of 0.0001346. The p-value is very small, and consequently the null hypothesis is rejected. It is concluded that the probability of germinating, for the seeds in the sack, is below 0.85.

The mallards, fruit flies and seeds examples cover the three cases with respect to the alternative hypotheses: 'greater than' (mallards), 'different from' or two-sided alternative (fruit flies), and 'less than' (seeds).

Using statistical software: The command

R: exact test and binomial table

binom.test(3,10,0.85,alternative=c('less'))

performs the test for Example 5.3. The following commands produce Table 5.4 and the plot for the binomial distribution

```
n<-10; p<-0.85; x<-seq(0,n,by=1)
px<-dbinom(x,n,p);table<-cbind(x,px);table;
plot(x,px,'h')
```

5.6   Steps in the testing hypotheses procedure

A review of the 3 previous examples (mallards, fruit flies, and seeds), shows the following steps:

1. State very clearly a research question or a scientific hypothesis.

2. Identify a quantity that describes the population and is related to the scientific hypothesis or the research question. In all 3 examples, the quantity of interest was the population proportion. In general, that quantity is called a 'parameter.'

3. Write the null and alternative statistical hypotheses in terms of the parameter of interest.

Steps in hypotheses testing

4. Fix a value of α, the probability of Type I error.

5. Collect data through observation or experimentation.

6. Summarize the data through a quantity that in general is called 'statistic.' In the case of the three examples, it was the 'number of successes in the sample' (or equivalently, the 'sample proportion': the number of successes divided by the sample size).

7. Calculate the p-value: the probability that the statistic takes the observed value, or a more extreme one, when the null hypothesis is true. In this case, the p-value is the probability of getting the observed number of successes, or a more extreme one, when $H_o : p = p_o$ is true.

8. Decide if the p-value is small or large (the pre-fixed value of α serves as a reference).

9. Decide, based on the p-value, if the null hypothesis H_o is rejected or not rejected.

10. The decision about H_o is interpreted in the context of the problem.

The process of testing a statistical hypothesis is explained also in Figure 1.2 of Chapter 1. In Chapters 3, 10 and 11 other cases of hypothesis testing are studied. The different cases might differ in the type of parameters used to write the statistical hypotheses and the distributions of reference used to calculate the p-values. In Chapter 3, an empirical distribution was created by randomization in order to calculate the approximated p-value. In the three examples seen in this chapter (mallards, fruit flies, and seeds) the binomial distribution was used because the statistical hypotheses were about a proportion in the population. In Chapters 10 and 11, other distributions will be used to calculate the p-value. However, all hypothesis testing cases have in common the 10 steps listed above.

5.7   Power of a test

The power of a test is the probability of rejecting the null hypothesis when it is NOT true. This definition is valid for all the tests of hypotheses. In this book, the concept of power will be introduced in the context of the exact test for proportions, as in Seier & Liu (2011).

Revisiting Example 5.1 Female mallards and green bread

Assume that the study with the mallards is still in the planning stage. Before going to the pond and collecting the data, we need to design a decision rule describing when it is that the null hypothesis will be rejected: 'We will reject the null hypothesis of 'no preference,' if among the 10 female mallards,____ pick the the green bread first.' How should we fill in the blank? What would be a good decision rule? Well, it depends on what risks we want to take. In particular, it depends on the value we want to allow the probability of Type I error to have.

Consider the null and alternative statistical hypotheses

The null and alternative hypotheses are:

$$H_0 : p = 0.5 \qquad \text{and} \qquad H_a : p > 0.5$$

A decision will be made about the null hypothesis based on the results for a sample of size 10.

Deciding the value of α and writing the decision rule

Since the alternative hypothesis is one-sided ($>$), the natural thing is to define the decision rule as: **The null hypothesis will be rejected if _____ mallards or <u>more</u> pick the green bread first.**

$\alpha=$ P(Type I error) = Probability (rejecting H_o when it is true). Thus, α could be written as:

α=P(_____ mallards or more pick the green bread first when p=0.5)

We will fill that blank so that the value of α is kept at a desirable level. In Figure 5.4, there is a binomial probability table for $n = 10$ and several values of p. To set the value of α, we need to focus on the column '$p = 0.5$' because when the null hypothesis is true, p has the value 0.5. Assume that we want α to be as close as possible to 0.05 (we cannot make it exactly equal to 0.05 in this case, but we want to be close to it). Thus, looking at the column $p = 0.5$ (corresponding to the null hypothesis): Where should you trace a line to define the decision rule? Start adding probabilities from the bottom of the column, until a sum that is as close as possible to 0.05 is achieved. The decision rule:

'reject Ho when 8 or more ducks go for the green bread first'

sounds like a good idea because that rule assigns the value 0.0547 to α, which is very close to the 0.05 target.

x	p=0.5	p=0.6	p=0.7	p=0.8	p=0.9	p=0.99
0	0.000977	0.000105	0.000006	0.000000	0.000000	0.000000
1	0.009766	0.001573	0.000138	0.000004	0.000000	0.000000
2	0.043945	0.010617	0.001447	0.000074	0.000000	0.000000
3	0.117188	0.042467	0.009002	0.000786	0.000009	0.000000
4	0.205078	0.111477	0.036757	0.005505	0.000138	0.000000
5	0.246094	0.200658	0.102919	0.026424	0.001488	0.000000
6	0.205078	0.250823	0.200121	0.088080	0.011160	0.000002
7	0.117188	0.214991	0.266828	0.201327	0.057396	0.000112
8	0.043945	0.120932	0.233474	0.301990	0.193710	0.004152
9	0.009766	0.040311	0.121061	0.268435	0.387420	0.091352
10	0.000977	0.006047	0.028248	0.107374	0.348678	0.904382

Figure 5.4: Binomial probability table for $n = 10$

The power

Power of a test is the probability of rejecting the null hypothesis when it is NOT true. In other words, the power of a test is the probability that the test finds out that the null hypothesis is not true when it is not true.

Power is the probability of rejecting H_o when it is NOT true

What is the power of the test or decision rule 'reject if 8 or more mallards pick the green bread first'? We need to calculate the probability that 8 or more mallards go for the green bread first when p has values other than the one specified in the null hypothesis.

If p is not 0.5, it could be 0.6, 0.7, 0.9 or any other value larger than 0.5. The power can be calculated for each one of those values, from the table in Figure 5.5, by simply adding the probabilities below the horizontal line in the appropriate column of the table. The values for power were calculated (for $p=$ 0.6, 0,7, 0.8, 0.9 and 0.99) from the table in Figure 5.5 and plotted in Figure 5.6a.

x	p=0.5	p=0.6	p=0.7	p=0.8	p=0.9	p=0.99
0	0.00098	0.00010	0.00001	0.00000	0.00000	0.00000
1	0.00977	0.00157	0.00014	0.00000	0.00000	0.00000
2	0.04395	0.01062	0.00145	0.00007	0.00000	0.00000
3	0.11719	0.04247	0.00900	0.00079	0.00001	0.00000
4	0.20508	0.11148	0.03676	0.00551	0.00014	0.00000
5	0.24609	0.20066	0.10292	0.02642	0.00149	0.00000
6	0.20508	0.25082	0.20012	0.08808	0.01116	0.00000
7	0.11719	0.21499	0.26683	0.20133	0.05740	0.00011
8	0.04395	0.12093	0.23347	0.30199	0.19371	0.00415
9	0.00977	0.04031	0.12106	0.26844	0.38742	0.09135
10	0.00098	0.00605	0.02825	0.10737	0.34868	0.90438

α =0.0547

0.6778
power when p=0.8

Figure 5.5: Binomial probability table for $n = 10$ and decision rule 'Reject H_o if $X \geq 8$'

It is understandable that the farther the true value of p is from the value specified in the null hypothesis, the higher the probability of noticing that H_o is not true. If we were not limited to the binomial table in Figure 5.5, but we had access to a computer and were able to calculate many more columns of the table for other values of p, $0.5 < p < 1$, the power plot would look like the one in Figure 5.6b. Power plots are used when trying to decide the sample size while planning a study to test a hypothesis.

Power and sample size

Software such as MINITAB or **R** can be used to produce power plots for different values of n. The researcher can then select the number n of individuals to include in the experiment so that he/she feels comfortable with the power of the test. As an example, Figure 5.7 compares the power of the test when $n = 10$ and $n = 20$. It can be seen that if more subjects are involved in the experiment, the test becomes more powerful.

The researcher might be interested in a particular value of p for which high power is desired. For example, in the mallards case, we might not be

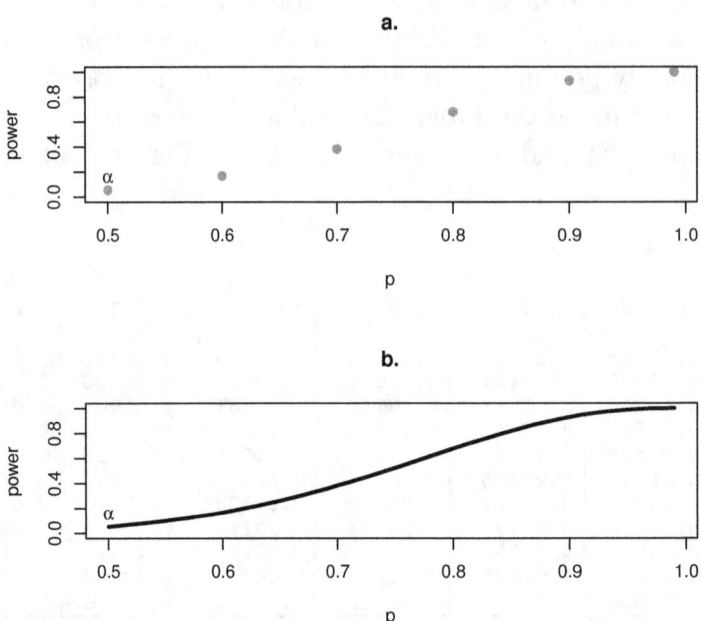

Figure 5.6: Power for the test for $H_o : p = 0.5$ against $H_a : p > 0.5$ when $n = 10$

too interested in having a high probability of rejecting the null hypothesis if the green bread was only slightly preferred by the female mallard population. However, we would like the probability of rejecting the null hypothesis to be high if there was a strong preference in the female mallard population for the green bread, for example if $p = 0.8$. If that were the case, the researcher would like to have a sample large enough such that the power, when $p = 0.8$, **Effect size**   is large. Looking at Figure 5.7, if the researcher wants the power of the test to be over 0.9, when $p = 0.8$, he/she would find it convenient to work with a sample of size 20 instead of size 10. The difference between the value of p in the null hypothesis ($p = 0.5$) and the value of p for which the researcher would like to reject H_o with high probability, if that value of p were true, is called the **effect size**. In this example, the effect size would be $0.8 - 0.5 = 0.3$.

Power and Type II error

Power and Type II error   Look at the binomial table in Figure 5.5. Which probability values would we need to add in order to calculate β=**P(Type II error)**=**P(not rejecting Ho when it is false)**? For the decision rule defined in Figure 5.5, the null

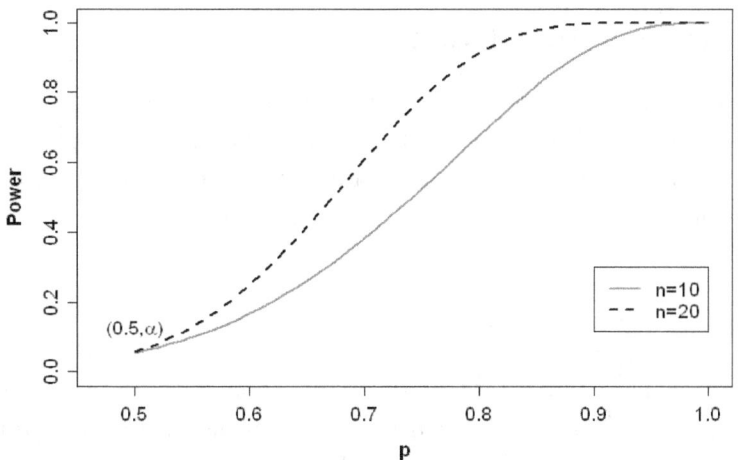

Figure 5.7: Power for the test for $H_o : p = 0.5$ against $H_a : p > 0.5$ when $n = 10$ and $n = 20$

hypothesis is NOT rejected when the number of mallards who pick the green bread first is below 8. Thus, in order to calculate β we need to add the values ABOVE the horizontal line corresponding to the decision rule. To practice, calculate β when p=0.8; the answer should be 0.3222. It becomes clear that **power**$=1 - \beta$

Look at the table in Figure 5.5 again. What happens if the line is moved down? (i.e., if the decision rule was to reject Ho if 9 or more mallards pick the green bread first). Will α increase or decrease? Will β increase or decrease? If α is reduced, the power decreases and β increases. Working with a given sample size n, making α very small might cause β to be too large. When the probability of Type I error is reduced, the probability of Type II error increases.

Look at Figure 5.5 again. What happens if the line is moved up? (i.e., if the decision rule was to reject Ho if 7 or more mallards pick the green bread first). Will α increase or decrease? Will the power of the test increase or decrease? Will β increase or decrease? For a given sample size n, when α increases, so does the power. Thus, $\beta = 1 - Power$ decreases.

5.8 Exercises

5.8.1 Review questions

All the review questions in this chapter refer to the following story:
Assume that 10% of the population of a country has a given trait. Somebody claims that, in an isolated region, the trait is more common than in the general population. The statistical hypotheses are:

$$H_o : p = 0.1$$
$$H_a : p > 0.1$$

where p represents the proportion of people in that region who have the trait. You go to that region and select a random sample of 20 individuals. Assume that you are asked to work with $\alpha = 0.04317450$. The binomial probabilities for $n = 20$ and some selected values of p are provided in Table 5.4.

1. Assume that in the sample of 20 individuals, 4 have the trait. Use Table 5.4 to calculate the p-value.
   A)0.089779   B)0.031921   C) 0.132953   D)0.956826

2. Based on the calculated p-value, what is your decision about the null hypothesis H_o?
   A) Reject H_o B) Do not reject H_o C) I can't decide

3. What is the correct interpretation of your decision in question 2?
   A) The data do not constitute enough strong evidence against the hypothesis that in this region 10 % of the population have the trait, the same as in the general population of the country.
   B) The percent of people with the trait in that region is definitely and exactly 10%.
   C) Data constitute strong evidence in favor of the claim that in this region the percent of people with the trait is over 10%.
   D) We have proved that exactly 20% of the population in the region have the trait.

4. Fill in the blank in the following decision rule so that $\alpha = 0.04317450$:
   Reject the null hypothesis $H_o : p = 0.1$ if _____ or more of the 20 individuals in the sample have the trait. Trace a horizontal line in Table 5.5 to represent that decision rule.

Table 5.4: Binomial table for $n = 20$ and selected values of p

x	$p = 0.1$	$p = 0.3$	$p = 0.5$
0	0.12158	0.00080	0.00000
1	0.27017	0.00684	0.00002
2	0.28518	0.02785	0.00018
3	0.19012	0.07160	0.00109
4	0.08978	0.13042	0.00462
5	0.03192	0.17886	0.01479
6	0.00887	0.19164	0.03696
7	0.00197	0.16426	0.07393
8	0.00036	0.11440	0.12013
9	0.00005	0.06537	0.16018
10	0.00001	0.03082	0.17620
11	0.00000	0.01201	0.16018
12	0.00000	0.00386	0.12013
13	0.00000	0.00102	0.07393
14	0.00000	0.00022	0.03696
15	0.00000	0.00004	0.01479
16	0.00000	0.00001	0.00462
17	0.00000	0.00000	0.00109
18	0.00000	0.00000	0.00018
19	0.00000	0.00000	0.00002
20	0.00000	0.00000	0.00000

5. Use the binomial table to find the power of the test when the true value of p is 0.3 (i.e., when 30% of the population of that remote region has the trait) and α=0.04317450. Also calculate the power when the true value of p is 0.5. Does it make sense that the power when $p = 0.5$ is larger than when $p = 0.3$? Why?

6. What is the best description of Type I error in this example?
A) To conclude that in that region the percent of the population with the trait is larger than 10% when it actually is 10%.
B) To conclude that in that region the percent of the population with the trait is 10% when actually it is larger than 10%.
C) To make a mistake in calculating the p-value.
D) To conclude that the percent of the population with the trait is 10% when it actually is 20%.

7. What is the best description of Type II error in this case?
A) To conclude that in that region the percent of the population with the trait is larger than 10% when actually it is 10%.
B) To conclude that in that region the percent of the population with the trait is 10% when actually it is larger than 10%.
C) To make a mistake in calculating the p-value.
D) To conclude that the percent of the population with the trait is 20% when it actually is 10%.

8. Check your answer to question 1 by performing the test of hypothesis using software.
If using **R**, type: **binom.test(4,20,p=0.1,alternative=c('greater'))**

If using MINITAB, select STAT>Basic Statistics> 1 proportion

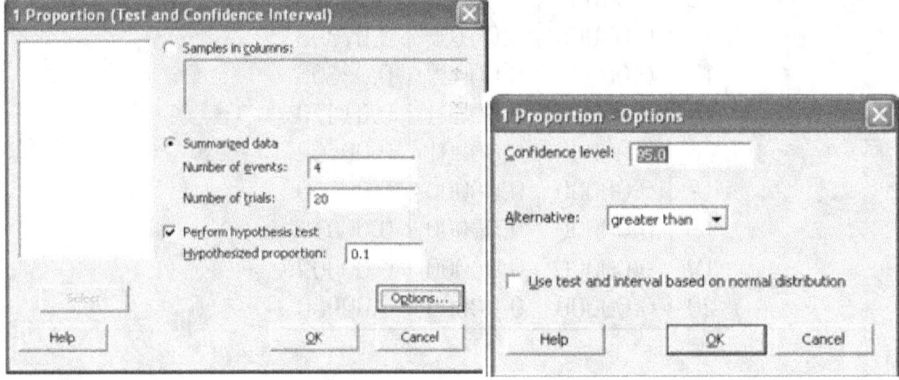

5.8.2 Exercises for discussion or homework

1. Usually at most 3% of the population is affected by a given disease. Somebody makes the hypothesis that in a region the % of people with the disease is way over 3% and that there is an epidemic there. Write the null and alternative statistical hypotheses for this story in terms of p, the proportion of people with the disease in that region. Describe the two types of error in the context of this story.

2. Nesting success for birds is defined as 'the % of nests that success-fully hatch at least one egg.' Nesting success varies across species, and it also depends on environmental conditions, predators, etc. Experts think that the nesting success necessary to maintain a mallards population is 15%. There are suggestions that the mallard population might

be decreasing. A test for $H_o : p = 0.15$ vs. $H_a : p < 0.15$ will be done with $\alpha=0.048$. A random sample of 30 nests is selected, the nests are observed and only in 3 of the nests the mallards are successful at nesting. Use software to calculate the p-value (notice that the alternative hypothesis is < 'less than'). State your conclusion about the null hypothesis and interpret it.

3. In the year 2002, approximately 22.5% of the U.S. population were currently smokers. We wonder if in Tennessee the proportion of current smokers was also 0.225 or if it was higher than that. Write the null and alternative hypotheses. Assume that in that year, a survey was conducted in Tennessee, and that a simple random sample of 2466 individuals was selected. Out of the 2466, 611 were classified as 'current smokers.' Use software to conduct the test. What is the p-value? Do you reject or not reject H_o? What is your answer to the research question?

4. It is known that for 'quarters' the probability of heads is approximately 1/2. We wonder if the same happens with dimes. Write the null and alternative hypotheses. What type of alternative hypothesis is that? One-sided? Or two-sided? Take one dime and toss it 25 times. How many heads did you obtain? Use the binomial table for $n = 25$ and $p = 0.5$ (Table 5.2) to calculate the p-value. What is your conclusion for that particular coin?

5. Consider that you are in the planning stage of a little experiment with a hypothetical type of coin for which you suspect that the probability of heads is higher than 0.5. In this case the alternative hypothesis is one-sided. This experiment is going to be repeated independently by many people, so you want to define an easy rule for when to reject the null hypothesis such as :'Reject H_o if more than ____ heads happen.' Use Table 5.2 (Binomial $n = 25$, $p = 0.5$), and decide where to trace the line for rejecting H_o so that α is as close as possible to 0.05. In this case, work with $\alpha = 0.05387$. Fill in the blank in the definition of the rule. Now analyze how powerful this test would be to detect a slight bias of the coin toward heads, such as when $p = 0.6$. Use software to produce the column of binomial probabilities for $n = 25$ and $p = 0.6$. Calculate the power of the test or decision rule you just defined. Is that power high enough for you? What would you do in case you want to increase the power of the test when the true probability of heads is not 0.5 but 0.6?

Note: This exercise was about coins; however, the issues discussed would be valid for any other type of problem in which you are trying to plan a study in which the statistical hypotheses are $H_o : p = 0.5$ and $H_1 : p > 0.5$.

6. A case that has become a classic in introductory statistics books is that of an experiment on therapeutic touch (TT) conducted by Emily Rosa as a science fair project when she was about 10 years old, and later published in a medical journal (Rosa et al., 1998). Videos and other accounts of the experiment are available in the web. She worked with a total of 21 TT therapists. The therapist was behind a screen with only the hands exposed. The experimenter passed her hand, without actually touching, over one of the hands (determined at random) of the therapist, who had to say if it was the right or the left. The question is whether the individual can actually feel the supposed 'energy' transmitted by the experimenter, which would lead to the identification of the correct side. The null hypothesis is that therapists cannot really feel any energy, and that they would be just guessing at random between right and left, thus $H_o : p = 0.5$. If the therapists could actually feel something, we would expect the proportion of correct identifications to be higher than that. Write the H_a. A total of 280 trials were done, in 122 of them the correct hand was identified. Use software to perform the test, report the p-value, and interpret the results.

7. In the table in Figure 5.5 calculate the power when $p = 0.6$. Why do you think it is so low?

8. In a certain college, the success rate in a typical introductory statistics course (proportion of students obtaining a C or better grade) is 70% of the students registered at the beginning of the semester or $p = 0.7$. Somebody creates a new method to teach the course and claims that the success rate will be higher. Write the null and alternative hypotheses. An experiment is conducted for which 30 students are randomly recruited. Prepare a binomial table for $n=30$ using software. Trace a horizontal line in that table (as was done in the table in Figure 5.5) defining a decision rule to know when to reject the null hypothesis, in such a way that α is the closest it can get to 0.05 (it could be larger or smaller than 0.05). Calculate the power of the test for $p=0.75$, 0.80, 0.85, 0.9, and 0.99. Plot the power as was done in Figure 5.6. What would be your decision about H_o if from the 30 students registered, 24 get a C or better at the end of the course? What is your conclusion

about the new method of teaching? If the study is going to be repeated, how could you improve the power of the test maintaining a more or less similar value for α?

9. This exercise is inspired by Exercise 6, but it includes some variations. Assume that there is a person who claims to have the ability of perceiving the therapeutic touch described in Exercise 6. You plan to conduct an experiment to find out if his/her claim is false. Again, the statistical hypotheses are $H_o : p = 0.5$ and $H_a : p > 0.5$. You are thinking of doing 50 trials. The 50 trials will be done by the same person and our conclusions will be for that person alone. We can use software to produce a binomial table for $n = 50$ and 3 values of p : 0.5, 0.7, and 0.9, that would represent 3 different scenarios with respect to the capacity of feeling the TT: 'no capacity at all,' 'some limited capacity,' and 'he/she definitely feels the touch.' The commands in **R** that would prepare such a table are:

```
x<-seq(0,50,by=1) ; px5<-dbinom(x,50,0.5) ;
px7<-dbinom(x,50,0.7); px9<-dbinom(x,50,0.9) ;
table<-cbind(x,px5,px7,px9); table
```

We want to define a decision rule that says 'Reject the null hypothesis if _____ , or more, correct identifications are done.' In order to fill in the blank, the value of α needs to be decided. We would need to start adding probabilities starting from the bottom of the column for $p = 0.5$ until we get as close to the value 0.05 as possible, a task that can sound a little daunting. This is when cumulative probabilities **cumulative** $P(X \leq x)$ become handy. If we replace **dbinom** by **pbinom** in the **R** **probability** commands above, we will get the probabilities in each column added up to every value of x. The rows for $x = 30$ and $x = 31$ read as follows:

```
       Cumulative binomial probabilities
  x    p=0.5         p=0.7          p=0.9
 30 9.405398e-01 8.480260e-02 2.368599e-08
 31 9.675457e-01 1.405599e-01 1.396909e-07
```

In the binomial distribution, when p=0.5, $P(X \leq 30) = 0.9405398$, that is the closest we can get to 0.95. Thus, $P(X > 30)$=1-0.9405398= 0.0594602, and that will be the value of α we will work with. So, now we have the decision rule 'Reject the null hypothesis if 31 or more correct identifications out of 50 are done.' We trace a line after $x = 30$ in the binomial table we produced.

Calculate the power of the test when $p = 0.7$ and when $p = 0.9$. Would the 50 trials be enough? In other words: Are you satisfied with the power of the test? (You do not actually need to produce the binomial table if you don't want to; the information given here is enough to calculate the power). Plot the values of the power for $p = 0.7$ and $p = 0.9$ in Figure 5.7. Calculate and interpret the probability of Type II error when $p = 0.7$ and $p = 0.9$.

5.8.3   Small data analysis projects

1. Think of a situation involving a categorical variable, for example hair color, being left handed, having had the flu during the past year, having ever broken a bone, or being Rh+. Write the null and alternative statistical hypotheses about the proportion of a category in a population of interest. Collect data through observation (random samples, please). Perform a test of hypothesis using the binomial distribution and state your conclusion.

2. Do a small project to test the hypothesis that 90% of the tomato seeds from a given brand germinate against the alternative hypothesis that the germination rate is below 0.9. Tomato seeds could be substituted by other type of seeds. You can buy 20 seeds, plant them, see how many of them germinate, and arrive at a conclusion about the null hypothesis.

References

1. Rosa, L., Rosa, E., Sarner, L., Barrett, S., (1998) A Close Look at Therapeutic Touch. *JAMA* **279**: 1005-1010.

2. Seier, E. and Robe, C. (2002) Ducks and Green - An Introduction to the Ideas of Hypothesis Testing. *Teaching Statistics* **24**: 82-86

3. Seier, E. and Liu, Y. (2011) An Exercise to Introduce Power. *Teaching Statistics*, first published online 4 APR 2011 DOI: 10.1111/ j.1467-9639.2011.00464.x.

Chapter 6

Conditional probability

In this chapter we focus on the calculation of conditional probabilities in the context of diagnosis. Important concepts in the application of statistics to medicine, such as false positive, false negative, specificity, sensitivity, and positive and negative predictive value will be discussed.

In Section 6.3, conditional probabilities are calculated from two-way tables. Some of the tables come from surveys, others are constructed for hypothetical populations based on information available in the medical and public health literature. In Section 6.4 reverse conditional probabilities are calculated with Bayes Rule in two forms: using probability trees and directly with the formula. We recommend using probability trees in class, as in Section 6.4.1. Section 6.4.2 is the classical presentation of the rationale of Bayes Rule; it is for those who want to understand where the formula applied in the probability tree comes from.

Appendix 1 of the book explains how to build two-way tables for hypothetical populations based on information about specificity and sensitivity of a medical test and the prevalence of the disease or condition in the population.

6.1   What is conditional probability?

Assume that a nurse working at a health care center knows that approximately 25% of adults have a cholesterol level above the threshold of 240 mg/dL. A man walks in (consider him a randomly selected man from the entire adult population), and the nurse thinks to herself: 'the probability that he has high cholesterol is 0.25.' The nurse measures the blood pressure of the patient and finds it elevated. She knows that there is an association between high cholesterol and high blood pressure. Given that the man has high blood

**Idea of
conditional
probability
through
an example**

pressure, she re-evaluates the situation and wonders if the probability that he has high cholesterol is greater than 0.25. This type of situation happens frequently. Under the absence of additional information, the probability of an event A is a given number. However, the occurrence of another event B might change the probability of A. This is what conditional probability is all about.

The notion of independence was introduced in Section 4.5 of Chapter 4. Figure 4.4 depicts two urns. In one urn, color and size of the balls are independent, and in the other urn they are not independent. Color and size are obviously not independent for the urn in Figure 6.1. There are more red balls among the large balls than among the small balls. When randomly selecting a ball from that urn, the probability of obtaining a red ball is $P(red) = 5/10 = 1/2$. Now assume that a ball is picked by a blindfolded person. The person perceives that the ball is large, but does not see the color. What is the probability that the ball is red? In this case $P(red/large) = 3/5$ because all the small balls are ruled out and only the large balls are considered. There are 5 large balls, 3 of which happen to be red. Thus, the information about the size of the ball has changed the probability of the ball being red.

This is the most simple way of calculating conditional probabilities: reduce the universe according to the information given, and calculate the probabilities in that restricted universe.

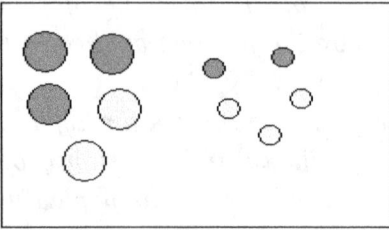 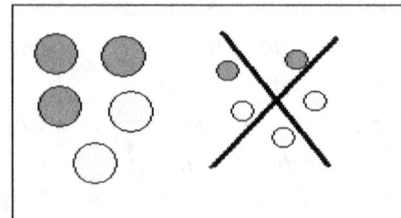

Figure 6.1: Urn before and after knowing that the selected ball is large

**Definition of
conditional
probability**

Conditional probability is formally defined as

$$P(B/A) = \frac{P(A \cap B)}{P(A)} \tag{6.1}$$

Consider the urn for which $P(red/large)$ was calculated intuitively. Now it can be calculated using the expression (6.1):

$$P(red/large) = \frac{P(red \cap large)}{P(large)} = \frac{3/10}{1/2} = \frac{3}{5}$$

$P(red \cap large) = 3/10$ because there are three balls that are red AND large. $P(large) = 1/2$ because 5 out of 10 balls are large.

In the urn example, the *probability of red* changed when it was known that the ball was large, $P(red/large) \neq P(red)$. This is an indication that color and size are NOT independent. Two events A and B are said to be independent if

$$P(A \cap B) = P(A) \times P(B) \qquad (6.2)$$

In general,

$$P(A \cap B) = P(A) \times P(B|A) \qquad (6.3)$$

Independent events

Expression (6.3) means that the probability that two events A and B occur is equal to the probability that one of them occurs multiplied by the probability that the second occurs, given that the first one has already occurred. For example, the probability of being exposed to pollen and getting an allergic reaction to it could be understood as the probability of being exposed to pollen multiplied by the probability of getting an allergic reaction given that there has been exposure to pollen.

Probability that both A AND B happen

6.2 Decisions and consequences

The following is an example of how conditional probabilities are not only easy to calculate, but quite useful for evaluating decision or diagnosis rules.

Revisiting Example 2.8.3

The Khan data set (Desper, Khan & Schaffer, 2004) contains the intensity of 2308 genes for 83 patients with one of four varieties of small, round blue cell tumors. Figure 2.23 compares the level of expression of one particular gene for four varieties or sub-types of a tumor. From Figure 2.23, it is clear that the gene tends to be expressed with more intensity in tumors of varieties 1, 3, and 4 than in tumors of variety 2 (Burkitt's lymphomas or BL). In the dotplots in Figure 6.2, types 1, 3 and 4 have been grouped together. Assume that the objective is to identify whether the tumor is a Burkitt's lymphoma or not, based on the intensity of this sole gene. Diagnoses based on microarrays are generally more complex and based on the intensities of several genes. The one-gene example included here is a simplified version designed to help the student understand some concepts and introduce some terminology.

One gene & 4 subtypes of tumor

To set the threshold at *intensity* $= -1$ seems a sensible rule. When a new patient comes with a small, round blue cell tumor, a microarray is prepared with the patient's tissue. If the intensity of that particular gene is below -1, the tumor will be identified as a Burkitt's lymphoma. If the intensity of this particular gene is above -1, it will be thought that the tumor is not BL but of one of the other sub-types.

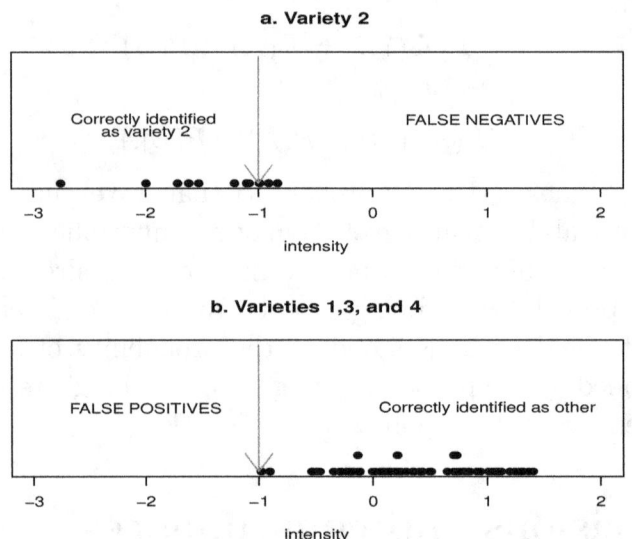

Figure 6.2: Diagnosis based on the expression of one gene

The way a
decision rule
is defined
has an impact
on the
probability
of each type
of error

It is a recommended practice to define decision rules and diagnosis methods with one data set ('training data set') and evaluate them with another data set ('test data set'). However, here we will use the data for the 83 patients at once. In the first dotplot there are 11 patients with Burkitt's lymphoma, and for 3 of them the intensity of the gene is above the value -1. Thus, applying the decision rule with the threshold at -1, they would be classified as not having Burkitt's lymphoma, when actually they do. Those three cases will be called 'false negatives.' If one of the 83 patients was selected at random, the probability of a 'false negative' will be the conditional probability:

P(false negative)=P(diagnosed as 'not BL'/the tumor is BL)=3/11

Notice also that 8 of the 11 patients with Burkitt's lymphoma are correctly diagnosed. In the second dotplot, there is a total of 72 patients and all of them are correctly classified as 'other,' there are no 'false positives.'. Thus,

P(false positive)=P(diagnosed as 'BL'/the tumor is not BL)=0

If the arrow is moved to the right of −1 in Figures 6.2a and 6.2b, the number of false negatives will diminish. However, some false positives will appear. **When the probability of one type of error decreases, the probability of the other type of error increases.**

6.3 Conditional probabilities from 2-way tables

Example 6.1: Smoking and marijuana

Smoking
& Marijuana

Table 6.1 displays information for 4304 individuals, 18 years or older, interviewed during an illegal drug, tobacco, and alcohol survey in the Midwest in 1996. The respondents have been classified according to their answers to the following questions:

- Have you smoked more than 100 cigarettes in your lifetime? If the answer is 'YES,' the individual is classified as an 'Ever smoker;' if the answer is 'NO,' the individual is classified as 'Never a smoker.'

- Have you used marijuana at least once in your life?

Two-way table
or joint
frequency
distribution

Table 6.1: Cigarette smoking and marijuana

| | Has ever used marijuana? | | |
Is or has been a smoker	YES	NO	Total
YES	722	1207	1929
NO	397	1978	2375
Total	1119	3185	4304

Table 6.1 describes the **joint frequency distribution** of cigarette smoking and marijuana use. Two-way tables are commonly produced in the analysis of survey data. To do 'cross-tabulation' is to combine the answers of two questions in the survey in order to produce this type of table.

Assume that one individual is randomly selected from the 4304 in Table 6.1. What is the probability that the individual is or has been a smoker AND has used marijuana at least once in his/her life? There are 722 individuals who satisfy both conditions, out of a total of 4304. Thus, P(smoker AND marijuana)=722/4304 (or $P(S \cap M) = 722/4304$).

Look at the totals of rows and columns in Table 6.1. A **marginal frequency distribution** is the distribution of only one of the variables ignoring the other variable. What is the probability that a randomly selected individual has used marijuana at least once in his/her life? This question is answered by looking only at the variable marijuana and the totals of each column. There are 1119 individuals out of the 4304 who have used marijuana at least once, thus P(marijuana)=1119/4304=0.2599907 $\sim$ 0.26. What is the probability that a randomly selected individual is or has been a smoker? There are 1929 smokers out of a total of 4304 individuals, thus P(smoker)=1929/4304.

Conditional probabilities

Conditional probabilities can also be calculated from Table 6.1. For example: What is the probability that a randomly selected person has used marijuana given that the person is or has been a smoker? Focus on the first row of the table, where the smokers are located, while ignoring those who have never been smokers, as in Figure 6.3. There are only 1929 'ever smokers,' among them 722 have used marijuana. Thus, P(marijuana/ever smoker)=722/1929 =0.3742872 $\sim$ 0.37.

Table 6.1: Cigarette smoking and marijuana

Has ever been an smoker	Has ever used marijuana?		
	YES	NO	Total
YES	722	1207	1929
~~NO~~	~~897~~	~~1978~~	~~2375~~
Total	1119	3185	4304

Figure 6.3: Looking just at the 'ever smokers' in Table 6.1

The conditional probabilities in the other direction can also be calculated. For example: P(ever smoker/ marijuana)= 722/1119=0.645219 $\sim$ 0.65, because there are only 1119 individuals who have used marijuana and 722 of them are or have been smokers. A two-way table describes how two categorical variables vary jointly (or the 'joint distribution' of the variables) in either a population or a sample. In a two-way table, marginal frequencies, probabilities, AND conditional probabilities in either direction can be calculated.

Example 6.2: Trisomy 21 and nasal bone

Trisomy 21

Down syndrome is associated with an error in meiosis that results in 3 copies of chromosome 21 (Trisomy 21). There is a definite test for Trisomy 21 in the fetus, but it is an invasive procedure called amniocentesis. Amniocentesis can produce infections and miscarriages. Thus, pregnant women usually prefer to have other tests or screenings before deciding to have an amniocentesis. Fan & Levine (2007) used Bayes' theorem, also called Bayes Rule, to be studied in Section 6.4, to make a decision. Some pregnancies are considered of high risk for Trisomy 21 due to the age of the mother and the results of the triple marker test, which is routinely applied to all pregnant women. For a type of high-risk pregnancies, the probability that the fetus has Down syndrome is 1/80. There is a non-invasive exam that uses ultrasound and focuses on the presence or absence of the nasal bone. The ultrasound exam has two possible results:

Abnormal marker (AM): absence of nasal bone = **TEST +**

Normal marker (NM): presence of nasal bone = **TEST −**

There is historical data of all the fetuses in which the ultrasound test was applied and, after birth, verification of Down syndrome was recorded. Half of the fetuses with Down syndrome showed absence of the nasal bone (abnormal marker) in the ultrasound. All fetuses without Down syndrome got NM (normal marker) in the ultrasound test. All this information is provided by Fan and Levine (2007). Rossman and Short (1996) suggest using two-way tables to facilitate the understanding of conditional probabilities in introductory statistics courses. We prepared Table 6.2 using the information in Fan and Levine (2007) (please see warning at the end of Section 6.3.3). Instructions on how to prepare such a table are included in Appendix 1. Table 6.2 displays the information for 160,000 hypothetical fetuses in the high-risk group. We will use this example to introduce some conditional probabilities that have special names in the medical literature. Consider the ultrasound exam as a test for Down syndrome or Trisomy 21 and the abnormal marker (absence of nasal bone) as a '+' (positive) in that test.

Table 6.2: Down syndrome and nasal bone in ultrasound

	Normal marker (nasal bone present) −	Abnormal marker (no nasal bone) +	Row total
Down syndrome	1000	1000	2000
No Down syndrome	158000	0	158000
Column Total	159000	1000	160000

6.3.1 False positives and false negatives

A false positive happens when the test is positive but the person does not have the disease or condition.

P(false +)= P(test + / individual does NOT have the disease or condition)

false +

In the case of Example 6.2, a false positive would be 'abnormal marker' (absence of nasal bone) when the fetus does not have Down syndrome. What is the probability of a false positive? There is a total of 158,000 fetuses in the 'No Down syndrome' row, none of them tested positive. Thus,

P(false +)=P(abnormal marker/No Down syndrome)=0/158000=0

false −

A false negative happens when the test is negative but the person has the disease or condition.

P(false −)=P(test − / individual has the disease or condition)

In the case of the example, a false negative would be that a normal marker is observed in the ultrasound (nasal bone is visible) but the fetus has Down syndrome. What is the probability of a false negative? There are 2,000 fetuses in the 'Down syndrome' row, and for 1,000 of them the test was negative. Thus,

P(false −)=P(normal marker/Down syndrome)=1000/2000=0.5

In Example 6.2, there are no false positives but the probability of false negative is 1/2.

6.3.2 Sensitivity and specificity

sensitivity

The sensitivity quantifies the ability of the test to find out that the person has the disease or condition when he/she has it.

Sensitivity=P(test +/ individual has the disease or condition)

What is the sensitivity of the ultrasound test? In Table 6.2, there are 2,000 fetuses with Down syndrome and 1,000 of them tested positive. Thus,

Sensitivity = P(test +/Down syndrome)=1000/2000=0.5

The ultrasound exam only detects 50% of the Down syndrome cases.

specificity

The specificity of a medical test evaluates how well the test performs in identifying who DOES NOT have the disease or condition.

Specificity = P(test −/ individual does NOT have the disease)

There are 158,000 fetuses that do not have Down syndrome in Table 6.2, and all of them tested negative; that is, the nasal bone was visible. Thus,

Specificity of the test=P(test −/No Down syndrome)=158000/158000=1.

The ultrasound test has a sensitivity of only 1/2 but a specificity of 1. Notice that:

$$\text{Sensitivity} = 1 - \text{P(false } -)$$
$$\text{Specificity} = 1 - \text{P(false } +)$$

6.3.3   PPV and NPV

When a person tests positive for a disease, it does not mean necessarily that the person has the disease; it might be a false positive. A very interesting probability for those who test positive is the one called **positive predictive value** or PPV.   PPV

PPV=P(has the disease or condition/ test +)

Notice that in the high-risk group depicted in Table 6.2, there is a total of 1000 fetuses that tested positive for Down syndrome in the ultrasound ('abnormal marker' or absence of nasal bone), all of them had Down syndrome. The probability of having the disease or condition given that one has tested positive, is called the **positive predictive value** or PPV. Thus, for the ultrasound test:

PPV = P(Down/abnormal marker)=1000/1000=1.

In a similar way, we can calculate the probability of not having the disease given that the result of the test was negative. That probability is called **negative predictive value** or NPV: NPV

NPV=P(does not have the disease or condition/ test −)

In Table 6.2 there are 159,000 fetuses that tested negative, among them 158,000 did not have Down syndrome. Thus:

NPV = P(no Down/ normal marker)=158000/159000=158/159

From the patient's perspective, it is important to calculate 1-NPV. This represents the probability of having the condition or disease given that the result of the test was negative. In Example 6.2, 1-NPV= P(Down/normal marker)=1/159.

What information is necessary?

All the conditional probabilities of interest in one direction or the other were calculated from Table 6.2. What information was used to build that table? Now that you are familiar with the new vocabulary, look again at the information given in Fan & Levine (2007):

1. In this specific high-risk group, the probability of Down syndrome is 1/80. This is the 'prevalence' of the condition, the proportion of individuals in the population or sub-population of interest that have the condition under study.

2. Half of the fetuses with Down syndrome show absence of the nasal bone (abnormal marker) in the ultrasound. This is telling us that the sensitivity, P(test +/ Down), of the ultrasound test is 1/2.

3. All babies without Down syndrome get NM (normal marker) in the ultrasound test. This is telling us that P(normal marker/No Down)=1, this is P(test −/ the condition or disease is not present) or the 'specificity' of the test.

The three pieces of information were **prevalence**, **sensitivity**, and **specificity**. All three were necessary to build Table 6.2 to calculate the conditional probabilities of interest such as PPV, NPV and the probability of false positive and false negative. In Appendix 1, guidelines are given to build a two-way table based on prevalence, sensitivity, and specificity.

Warning

Warning: There is a fundamental difference between Table 6.1 and Table 6.2. Table 6.1 was prepared with real data from a survey and therefore it will be used in hypothesis testing in Chapter 10. Table 6.2 was prepared for a hypothetical population using valid information and can be used to calculate conditional probabilities. However, we should never use hypothetical tables to test hypotheses or to build confidence intervals. We have learned in Chapters 3 and 5 that in hypothesis testing and confidence interval estimation, the sample size matters. In tables such as 6.2 or 6.3, the sample size is arbitrary and hypothetical as it is explained in Appendix 1. If we knew how many babies were actually involved in the study of ultrasound and Trisomy 21, then we could proceed with hypothesis testing or confidence interval estimation.

6.4   Bayes Rule

Sometimes we receive information in the form of the probability of an event A, $P(A)$, and a conditional probability for an event B, $P(B/A)$, but what we really want to know is $P(A/B)$. In medical studies, the sensitivity P(+/disease) and specificity P($-$/no disease) of a test might be known, as well as the prevalence of the disease in the population or 'pre-test probability of disease,' P(disease). The question is: How do we calculate the positive predictive value PPV=P(disease/test +) or 'post-test' probability of disease? The tool to solve the problem is Bayes Rule. There are two ways of organizing the calculations for the application of Bayes' theorem: applying the formula directly (as in Section 6.4.2), or sketching a probability tree (as in Section 6.4.1). There is also the alternative of converting the information into a two-way table for a hypothetical population as in Table 6.2. Using any one of these three strategies, the same result is obtained.

6.4.1 Probability trees

Probability trees are useful to organize the information of random processes that involve several stages. In particular, they offer an easy way of organizing the calculation of reverse or inverse conditional probabilities. In the examples that follow, notice that the branches coming out from the first node of the tree refer to the possible states of the individual. The probabilities are the pre-test probabilities that depend on the prevalence of the disease or condition in the population. The second set of branches corresponds to the result of the test and the probabilities assigned come from the sensitivity and specificity of the test.

Revisiting Example 6.2

Trisomy 21

The question of interest is: For a fetus in the high-risk group, what is the probability of Down syndrome, given that the ultrasound gave a normal marker? This is a reverse or inverse conditional probability question. In the medical literature, this is known as a 'post-test' probability. The probability of getting a normal marker in the ultrasound when the fetus has Down syndrome is 1/2, but what is the probability that the fetus has Down syndrome given that the ultrasound showed a normal marker? In other words, P(NM/DS)=0.5, but what is P(DS/NM)?

The information given by Fan & Levine (2007) is now used to build the probability tree in Figure 6.4:

1. Among this type of high-risk pregnancies, the probability of the fetus having Down syndrome is 1/80, thus the probability of NOT having Down syndrome is 79/80. Those two values go in the branches that come out from the first node.

2. Among fetuses with Down syndrome, half show a 'normal marker' in the ultrasound and half show the 'abnormal marker' or test +. Thus, in each one of the two branches (AM and NM) that come out from the node for Down syndrome, we write 1/2.

3. All fetuses without Down syndrome show the 'normal marker' in the ultrasound, and none shows the 'abnormal marker' or test +. Hence, in the second set of branches that come out from the node for 'No Down,' the branch for abnormal marker (test +) has probability 0 and the branch for normal marker (test−) has probability 1.

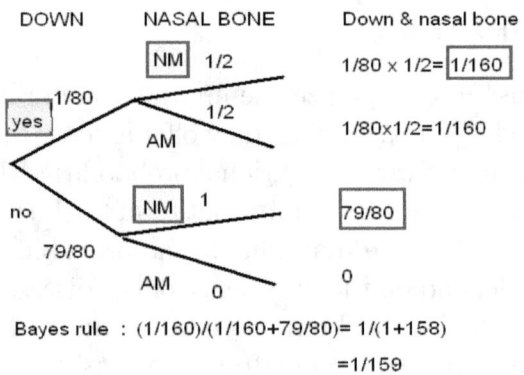

Figure 6.4: Probability tree for the Trisomy 21 example

In the first set of branches (those coming out from the first node) the probabilities add up to one. In the second set of branches (corresponding to the results of the test), the probabilities in each subset (branches coming from the same node) also add up to 1. Now we multiply the probabilities along each sequence of branches. The values at the end are probabilities of two events happening. For example P(Down AND normal marker)$=\frac{1}{80} \times \frac{1}{2} = \frac{1}{160}$. Those probabilities are the joint probabilities.

In order to get the reverse probabilities such as P(Down/Normal marker), consider only the branches that have 'normal marker' in them. There are two such branches, and the probabilities of those two branches are: 1/160 and 79/80. Now the universe is reduced to these two branches, and the total probability is $\frac{1}{160}+\frac{79}{80}$. To calculate the probability P(Down/Normal marker), pick the branch in which the fetus HAS Down syndrome. Divide the probability of that branch by the sum of the probabilities of all the branches that have 'normal marker' in them. Thus, P(Down/Normal marker)$=\frac{1/160}{1/160+79/80} = 1/159$.

Considering the age of the mother and the results of the triple marker blood test, the probability of the fetus having Down syndrome was 1/80 (this is the 'pre-test' probability). However, after receiving the result 'normal marker' in the ultrasound, the probability has been reduced to 1/159. The calculations done in the probability tree are actually the implementation of the Bayes Rule formula in expression (6.4) and explained in Section 6.4.2.

In the exercises section (Exercise 7) you will be asked to rework this case for another population for which the prevalence of Trisomy 21 is 92/100000.

Probability tree for the general case

Bayes' theorem is not applied only in the medical diagnosis context, but to many fields in which there are sequences of events and reverse probabilities are needed. In Example 6.2, the first node had only two branches: Down or no Down. However, this is not always the case. For example, if we are studying lung cancer with regard to smoking status, and smoking status goes in the first node, we would need to consider three possibilities: never a smoker, former smoker, and current smoker. In general, there could be k possibilities: A_1, A_2, ...,A_k. In the second set of nodes, there are always two possibilities: the event B happens or does not happen ($\bar{B}$). In Example 6.2, the two possibilities are 'test positive' or 'test negative.' In the smoking and lung cancer case, B would be to develop lung cancer and $\bar{B}$, not to develop lung cancer.

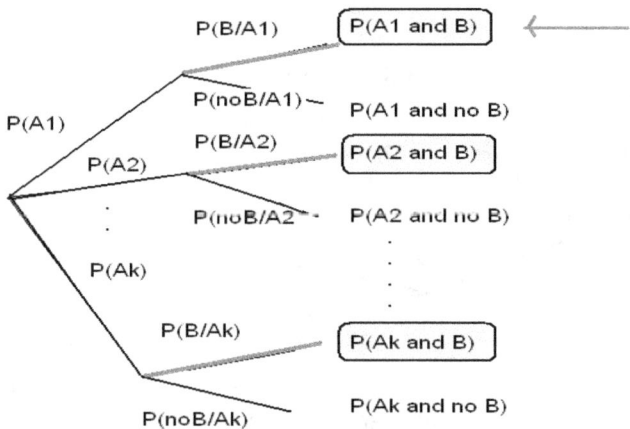

Figure 6.5: Tree for the general case of Bayes rule

A probability tree for the general case would look like the one in Figure 6.5. In the first node, there are k possible states: A_1, A_2,..., A_k. The probability for each one of these states is: $P(A_1)$, $P(A_2)$,..., $P(A_k)$. In the second set of nodes, there are two possibilities: B and *not* B. In the second set of branches, the probabilities are conditional probabilities of the type $P(B/A_i)$ and $P(\bar{B}/A_i)$. Multiplying the probabilities along each branch we obtain, according to expression (6.3), the probabilities that both things happen. For example, $P(A_1 \text{ and } B) = P(A_1) \times P(B/A_1)$ is the probability that both events A_1 and B happen. To calculate the reverse conditional probability $P(A_1/B)$, we take the probability at the end of the branch that has A_1 and B, $P(A_1 \text{ and } B)$, and divide it by the sum of the probabilities at the end of all the branches that have B on them, since B is the condition given. Bayes Rule in equation (6.4) is used to calculate reverse conditional probabilities and is a theorem proved by Thomas Bayes in the 18th century:

$$P(A_i/B) = \frac{P(B/A_i) \times P(A_i)}{\Sigma_1^k P(B/A_i) \times P(A_i)} \qquad (6.4)$$

Example 6.3 Melanoma

Soon et al. (2003) report that the sensitivity of a digital dermoscopic test for melanoma is 93% and specificity is 92.75%. We can prepare a tree to write this information. Sensitivity and specificity are conditional probabilities:

Sensitivity: P(test +/ melanoma)=0.93
Specificity: P(test−/no melanoma)=0.9275

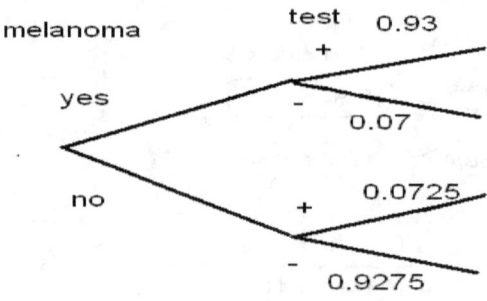

Figure 6.6: Probability tree for Example 6.3

Sensitivity and specificity are properties of the test, thus the probabilities are written in the second set of nodes (those corresponding to the results of the test), remembering that the probabilities in each subset of branches add up to 1. However, the probabilities for the first set of branches are missing in

Figure 6.6. We cannot proceed unless the prevalence of the disease is known. The pre-test probability of a disease can vary for different age groups and other types of sub-populations. The proportion of people getting melanoma varies around the world depending on skin color and exposure to the sun, among other factors. Other variables, such as age, can also play a role. Assume that there is a sub-population for which the pre-test probability of having melanoma is 0.001 or 1/1000. The tree in Figure 6.7 can be used to calculate P(melanoma/test+). Since we already know that the test is positive, we only have to worry about sequences of branches that have a 'test +' label:

$$P(\text{melanoma/test } +) = \frac{0.00093}{0.00093 + 0.0724275} = 0.01266512$$

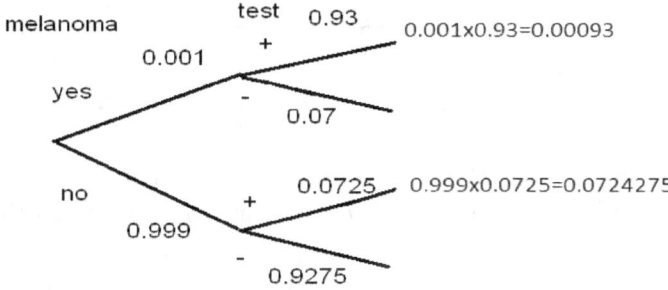

Figure 6.7: Probability tree for Example 6.3 assuming P(melanoma)= 1/1000

The post-test probability P(disease/test +) or PPV depends not only on the sensitivity and specificity of the test but on the prevalence of the disease, or pre-test probability, in the sub-population to which the individual belongs.

6.4.2 Bayes Rule formula *

Bayes Rule, formula (6.4), is named after Thomas Bayes (c.1702−1761), who solved the problem of the 'reverse' probability. His method was published after his death. Fan & Levine (2007) used Bayes's theorem to solve a decision problem. Bayes Rule was already applied to Examples 6.2 and 6.3, doing the calculations with the help of probability trees. The justification of the formula will be discussed using Example 6.2.

Revisiting Example 6.3 Trisomy 21

Consider the population of fetuses in the high-risk group for Down Syndrome. The population is divided in two parts, those who have it and those that do not have it: DS (Down Syndrome) and No DS (No Down Syndrome) as shown in Figure 6.8. The probabilities are: P(DS) = 1/80 and P(no DS)=79/80. On top of that partition we place an event, that in this case is the result

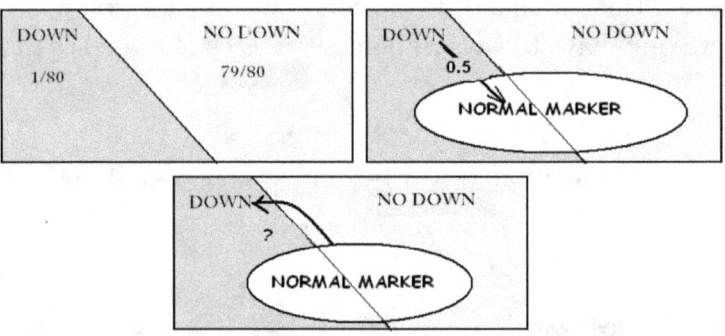

Figure 6.8: Bayes rule for Trisomy 21 example

'normal marker' from the ultrasound test. The Bayes Rule formula results from a repeated use of the definition of conditional probability and the fact that the event 'normal marker' is composed of two parts.

Assume that the ultrasound is conducted for a woman in the high-risk group, and the test shows presence of the nasal bone (normal marker or test negative). What is the probability that the fetus has Down Syndrome? P(Normal marker/Down Syndrome) is known, but the question is P(Down Syndrome/Normal marker) or P(DS/NM). In order to calculate P(DS/NM), we will write it in terms of quantities that are already known: P(DS)=1/80, P(no DS)=79/80, P(AM/DS)=0.5, P(NM/DS)=0.5, P(NM/no DS)=1, and P(AM/ no DS)=0. According to expression (6.1),

$$P(DS/NM) = \frac{P(DS \cap NM)}{P(NM)}.$$

The values of $P(DS \cap NM)$ or $P(NM)$ are not known, but expression (6.1) can be used again:

$$P(NM/DS) = \frac{P(DS \cap NM)}{P(DS)}.$$

Thus, $P(DS \cap NM) = P(NM/DS)P(DS) = \frac{1}{2} \times \frac{1}{80} = \frac{1}{160}$.

The numerical value of the numerator has been found, now the denominator needs to be calculated. The event 'normal marker' is composed of two mutually exclusive events (see Figure 6.8), 'Normal marker and Down syndrome' and 'Normal marker and no Down syndrome,' so each probability can be calculated separately and added later:

$$P(Normal\ marker) = P(NM \cap DS) + P(NM \cap noDS).$$

Since $P(NM \cap DS) = P(DS \cap NM) = P(DS)P(NM/DS) = \frac{1}{80} \times \frac{1}{2} = \frac{1}{160}$, and $P(NM \cap noDS) = P(noDS)P(NM/noDS) = \frac{79}{80} \times 1 = \frac{79}{80}$,

$P(DS/NM) = \frac{P(DS)P(NM/DS)}{P(DS)P(NM/DS) + P(noDS)P(NM/NoDS)}$.

Thus, $P(DS/NM) = \frac{(1/80) \times (1/2)}{(1/80) \times (1/2) + (79/80) \times 1} = \frac{1/160}{159/160} = 1/159$.

For fetuses in the group with high-risk for Down syndrome, the probability of Down syndrome is 1/80, before the ultrasound. It goes down to 1/159 after the ultrasound outcome is known to be 'normal marker' (visible nasal bone). This is the same result obtained with the probability tree in Figure 6.4.

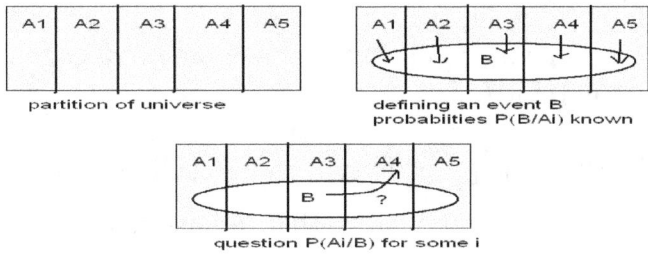

Figure 6.9: Bayes rule in general

In general, Bayes rule is applied to the situation described by Figure 6.9, which can be summarized as follows:

1. There is a partition of the universe with respect to one criteria. The subsets are: $A_1, A_2, ..., A_k$ and the probabilities $P(A_1)$, $P(A_2)$,...,$P(A_k)$ are known.

2. There is another criterion that defines the event B.

3. The probabilities $P(B/A_1)$, $P(B/A_2)$,..., $P(B/A_k)$ are known.

4. The conditional probability $P(A_i/B)$, for some value of $i = 1, ..., k$, is calculated using expression (6.4).

In the example on high cholesterol and high blood pressure used as motivation at the beginning of this chapter, the nurse modified her first idea about the probability that the patient had high cholesterol based on the result of the blood pressure measurement. In Example 6.2, the probability that a fetus has Down syndrome when the pregnancy is in the high-risk group was 1/80, but it changed to 1/159 once it was known, from the ultrasound exam, that the nasal bone was present. There is a way of doing statistical inference called the Bayesian approach, in which the idea the scientist has about the possible values of a parameter is represented by a 'prior' probability distribution. Based on that prior distribution and the data, a 'posterior' distribution for the parameter is calculated. The Bayesian approach is used in medical research and in many other fields of application, but it is beyond the scope of this book.

6.5   Exercises

6.5.1   Review Questions

1. Assume that you will select a person at random from the sample described in Table 6.1.

    (a) What is the probability that he/she has ever been a smoker?

    (b) What is the probability that he/she has ever been a smoker and has used marijuana?

    (c) What is the probability that he/she has used marijuana given that he/she is a smoker?

2. Use Table 2.18 (in Chapter 2) to answer the following question. In some places, only one hip is examined for osteoporosis in order to save costs. Assume that a white woman, 50 years or older, is selected at random from the population.

    (a) What is the probability that she has osteoporosis in the right hip but not in the left?

    (b) What is the probability that she has osteoporosis in one hip but not in the other?

    (c) What is the probability that she has osteoporosis in the right hip given that she has osteoporosis in the left hip?

    (d) What is the probability that she has osteoporosis in both hips?

3. Ongut et al. (2006) evaluate the ICT (immunochromatographic test) to detect tuberculosis. From 72 patients with active pulmonary tuberculosis, 53 tested positive and 19 tested negative. The 54 controls tested negative.

   (a) Based on the information given, fill in the blank cells in the two-way table:

	Test +	Test −	Total
Active Pulmonary TB			72
Controls (No TB)			54
Total			126

   (b) Calculate the sensitivity of the test: P(test + / has TB)=

   (c) Calculate the specificity of the test: P(test − / does not have TB)=

   (d) Calculate the probability of a false positive: P(Test +/does not have TB)=

   (e) Calculate the probability of a false negative: P(test −/ has TB)=

   Be aware that we don't ask you to calculate reverse probabilities, or post-test probabilities, of the type P(has TB/test +), because there is not enough information. The information given is the result of an experiment with 72 TB patients and 54 controls, but no information about the prevalence of TB in the population is available.

4. Traditionally, the Papanicolau smear test is used for the detection of cervical cancer in women. In Nature Reviews Cancer **7**, p. 893 (December 2007) the following information is found: 'The sensitivity of the Pap test is 55.4%, the specificity is 96.8%.' Writing sensitivity and specificity as conditional probabilities: P(Pap +/cancer)= 0.554 and P(Pap−/no cancer) =0.968. Calculate the probability of a false positive and a false negative.

5. Use the probability tree in Figure 6.6 for Example 6.3, but now work with a sub-population for which the pre-test probability of having melanoma is 0.04. Calculate the PPV and NPV.

6. HIV is the virus that causes AIDS, and ELISA (Enzyme Linked Immuno-Sorbent Assay) is a test for HIV. Table 6.3 was constructed by Rossman and Short (1995), for one million hypothetical individuals, based on the

specificity and sensitivity of the ELISA test and the prevalence of HIV among the U.S. population provided by Gastwirth (1986). Use Table

Table 6.3: HIV and the ELISA test

	Test Positive	Test Negative	Row total
Carry HIV	4885	115	5000
No HIV	73630	921370	995000
Column Total	78515	921485	1000000

6.3 to calculate: sensitivity, specificity, probability of false negative, probability of false positive, PPV, NPV and the probability of having HIV despite the result of the ELISA test being negative.

6.5.2   Exercises for discussion or homework

1. Snee (1974) reports the following table corresponding to hair and eye color for 592 students:

Table 6.4: Hair and eye color for a population of students

Eye color	Hair color				
	Black (B)	Brown (E)	Red (R)	Blond (L)	Total
Brown (N)	68	119	26	7	220
Blue (U)	20	84	17	94	215
Hazel (Z)	15	54	14	10	93
Green (G)	5	29	14	16	64
Total	108	286	1	127	592

Source: Snee (1974)

Assume that a person will be selected at random from that population. The events B, E, R, and L are defined with respect to hair color and the events N, U, Z, and G are defined with respect to eye color.

(a) Explain in words the meaning of the following notation: B, $G \cap R$, $U \cup L$.

(b) Calculate the following probabilities: $P(R)$, $P(B')$, $P(G \cap R)$, $P(U \cup L)$, $P(U \cap L)$, $P(L|U)$.

2. Consider two events A and B such that $P(A) = 0.5$, $P(B) = 0.6$, and $P(A \cap B) = 0.2$. Calculate $P(B|A)$. Is $P(B|A)$ equal to $P(B)$? Are A and B independent? Are they mutually exclusive?

3. In Egwaga et al.(2007), the following information is reported: A test for HIV based on sputum was developed and compared with the assumed conclusive results of the blood serum test. The sputum test is cheaper and easier to apply and could be of practical use in developing countries where people are also tested for tuberculosis. The authors report that the sensitivity of the sputum HIV test (if applied on the same day of sputum collection) was 94.7%, and the specificity was 92.9%. Write sensitivity and specificity in the form of conditional probabilities in the context of the problem:

   P(/   )=0.947
   P(/   )=0.929

   What is the probability of a false positive? What is the probability of a false negative?

4. Traditionally the Papanicolau smear test is used for the detection of cervical cancer in women. Currently, a DNA test for Human Papilloma virus (HPV), the main cause for cervical cancer, also exists. In Nature Reviews Cancer 7, 893 (December 2007) the following information is found: 'The sensitivity of HPV testing was 94.6%, whereas that of Pap testing was 55.4%. The specificity was 94.1% for HPV testing and 96.8% for Pap smears.' What is the probability of a false negative in the DNA test for HPV? Compare it to the probability of a false negative for the Pap smear test. **Exercises for discussion or homework in Chapter 6**

5. Only 1 in 1000 adults is afflicted with a rare disease for which a diagnostic test has been developed. The test is such that when an individual actually has the disease, a positive result will occur 99% of the time, whereas an individual without the disease will show a positive result (false positive) only 2% of the time. A randomly selected individual is tested and the result is positive, what is the probability that the individual has the disease? Sketch a probability tree in order to calculate the probability.

6. Consider a small variation to the previous problem: the disease is not so rare and 5% of the population is affected by it. What is the probability that a person has the disease given that he/she has tested positive?

7. Consider a variation to Example 6.2. Instead of considering the high-risk group, work with a population for which the prevalence of Trisomy 21 is much lower, P(Down syndrome)=92/100000. Of course, the sensitivity and specificity of the test remain the same. Calculate the probability of false positive, probability of false negative, PPV and NPV.

References

Desper, R., Khan, J. and Schaffer, A.A. (2004) Tumor classification using phylogenetic methods on expression data. *Journal of Theoretical Biology* **28**: 477-96.

Egwaga, S.M., Chonde, T.M., and Matee, M.I. (2007) Low specificity of HIV-testing on sputum specimens kept at ambient temperatures for 4 to 7 days: a blinded comparison. *BMC Clinical Pathology* **7**:8 doi:10.1186/1472-6890-7-8

Fan, J.J. and Levine, R.A. (2007) To Amnio or Not to Amnio: That is decision for Bayes *Chance* **20**: 26-32.

Gastwirth, J.L (1987) The Statistical Precision of Medical Screening Procedures: Application to Polygraph and AIDS Antibodies Test Data, *Statistical Science* **2**: 213-222.

Rossman, A.J. and Short, T.H. (1995) Conditional Probability and Education Reform: Are They Compatible? *Journal of Statistics Education* **3**:2

Ongut, G., Ongunc, D., Gunsere, F., Ogus, C., Donmez, L., Colak, D. and Gultekin, M. (2006). Evaluation of the ICT Tuberculosis test for the routine diagnosis of tuberculosis. *BMC Infectious Diseases* **6**: 37 doi:10.1186/1471-2334-6-37.

Snee, R. D. (1974) Graphical display of two-way contingency tables. *The American Statistician* **28**: 9-12.

Soon, S., McCall, C. and Chen, S. (2003) Computerized Digital Dermoscopy: Sensitivity And Specificity Aren't Enough, *Journal of Investigative Dermatology* **121**: 214-215.

Chapter 7

Discrete distributions

The binomial probability distribution was introduced in Chapter 4. This chapter furthers the discussion on discrete probability distributions including the concept of 'expected value.' Some other discrete distributions useful in the life sciences, such as the geometric distribution and the Poisson distribution, are also introduced.

7.1 Discrete random variables

A random experiment is an experiment for which all the possible **outcomes** are known, but we do not know which outcome will actually happen. The **sample space** (S) is the set of all the possible outcomes of the random experiment. The outcomes are also called **simple events**.

Sample space: all the possible outcomes

Example 7.1 Rolling two dice

Consider the random experiment of rolling one die twice or rolling two distinguishable dice. Throughout the book we are assuming that dice are cubes and that they are 'fair' in the sense that all the six faces are equally likely to come up on top when the die is rolled. The sample space has 36 possible outcomes that can be written as pairs of numbers. The first element of the pair is a number from 1 to 6, the outcome of the first roll. The second element of the pair, the outcome of the second roll, is also a number from 1 to 6. The sample space is:

S={(1,1),(1,2),(1,3),(1,4),(1,5),(1,6) (2,1),(2,2),(2,3),(2,4),(2,5),(2,6)
(3,1),(3,2),(3,3),(3,4),(3,5),(3,6) (4,1),(4,2),(4,3),(4,4),(4,5),(4,6)
(5,1),(5,2),(5,3),(5,4),(5,5),(5,6) (6,1),(6,2),(6,3),(6,4),(6,5),(6,6)}

**Defining
a random
variable:
each outcome
is assigned
a number**

Assume that only the sum of the two values is of interest, and not the detailed information on the outcome of the first and second rolls. In this case, each element of the sample space is assigned a number that is the sum of the two elements of the pair. For example, the value 4 is assigned to the pairs (1,3), (3,1) and (2,2). Each element of the sample space is assigned a real number. In mathematics, this process is known as 'defining a random variable.' In this example, the variable X is being defined as:

X : *sum of the two faces that come up when rolling a die twice.*

X can take the values 2, 3, 4, 5, 6, 7, 8, 9, 10, 11, 12. We might be interested in another aspect of the outcome when rolling a dice twice. Other random variables that can be defined over the same sample space are:

Y: *the maximum value when a die is rolled twice.*
Z: *absolute value of the difference between the two values.*

**Discrete random
variables take
only integer
values**

What are the possible values for Y? What are the possible values for Z? Random variables that take only integer values are called **discrete random variables**. Discrete random variables are naturally associated with counting processes such as:

1. Number of worms per cubic decimeter of soil

2. Number of animals killed per mile of road (i.e. road kills)

3. Number of chirps of a cricket per minute

4. Number of siblings who inherit a given trait in a family with four children

These examples correspond to two different counting situations. The first three are related to counting the number of times that a given event happens per unit of space, length, or time. In theory, there is no limit for the values of the variable *number of....* In the fourth example we are counting the number of 'successes' in n 'replicates' of a random experiment, the maximum possible value for X: *number of successes* is 4 (the value of n in the example). The Poisson distribution could be used as a model in the first three examples and the binomial distribution for the last one.

7.1.1 Mass or probability function

The behavior of a discrete random variable is described by its distribution, that is, which values the random variable can take and the probability of each one of those values. The 'probability function,' also called 'mass function,' is a rule that assigns a probability to each one of the possible values of the variable.

Revisiting Example 7.1

When rolling two dice, all the 36 elementary outcomes in the sample space S are equally likely. Thus, in order to calculate the probability of each value of the random variable X: *sum of the two values*, it is enough to count how many outcomes produce each one of the values 2, 3,..., 12. Table 7.1 and Figure 7.1 display the probability distribution of the variable X.

> The 'mass' or 'probability' function assigns a probability to each possible value of the variable

Table 7.1: Distribution of X: *sum of values in two rolls of a fair die*

x	2	3	4	5	6	7	8	9	10	11	12
$p(x)$	$\frac{1}{36}$	$\frac{2}{36}$	$\frac{3}{36}$	$\frac{4}{36}$	$\frac{5}{36}$	$\frac{6}{36}$	$\frac{5}{36}$	$\frac{4}{36}$	$\frac{3}{36}$	$\frac{2}{36}$	$\frac{1}{36}$
CDF	$\frac{1}{36}$	$\frac{3}{36}$	$\frac{6}{36}$	$\frac{10}{36}$	$\frac{15}{36}$	$\frac{21}{36}$	$\frac{26}{36}$	$\frac{30}{36}$	$\frac{33}{36}$	$\frac{35}{36}$	1

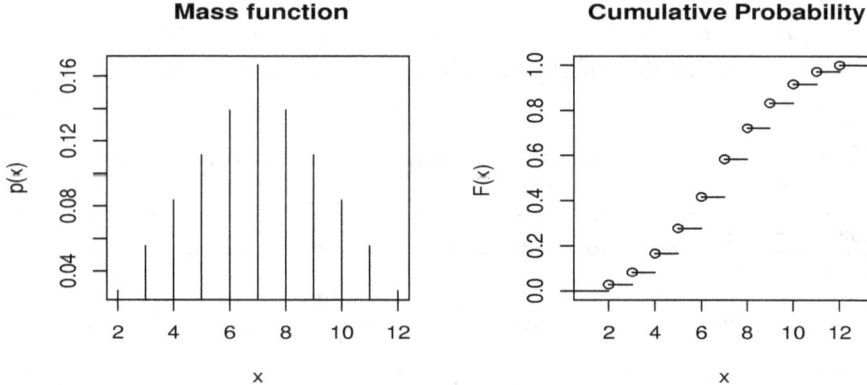

Figure 7.1: Probability or mass function $p(x)$ and CDF

The cumulative probabilities or Cumulative Distribution Function (CDF), $F(x) = P(X \leq x)$, is also included in both Table 7.1 and Figure 7.1. The

Cumulative probability

CDF for the value x is calculated by adding the probability for x and the probabilities for all the values of the variable X smaller than x. In Table 7.1, the value of CDF for $X = 4$, or $F(4)$, is calculated as $p(2) + p(3) + p(4) = \frac{1}{36} + \frac{2}{36} + \frac{3}{36} = \frac{6}{36} = \frac{1}{6}$. The CDF is useful to answer questions such as 'When we roll a die twice, what is the probability of obtaining a total of 4 points or less?'

7.1.2 Expected value and variance

Revisiting Example 7.1

Consider again the experiment of rolling a die twice or rolling two dice, one red and one green, once. Assume you are interested in the random variable X: *sum of the two faces*. If the random experiment of rolling the two dice is repeated many times, what will be the average value of X? The expected value of a random variable is the average value of the variable if the experiment were repeated an extremely large number of times. The expected value of a random variable is also known as the **mean** of the random variable. To calculate the expected value of a discrete random variable, multiply each possible value of the variable by its corresponding probability and calculate the sum of those products, as described by the following formula:

Expected value $E(X)$ or mean: multiply each value by its probability and add

$$E(X) = \sum x_i p(x_i) \tag{7.1}$$

For the experiment in Example 7.1, the expected value can be calculated by adding the values in the third row of Table 7.2. The value of the sum is $\frac{252}{36} = 7$. Figure 7.1 indicates that the distribution is symmetric. Notice that the value 7 is at the center of the distribution in Figure 7.1a.

Table 7.2: Calculating the mean and variance of a discrete random variable

x	2	3	4	5	6	7	8	9	10	11	12	Sum
$p(x)$	$\frac{1}{36}$	$\frac{2}{36}$	$\frac{3}{36}$	$\frac{4}{36}$	$\frac{5}{36}$	$\frac{6}{36}$	$\frac{5}{36}$	$\frac{4}{36}$	$\frac{3}{36}$	$\frac{2}{36}$	$\frac{1}{36}$	
$x \times p(x)$	$\frac{2}{36}$	$\frac{6}{36}$	$\frac{12}{36}$	$\frac{20}{36}$	$\frac{30}{36}$	$\frac{42}{36}$	$\frac{40}{36}$	$\frac{36}{36}$	$\frac{30}{36}$	$\frac{22}{36}$	$\frac{12}{36}$	$\frac{252}{36}$
x^2	4	9	16	25	36	49	64	81	100	121	144	
$x^2 \times p(x)$	$\frac{4}{36}$	$\frac{18}{36}$	$\frac{48}{36}$	$\frac{100}{36}$	$\frac{180}{36}$	$\frac{294}{36}$	$\frac{320}{36}$	$\frac{324}{36}$	$\frac{300}{36}$	$\frac{242}{36}$	$\frac{144}{36}$	$\frac{1974}{36}$

The variance and the standard deviation are indicators of the variability or dispersion of a random variable. The variance of a random variable is defined as the expected value of the squared distance from the values of the variable to the mean:

$$Var(X) = \sum [x_i - E(X)]^2 p(x_i) \qquad (7.2)$$

An alternative way of calculating the variance is expression (7.3). The variance can be written as the difference between the expected value of the square of the variable and the square of the expected value.

$$Var(X) = E(X^2) - [E(X)]^2 \qquad (7.3)$$

For Example 7.1, formula (7.3) becomes:

$$Var(X) = \tfrac{1974}{36} - 7^2 = 54.83333 - 49 = 5.833333$$

The standard deviation of the variable X is $\sqrt{5.83333} = 2.415229$.

7.2   Discrete uniform distribution *

Example 7.2 Lab technician performance

A lab technician frequently performs a technically demanding procedure in the lab. He has noticed that sometimes it takes two or even three attempts to complete it. If asked how frequently this happens, he says that one third of the times he completes it in the first trial, in one third of the times two trials are needed, and in one third of the occasions three trials are needed. The variable X is defined as *number of trials needed to complete the task successfully*. The variable X can take the values 1, 2, 3, all with the same probability of 1/3. The situation can be described by a probability model called the **discrete uniform distribution**; its mass function is plotted in Figure 7.2

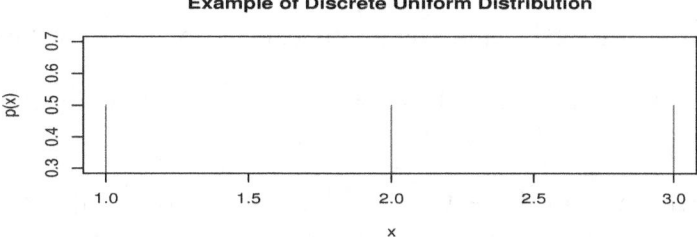

Figure 7.2: Discrete uniform distribution when X can take the values 1,2,3

In the discrete uniform distribution, the variable X can take a number of consecutive integer values, from a to b, and each one of them has the same probability. From a to b, including both extremes (a and b), there are $b-a+1$ different values. If there are $b-a+1$ values and all are equally likely, the probability of each one is $1/(b-a+1)$. In the lab technician example, we have $a=1$ and $b=3$. The variable X can take the values 1, 2, or 3, and all have the same probability 1/3. The mass function or probability function is in expression (7.4). Notice that $p(x)$ is written in terms of constants a and b, called the **parameters** of the distribution:

Mass function of the discrete uniform

$$p(x) = 1/(b - a + 1) \qquad\qquad x = a, a+1, ..., b \qquad (7.4)$$

The expected value or mean of the variable is at the center, equidistant from both extremes:

$$E(X) = \frac{a+b}{2} \qquad\qquad (7.5)$$

Expected value of a uniform random variable

In the example of the lab technician, if he has to perform the same task a very large number of times, he will need two trials on average to complete the task because $E(X) = 2$. The expected value can be calculated either with the general definition in (7.1)

$$E(X) = \sum x_i p(x_i) = 1 \times \tfrac{1}{3} + 2 \times \tfrac{1}{3} + 3 \times \tfrac{1}{3} = \tfrac{1}{3}[1 + 2 + 3] = \tfrac{6}{3} = 2$$

or with the specific formula (7.5) for the uniform distribution,

$$E(X) = \tfrac{1+3}{2} = 2.$$

7.3   Bernoulli distribution

Many questions in real life have a YES/NO answer. Some experiments have two possible outcomes: 'success' and 'failure'. Other experiments can have more outcomes but sometimes those outcomes can be grouped in two categories. Some examples are:

A person is tested for HIV and the result can be 'positive' or 'negative.'

A pea seed is planted. When the plant grows, the pea pods can be either green or yellow.

A hamster can have 4 different coat colors: white (albino), tan, black, or grey. If the question is how many albino hamsters are produced, the other three colors could be grouped together and the new categories would be 'albino' or 'not-albino.'

Each of these situations can be related to a Bernoulli experiment, a random experiment that has two possible outcomes, 'success' and 'failure,' with probabilities p and $1-p$, respectively. Which of the two outcomes is labeled as 'success' is arbitrary, provided that consistency is maintained. 'Success' does not necessarily mean what is colloquially called success. In the context of a Bernoulli experiment, success could be having a terminal illness. A variable X is defined as follows: $X = 1$ if the outcome is 'success' and $X = 0$ if the outcome is 'failure.'

The mass or probability function for a Bernoulli variable is described by equation (7.6) and Table 7.3. Notice that the expression for $p(x)$ includes the constant p, which is the **parameter** of the distribution. The value of the parameter (e.g. $p = 0.5$ or $p = 0.8$) indicates to which particular Bernoulli distribution we are referring. In Example 7.3, the distribution is Bernoulli with $p = 3/4$.

Parameter of a probability distribution

Table 7.3: Probability table for a Bernoulli variable

x	0	1
$p(x)$	$1-p$	p

$$p(x) = \begin{cases} 1-p & for \quad x = 0 \\ p & for \quad x = 1 \end{cases} \qquad (7.6)$$

Using expression (7.1), the expected value of a Bernoulli variable is equal to the parameter p because:

$$E(X) = 0 \times (1-p) + 1 \times p = p. \qquad (7.7)$$

Using expression (7.2) the variance is:

$$Var(X) = [0-p]^2(1-p) + [1-p]^2(p) = (1-p)(p)(p+1-p) = p(1-p). \quad (7.8)$$

The expected value and the variance are written in terms of the parameters of the distribution

Example 7.3 Color of a pea pod

In an experiment with peas, the probability of getting a green pea pod is $3/4$, and the probability of getting a yellow pea pod is $1/4$. Define 'success' as green and 'failure' as yellow. The variable X is defined as $X = 0$ if the pod is yellow and $X = 1$ if the pod is green. The expected value of the variable is $E(X) = p = 3/4$, meaning that if many pea pods are obtained, on average the value of X will be $3/4$. The variance is $p(1-p) = \frac{3}{4} \times \frac{1}{4} = \frac{3}{16}$.

7.4   Binomial distribution

The binomial distribution has already been studied in Chapter 4. Consider n independent replicates of a Bernoulli experiment and in each the probability of success is p. The random variable X: *number of successes in n independent replicates of a Bernoulli experiment* has a binomial distribution.

The probability or mass function for the variable X is:

$$P(X = x) = \binom{n}{x} p^x (1 - p)^{n-x} \qquad\qquad x = 0, 1, 2, 3, ..., n \qquad (7.9)$$

where $\binom{n}{x} = \frac{n!}{x!(n-x)!}$, and $x!$ represents the factorial of x that is calculated as the product $x(x-1)(x-2)...1$. The **parameters** of the binomial distribution are:

$$n : \text{number of trials}$$
$$p : \text{probability of success in each trial}$$

It is useful to know the relationship between the mean and the variance of a distribution and the parameters of the distribution. In the case of the binomial, the mean (or expected value) and variance of X are:

$$E(X) = np \qquad\qquad Var(X) = np(1 - p) \qquad (7.10)$$

Example 7.4: Genetic traits

When both parents are carriers of a recessive genetic trait, the probability that the offspring gets the trait is 1/4. A couple plans to have 4 children. What is the distribution of the variable X: *number of children with the genetic trait?*

The binomial distribution and genetic traits

Each child can be considered an independent Bernoulli experiment, with probability of 'success,' having the genetic trait, equal to 1/4. The probability of each value of X can be calculated by replacing n and p in equation (7.9) by 4 and 0.25, respectively:

$$P(X = x) = \binom{4}{x} 0.25^x (0.75)^{4-x} \text{ for } x = 0, 1, 2, 3, 4$$

Interpreting the expected value

The values of the mass or probability function of X are displayed in Table 7.4 and Figure 7.3. The distribution is skewed with the lower values of X having a larger probability. The expected value is $E(X) = 4 \times \frac{1}{4} = 1$. If a very large number of couples (with both parents carriers of the genetic trait) have 4 children, families will have on average one child with the trait.

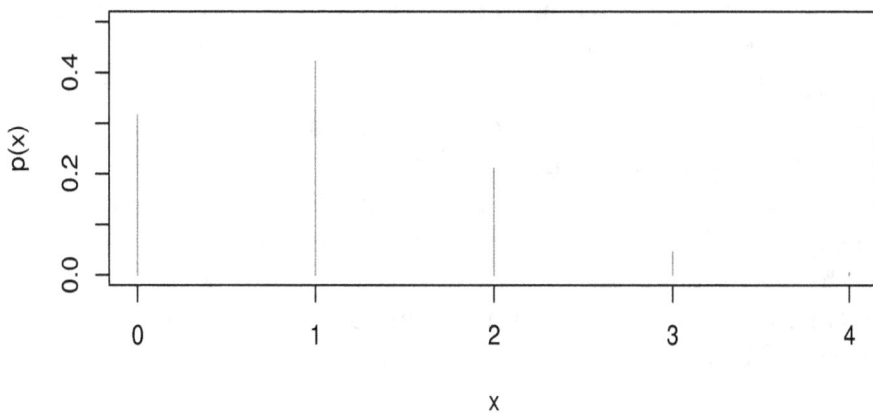

Figure 7.3: Binomial distribution with $n = 4$ and $p = 0.25$

Table 7.4: Binomial distribution, $n = 4$ and $p = 0.25$

x	p(x)
0	0.31640625
1	0.42187500
2	0.21093750
3	0.04687500
4	0.00390625

Using statistical software
A binomial table for $n = 4$ and $p - 0.25$ can be produced using statistical
software. In MINITAB, type the values 0 to 4 in a column and use CALC >
Probability distributions > Binomial.

To produce Table 7.4 and Figure 7.3 using R, type the following com-
mands:

MINITAB & R:
calculating
binomial
probabilities

```
x<-c(0,1,2,3,4)          ## values of the variable
px<-dbinom(x,4,0.25)     ## calculates binomial probabilities
table<-cbind(x,px)       ## creates a table
table                    ## displays the table
plot(x,px,'h',ylab='p(x)')  ## plots the distribution
```

7.5   Geometric distribution *

**Repeating a
Bernoulli
experiment until
the first success
is obtained**

Think of a Bernoulli experiment, but instead of having a fixed number of trials, the experiment is repeated until the first success happens. For example, a die is rolled until the first six is cast. The question is: How many trials are needed for the first '6' to be cast? The random variable X is defined as *the number of trials to get the first success*, and can take values 1,2,3,... What is the probability of each one of those values?

$X = 1$ means that the first roll was a '6.' Thus, $P(X = 1) = 1/6$.

$X = 2$ means that the first trial resulted in failure (no '6') and the second in success. Thus, $P(X = 2) = \frac{5}{6} \times \frac{1}{6}$.

$X = 3$ means that the first two trials resulted in failure and the third one was a success. Thus, $P(X = 3) = \frac{5}{6} \times \frac{5}{6} \times \frac{1}{6}$.

In general, if a number x of trials are needed to get the first success, it means that the first $x - 1$ trials resulted in failure and the xth trial was a success. Given that the probability of success in each trial is p and the probability of failure is $1 - p$, the probability of $x - 1$ failures and one success is given by expression 7.11. The only **parameter** of the distribution is p, the probability of success in each trial.

$$P(X = x) = p(x) = (1 - p)^{x-1}p \qquad\qquad x = 1, 2, ... \qquad (7.11)$$

The expected value is $E(X) = 1/p$. This means that if the probability of success is p, in the long run and on average, $1/p$ trials are needed to get a success.

Example 7.5 Looking for a rare trait

A researcher is looking for subjects who exhibit a rare trait in order to recruit them for a study. Assume that only 1% of the population has that trait. What is the probability that 8 individuals have to be examined in order to find one with the trait? Using equation (7.11):

$$P(X = 8) = (0.9)^7 0.1 = 0.04782969$$

Example 7.6 Keep trying until you succeed

A lab technician needs to repeat a delicate task in the lab. Only 70% of his attempts are successful, and he has to repeat the attempt until the task is done. Define the random variable X:*number of attempts until the task is accomplished.* The probability of needing 1, 2, or 3 attempts to complete the task can be calculated using the mass function in expression (7.11), replacing p by 0.7: $P(X = x) = (0.3)^{x-1}(0.7)$ for $x = 1, 2, 3$. If this task is performed very frequently, in the long run the average number of attempts needed to successfully complete it is $1/0.7 = 1.428571$. It is not a problem to have a non-integer expected value for a discrete random variable; expected values are long-term averages.

Example 7.7 A sequence of 10 A's

DNA is a sequence of the nucleotides C,A,T,G. In a given section of DNA, **Examples for** 19% of the nucleotides are A. To simplify the problem, assume independence **the geometric** between subsequent nucleotides. What is the probability that at a location **distribution** selected at random, a sequence of exactly 10 A's starts? Assume that the nucleotides previous to the location selected are not important. To get a sequence of exactly 10 A's means that a sequence of 10 A's is needed and that the 11th position is occupied by a different letter, i.e., AAAAAAAAAAĀ (where Ā= not A, which means C,G or T). The probability of that happening is $0.19^{10}0.81 = 0.00000004966164$.

An alternative version of the geometric distribution

Consider the variable Y: *number of failures before the first success.* The variable Y can take the value 0 because the first trial can result in success. The probability of y failures is $(1-p)^y$ and the probability of success in the next trial is p. Thus, the mass or probability function is:

$$P(Y = y) = p(y) = (1 - p)^y p \quad for \ \ y = 0, 1, 2, ... \qquad (7.12)$$

Using this version of the geometric distribution, we would calculate $P(y = 7)$ in order to solve the question in Example 7.5. The distribution, when $p = 0.1$, is shown in Figure 7.4.

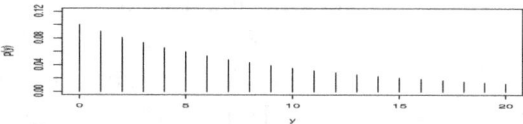

Figure 7.4: Probability distribution for the number of failures before first success when $p = 0.1$

Using statistical software:
In MINITAB, geometric probabilities can be calculated using the option CALC
> Probability distributions > geometric.

MINITAB &R:
geometric
distribution

In R, the command 'dgeom' uses equation (7.12). To solve Example 7.5, one
can write
dgeom(7,0.1)

7.6   Poisson distribution

Named after Simeon Denis Poisson, a French mathematician (1781-1840),
this distribution can be used as a model for a number of biological processes.
The variable X represents the number of events of a given type occurring on
non-overlapping units of space or time. Some examples are:

1. X: number of birds that come to a feeder per hour.

2. X: number of a given type of organisms per cubic decimeter of soil.

3. X: number of bacterial colonies per grid in a Petri dish.

4. X: number of children with cancer per municipality (Westermeier &
   Michaelis, 1995).

5. X: number of roadkills per unit length of road (Gomes et al., 2009).

Traditionally used to model the number of typing errors per page, the
Poisson distribution is currently used in bioinformatics in relation to nucleotide substitutions.

The mass function or probability function is:

$$p(x) = e^{\lambda} \frac{\lambda^x}{x!} \qquad x = 0, 1, 2, 3, ... \qquad (7.13)$$

The number e is the base of the natural logarithm, approximately 2.718282.
The factorial $x!$ is equal to $x \times (x-1) \times (x-2) \times .... \times 2 \times 1$ (keep in mind
that $e^0 = 1$ and $0! = 1$). The parameter λ represents the average number of
occurrences per unit of time or space. In Figure 7.5, notice how the value of
the parameter λ determines the shape of the distribution.

The mean of the distribution is $E(X) = \lambda$ and the variance is also equal
to λ. In the first three previous examples, the value of the parameter λ would
represent:

1. The mean and variance of the number of birds that come to the feeder per hour.

2. The mean and variance of the number of bugs per cubic centimeter of soil.

3. The mean and variance number of colonies per square of a grid superimposed on Petri dishes.

Example 7.8 Number of bacterial colonies

Assume that the average number of bacterial colonies per square in the grid is 10. What is the probability that in a given square there are no colonies at all? From Figure 7.5b it is clear that the probability is quite small:

$$P(X = 0) = e^{-10}\frac{10^0}{0!} = e^{-10} = 0.0000454.$$

If the average number of colonies per square in the grid was 4, then it would not be extremely unlikely to find a square with no bacteria at all (see Figure 7.5a) because:

$$P(X = 0) = e^{-4}\frac{4^0}{0!} = e^{-4} = 0.0183156.$$

Examples for the Poisson distribution

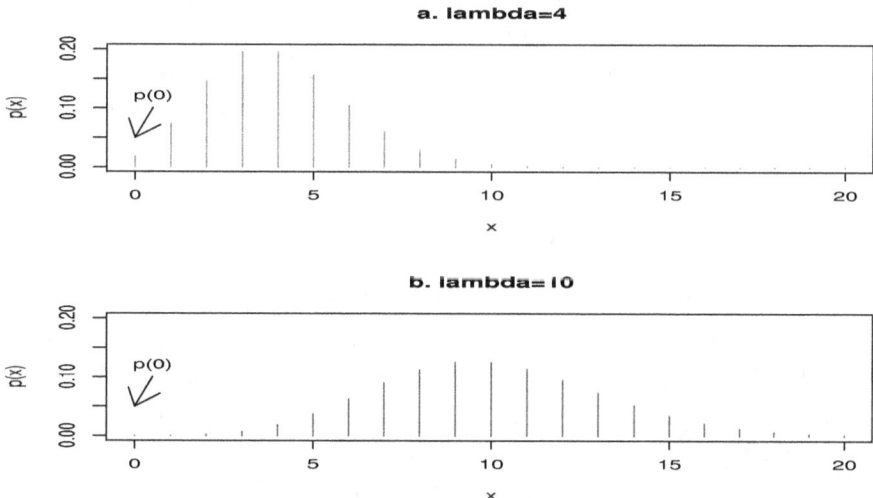

Figure 7.5: Poisson distributions with λ=4 and 10

What is the probability that a randomly selected square contains at least 1 colony? To calculate that probability directly, many terms would need to be added: $P(X \geq 1) = P(X = 1) + P(X = 2) + \ldots$ However, it would be much easier to remember the concept of 'complementary event' and calculate

$$P(X \geq 1) = 1 - P(0) = 1 - 0.0183156 = 0.981684.$$

Using statistical software:
In MINITAB, the values of x are typed in a column of the spreadsheet and then select CALC>Probability distributions> Poisson.
In R, enter the values for x and use the command 'dpois' to obtain the probabilities. For example, the distribution in Figure 7.5.a was obtained with the following commands:

MINITAB & R:
Poisson
probabilities

```
x<-seq(0,20,by=1)  ## creates values from 0 to 20
px<-dpois(x,4)     ## calculates the probabilities with lambda =4
plot(x,px, 'h', ylab='p(x)') ## makes the plot and labels the axes
table<-cbind(x,px) ## creates a table
table              ## displays the table on the screen
```

7.7   Poisson approximation to the binomial *

The parameters of the binomial distribution are n (number of trials) and p (probability of success in each trial), and the expected value is np. When n is very large and p is very small, the probabilities calculated with the binomial distribution are quite similar to those calculated with the Poisson distribution with parameter $\lambda = np$.

Example 7.9 Sickle cell anemia

Sickle cell anemia affects 1 out of 675 individuals in a given population. A random sample of 2000 individuals from that population is drawn. What is the probability that exactly 3 people in the sample have sickle cell anemia?

The number of people in the sample who have the condition has a binomial distribution with parameters $n = 2000$ and $p = 1/675 = 0.00148148$. A computer can be used to calculate $P(X = 3) = 0.224158$. Using **R**, the command *dbinom*$(3, 2000, 1/675)$ will calculate the probability. However, before computers were widely available, it was quite difficult to calculate expressions such as:

$$P(X = 3) = \binom{2000}{3} 0.00148148^3 (1 - 0.00148148)^{1997}.$$

The difficulty in performing some calculations was a strong motivator for finding expressions that have approximately the same value and are easier to calculate. One such approximation is the approximation of the Poisson to the binomial distribution. In the case of Example 7.9, it is easy to use the Poisson distribution with parameter $\lambda = 2000(0.00148148) = 2.96296$ and calculate $e^{-2.96296}\frac{2.96296^3}{3!} = 0.223990$. Table 7.5 displays the probabilities obtained with the Binomial(2000,3) and the Poisson(2.96296) distribution for X from 0 to 12; notice their similarity.

Table 7.5: Binomial(2000,0.000148148) and Poisson(2.96296) probabilities

x	Binomial	Poisson
0	0.051552	0.051666
1	0.152974	0.153084
2	0.226851	0.226790
3	0.224158	0.223990
4	0.166040	0.165918
5	0.098343	0.098322
6	0.048515	0.048554
7	0.020504	0.020552
8	0.007579	0.007612
9	0.002489	0.002506
10	0.000735	0.000743
11	0.000197	0.000200
12	0.000049	0.000049

7.8 Exercises

7.8.1 Review questions

1. When is it said that a random variable is 'discrete'?

2. What is the 'mass' or 'probability function'? Mention its two properties.

3. How is the expected value of a discrete random variable calculated?

4. What does 'the expected value of the variable X is 4' mean?

5. What are the parameters of the binomial distribution? What do they represent? What is the expected value of a variable X that follows a binomial distribution?

6. Both Poisson and binomial are distributions for X: *number of times something happens*. How do we decide which one to use?

7. What is the parameter of a Poisson distribution? What does it represent?

8. The cowbird, similarly to the cuckoo, places its eggs in the nests of other species. Based on information found in Preston (1948) and Norris (1947), the following table describes the number of cowbird eggs found in 112 nests from other species. The frequency distribution is displayed in Figure 7.6.

X(eggs)	0	1	2	3
Frequency(nests)	43	42	20	7

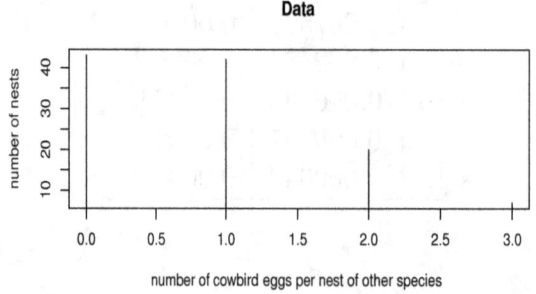

Figure 7.6: Frequency distribution for number of cowbird eggs in the nests of other species

Calculate the mean of the 112 observations from the table above, and round it to two decimals. Did you get 0.92 as the mean of the data? Assume that it is decided to use the Poisson distribution with parameter λ=0.92 as a probability model for the variable X: *number of cowbird eggs in nests of other species that live in the same habitat.* A nest is selected at random.

(a) What is the probability that the nest contains no cowbird eggs?

(b) What is the probability that the nest contains at least one cowbird egg?

(c) What is the probability that it contains more than 2 cowbird eggs?

9. A researcher is looking for subjects with a certain trait that is present in only 3% of the population. What is the probability that exactly 10 individuals would need to be examined in order to find the first person with the trait?

10. Identify the appropriate model (Geometric, Poisson or binomial) and the value of its parameter(s) for each one of the following cases:

(a) A random sample of 20 individuals will be selected from a very large population in which 20% of the individuals have a certain trait. Calculate the probability that 5 individuals in the sample have the trait.

(b) A region with an area of 1000 m^2 will be divided in quadrats that are one square meter each. In each quadrat you will count the number of nests of a given species. What distribution could be used to represent the behavior of the variable X:*number of nests per m^2*? If the mean number of nests per m^2 is 5, what will be the value of the parameter of the distribution?

(c) A biologist is searching a forest looking for frogs that exhibit a certain trait that is present in only 10% of the frog population. Calculate the probability that he has to examine 5 frogs in order to find the first one that has the trait.

7.8.2   Exercises for discussion or homework

1. Use the sample space S in Section 7.1 to prepare a probability table similar to Table 7.1 for the random variable Y: *Maximum value obtained when we roll a die twice.* Sketch the distribution as in Figure 7.1a. Is the distribution symmetric? Calculate and interpret $E(Y)$.

2. Assume that there is a random variable X that can take the values x=1,2,3,4,5 with probabilities proportional to their value. The mass function is $p(x) = x/15$.
a) Sketch the mass function. Is this a 'discrete uniform distribution?
b) Calculate $E(X)$.

3. An instructor routinely gives quizzes with 5 multiple choice questions (each question has 4 options: A,B,C,D) to the students. Assume the students answer the questions at random. What is an appropriate distribution to describe the distribution of the variable X:*number of questions correctly answered*? Prepare a probability table. Calculate and interpret the mean.

4. There is a disease associated with a mutation that is present in approximately 1/400 individuals. If a random sample of 400 individuals is taken:

   (a) What will be the 'exact' distribution for the variable X: *number of people in the sample who carry the mutation*?

   (b) What will be a distribution that can be used as an approximation to the distribution of X?

   (c) Use the approximated distribution in b) to calculate the probability that, in a sample of 400 individuals, nobody carries the mutation.

   (d) Use the approximated distribution in b) to calculate the probability that, in a sample of 400 individuals, exactly 2 people carry the mutation.

References

Gomes, L. , Grillo, C., Silva, C. and Mira, A. (2009) Identification methods and deterministic factors of owl roadkill hotspot locations in Mediterranean landscapes. *Ecological Research* 24: 355-370.

Norris, R. T. (1945) The cowbirds of Preston Frith. *The Wilson Bulletin* **59**: 83-103.

Preston, F.W. (1948) The cowbird (*M. Ater*) and the cuckoo (*C. Canorus*). *Ecology* **29**: 115-116.

Westermeier, T. and Michaelis, J. (1995) Applicability of the Poisson distribution to model the data of the German Children's Cancer Registry. *Radiation and Environmental Biophysics* **34**: 7-11.

Zvelebil, M. and Baum, J.O. (2008) *Understanding Bioinformatics*. New York: Garland Science.

Chapter 8

Probability and genetics *

Some of the exercises in the previous chapters have referred to genes or to sequences of nucleotides. In this chapter, more applications of probability will be discussed in the context of Mendelian genetics and DNA genetics. No new probability concepts will be introduced, just applications to the genetics environment of tools learned in previous chapters. Basic concepts of genetics will be introduced when necessary.

In the context of Mendelian genetics, a coin model will be used to explore the probabilities for the phenotype of the offspring given the genotype of the parents. Exercises on genetic disorders are included at the end of the chapter.

In the context of DNA genetics, we calculate the probability of specific sequences of letters such as palindromes. Also, the probability of matches, just by chance, when comparing two sequences will be calculated. Brief information about bioinformatics software is included.

The two parts of this chapter (Mendelian and DNA genetics) can be covered independently, one is not a prerequisite for the other. Neither are they prerequisite for the reminder of the book.

8.1 Phenotypes and genotypes

In chapters 4 and 7, when applying the binomial distribution, it was mentioned that the probability of a child getting a recessive trait is 1/4 when both parents are carriers. Why is that probability 1/4?

8.1.1   Chromosomes, loci, genes and alleles

Biology background: The cell nucleus contains chromosomes made of DNA and proteins. Chromosomes contain a large number of genes that are located at specific locations called **loci***. Each gene has a specific* **locus** *or location. Humans have 23 pairs of chromosomes and peas have 7 pairs of chromosomes. Peas have historical importance, because Gregor Mendel (1822-1884) used them to systematically perform experiments to understand the transmission of information from generation to generation. For the sake of simplicity, think of a chromosome as a strand of beads and the genes as the beads (a <u>very</u> simplistic model indeed). A gene contains information about a specific trait (color of unripe pod in pea plants, a mouse's coat).*

Genes and alleles

Genotypes and phenotypes

Nothing was known about chromosomes and genes at the time of Mendel, which makes his work even more amazing. Based on the appearance (**phenotype***) of plants he was able to make statements that are now understood to be based on the genetic nature* (**genotype***) of those plants. The word* **gene** *was first used in 1909, 25 years after Mendel's death. Chromosomes come in pairs with one element of the pair inherited from the father, and the other inherited from the mother. In the simplest examples, a gene responsible for a phenotype is present at the same location in each one of the two homologous chromosomes. For example, the chromosome in which the information about the color of the unripe pod is contained, has the same location in both homologous chromosomes.*

The two (or more) forms of a specific gene are called **alleles***, a name first introduced by the biologist Johannsen in 1926. The alleles determine the* **genotype***.*

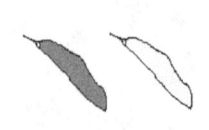

Example 8.1 Color of a pea pod

The gene responsible for pod color in peas has two alleles, green and yellow. The possible genotypes are 'green-green,' 'yellow-green' (or 'green-yellow'), and 'yellow-yellow.' The peas with the **genotype** *green-green* and *yellow-yellow* are called **homozygous**, the peas with different alleles are called **heterozygous**. Depending which color is dominant (in the case of the pea-pod, green is dominant), the genotype determines the **phenotype**, that is, the aspect of the pod visible to us. In Figure 8.1, green is represented by the dark shade and yellow by the light shade. If green is dominant, the presence of one green allele will cause the pod to be green. For the pod to look yellow, both alleles have to be yellow because yellow is recessive (non-dominant).

Which allele the offspring receives from the father and from the mother

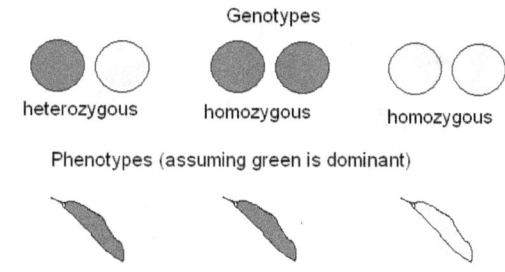

Figure 8.1: Genotypes and phenotypes

is independently and randomly determined. In the rest of the section, a coin model will be used to calculate the probabilities of the different genotypes and phenotypes depending on the nature of the parents (heterozygous or homozygous). Genes (with two different alleles) will be treated as fair coins or as chips with the two sides of different colors. Each side of the coin or chip will represent an allele. In genetics it is common to represent one allele (the recessive one) with the lower case version of a letter and the other allele (the dominant one) with the upper case version of the same letter. In the case of the pea pod color gene example, the alleles are green (B) and yellow (b). In the figures that follow, one coin or chip represents the mother and the other represents the father.

Which of the two alleles of a parent is copied to the offspring, is randomly determined

8.1.2   Genotype of parents

Heterozygous parents

Consider two parent plants, one of them (male) will contribute the pollen to fertilize the egg from the other plant (female). Both male and female are heterozygous, so their genotype is Bb, meaning they have one allele of each type (green and yellow). However, both plants have green unripe pods (the phenotype) because the dominant allele is green. Figure 8.2 describes the parents phenotype and their genotype.

Heterozygous parents

With heterozygous parents, all the combinations of green and yellow alleles are possible for the offspring. The allele inherited from each parent is determined at random, like flipping a fair coin. Figure 8.3 describes the 4 possible genotypes, the resulting phenotypes (assuming green is dominant over yellow), and their respective probabilities. Now it is clear why the probability of an offspring, from heterozygous parents, showing the recessive trait is 1/4 (look at the last column of the table in Figure 8.3). When both parents

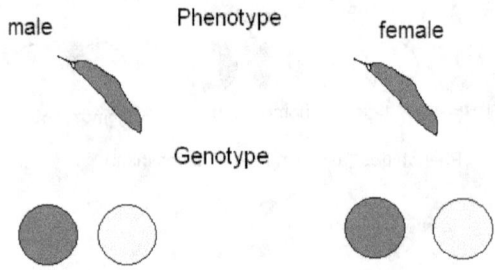

Figure 8.2: Heterozygous parents

Out-come	Allele copied from male	Allele copied from female	Alleles of the Offspring	Genotype	Phenotype dominant: green	Probability
1	●	●	● ● Homozygous	BB		1/4
2	●	○	● ○ Heterozygous	Bb		1/4
3	○	●	● ○ Heterozygous	Bb		1/4
4	○	○	○ ○ Homozygous	bb		1/4

Figure 8.3: Offspring of heterozygous parents

are heterozygous, the probability of the offspring showing the dominant trait is 3/4.

Probability trees are useful to organize the information of processes that involve two or more random experiments (in this case the allele inherited from the male and the allele inherited from the female) and calculate probabilities. To calculate the probabilities at the end, the probabilities of the successive branches are multiplied. Figure 8.4 displays the probability tree for the pea pod example. In biology, genotypes are usually represented using Punnett squares (not included here).

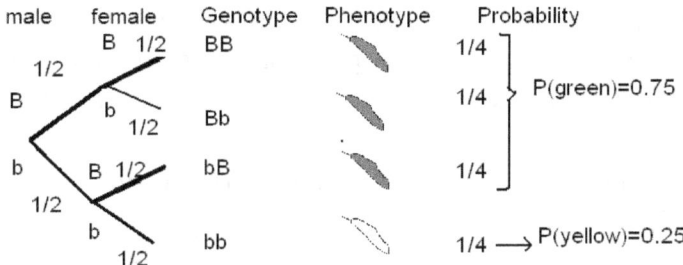

Figure 8.4: Probability tree for heterozygous parents

One heterozygous and one homozygous

Consider the case of one heterozygous and one homozygous parent. The heterozygous parent has both alleles. The homozygous parent can be homozygous for the dominant or the recessive allele.

One homozygous parent for the dominant allele

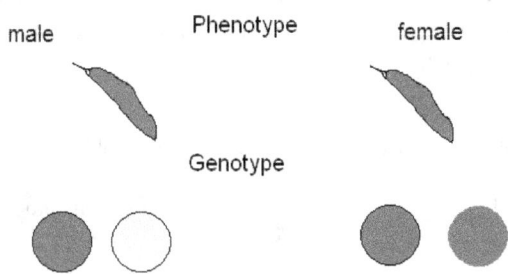

One parent is heterozygous and the other homozygous for the dominant trait

Figure 8.5: One heterozygous and one homozygous for the dominant allele

Figure 8.5 describes the situation in which the male is heterozygous and the female is homozygous for the dominant allele. As an exercise, prepare a table like the one in Figure 8.3 or a tree like the one in Figure 8.4. The only possible genotypes for the offspring are 'green-green' (BB) and 'yellow-green' (bB) because the only allele the female can pass on to the next generation is B. Each of the two possible genotypes (BB and Bb) has probability 1/2, that is the probability of green (or yellow) for the allele coming from the male. Since green is dominant, all the offspring will show the dominant phenotype, i.e., P(green pod)=1. Notice that the offspring can still have the yellow allele (actually, the probability of carrying the yellow allele is 1/2) but their phenotype will be green. If a disease is produced by a recessive allele and one of the parents does not have the recessive allele, none of the children will get the disease. However, the probability of carrying the recessive allele (that enables them to transmit it to their own offspring) is 1/2.

One homozygous parent for the recessive allele

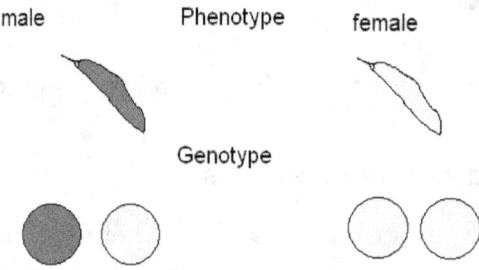

Figure 8.6: One heterozygous and one homozygous for the recessive allele

Figure 8.6 depicts the situation in which the male is heterozygous and the female is homozygous for the recessive allele, as depicted in Figure 8.6. Figure 8.7 shows the probability tree for this case. The probability of each phenotype (green or yellow) is 1/2.

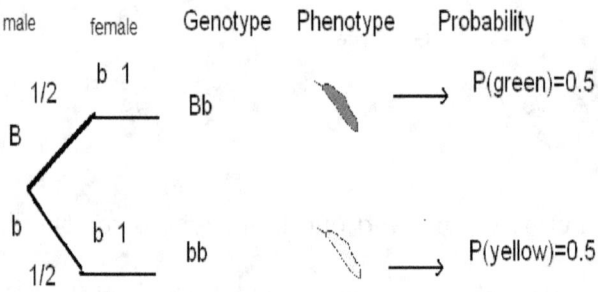

Figure 8.7: Offspring of one heterozygous and one homozygous for the recessive allele

Homozygous parents

Now consider the case of both parents being homozygous. If both parents are homozygous and of the same trait, either both are BB or both are bb, the offspring will be just like them. In genetics, this is called 'true breeding

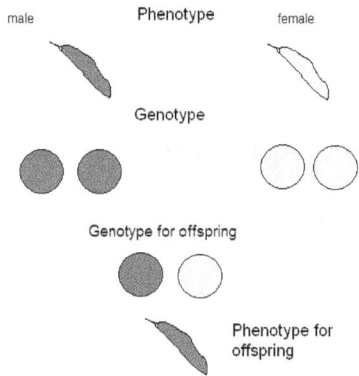

Figure 8.8: Two homozygous parents for different alleles

strains,' and continued breeding will stay the same. However, if the two
are homozygous for different alleles as in Figure 8.8, the offspring of these
parents will inherit a green allele from the male and a yellow allele from the
female. All will be heterozygous and will look green, if green is dominant.

**Both parents
are homozygous
but for different
alleles**

*Note: In biology, when there is doubt if an organism (for example a fruit-
fly or a plant) exhibiting the dominant phenotype is homozygous or heterozygous, it is crossed with another that is homozygous for the recessive trait.
This method is known as a 'test cross' or 'backcross.' Compare Figures 8.6
and 8.8, in both cases the female is homozygous for the recessive allele. However, in Figure 8.6 the male is heterozygous and in Figure 8.8 the male is
homozygous for the dominant trait. Figure 8.7 indicates that for the situation represented by Figure 8.6 (heterozygous male) the probability that the
offspring will show the dominant trait is 1/2 and the probability of the recessive trait is 1/2. In Figure 8.8, the male was homozygous and the offspring
will show the dominant trait with probability 1.*

8.2 DNA

Nucleotides

C

A

T

G

*Biology background: DNA, or deoxyribonucleic acid, is part of the mate-
rial which makes up chromosomes. DNA is a double helix in which double-
stranded DNA is formed by two very long strands of nucleotides that are
connected like a ladder. The nucleotides are represented by their bases C,
A, T, G (C=Cytosine, A=Adenine, T= Thymine, G=Guanine). Those 4
letters constitute the alphabet with which the genetic information contained
in DNA is written. In the double-stranded DNA, A pairs with T and C pairs
with G. This section deals with sequences of nucleotides. Genes are specific*

sections of a long DNA sequence, coding for proteins, transfer RNA (tRNA), or ribosomal RNA (rRNA)

8.2.1   Frequencies for each letter

One of the first questions that arises, when looking at a sequence of nucleotides, is how many of each one of C, A, T, G are present. A simple frequency table, like the ones introduced in Section 2.14, can be built to answer that question.

Example 8.2 GC content of Papilloma virus and BRCA1

The nucleotide frequency in one of the strands for the Human Papilloma virus 18 and the breast cancer gene BRCA1 gene are shown in Table 8.1. The sequences of nucleotides (data files *hpv18.doc* and *BRCA1BX255.doc*) were obtained from the NCBI (National Center for Biotechnology Information) website. To count how many times each one of G,C,A,T is present, bioinformatics software or Microsoft Word (using the FIND option) can be used. The % of G plus the % of C is known as the GC content.

Table 8.1: Nucleotide frequency in the HP virus 18 and the BRCA1 gene

Nucleotide	HPV 18		BRCA1-BX255925	
	Frequency	%	Frequency	%
G	1680	21.3822	31245	31.0220
C	1497	19.0531	29735	29.5227
A	2365	30.1005	19167	19.0302
T	2315	29.4642	20572	20.4251

Data bases with genome data usually report the GC content because DNA with a higher percentage of G and C is more stable than DNA with higher percentages of A and T. Table 8.1 shows that the HPV 18 has a GC content of about 40% while the BRCA1 gene has a GC content of about 60%.

8.2.2   Repeats of the same letter

In a genome, or part of a genome, a relatively long sequence of the same letter might be present, for example AAAAAAAAAA. What is the probability of finding a sequence of 10 A's if the four nucleotides were equally likely and the nucleotide in each position was independent of the others? For simplicity, assume that the question refers only to the sequence that starts on a particular location without concern for the nucleotide in the previous location, or that it is known that the previous location was occupied by a different letter. Calculate P(AAAAAAAAAA$\bar{\text{A}}$), where $\bar{\text{A}}$ represents a letter different from A. Since P(A)=0.25, P($\bar{\text{A}}$)=0.75. Thus, P(AAAAAAAAAA$\bar{\text{A}}$)= $(0.25)^{10} \times 0.75$=0.0000007152667. This is an application of the geometric distribution studied in the previous chapter. The probability of a repeat of exactly 4 A's is P(AAAA$\bar{\text{A}}$)=$(0.25)^4 \times 0.75$=0.002929688. Longer repeats of the same letter are less likely to happen than shorter repeats.

Note: In bioinformatics it is of interest to check if any sequence, not necessarily of the same letter, is repeated in different parts of a gene or genome. There is special software (e.g. MOLKIT) to look for those internal repeats. A dotplot, different from the one learned in Chapter 2, is used to visualize those internal repeats. Repeats of 'trinucleotides' are of special interest because a sequence of three nucleotides, such as CAG, translates into an aminoacid. Some neurological disorders are associated with the repeats of specific trinucleotides.

8.2.3   Specific sequences

Example 8.3 The TATA box

There are some 'words' or sequences of nucleotides that are of special interest. For example, the sequence TATAAA is found about 25 positions before a gene or a transcription starts (transcription regions are the sections of DNA in which the nucleotides code for amino acids, which are used to make proteins). These TATAAA sequences are referred to as a TATA box. Assuming equal frequency and independence between successive nucleotides, what is the probability of a given six-letter word such as TATAAA? There are 4^6 different six-letter words, TATAAA is only one of them, thus the probability that a randomly chosen sequence of 6 letters reads TATAAA is $1/4^6 = 0.0002441406$. Another way of solving the problem is to think of the probability of the specific letter in each location, for example P(T), P(A), etc. P(TATAAA)=0.25^6=0.0002441406.

Calculating the probability of a specific sequence

8.2.4   Palindromes

Other sequences of interest are palindromes. In everyday language 'palindromes' are words that read the same forward and backwards. For example, the name ANNA is a palindrome. In DNA, there are two strands or sequences

What is a palindrome in the DNA context?

of letters; one is read from top to bottom (5' to 3'), and the other is read from bottom to top (3' to 5' direction). These two strands are complementary because if there is a T in one strand, there is an A in the same position in the other strand. If there is a G in one strand there is a C in the other, T is linked to A and G is linked to C.

Example 8.4 The palindrome TGATCA

An example of a palindrome in DNA is shown in Figure 8.9. Read the column in the left from top to bottom (5' to 3') and the other one from bottom to top (3' to 5'). In both cases the sequence will read the same, TGATCA. In the DNA environment, TGATCA is a palindrome because its complementary sequence ACTAGT (read in the other direction) reads the same. To check if a sequence of nucleotides is a palindrome, check if the first letter is the complement of the last one. Then check if the second letter is the complement of the letter next to last, etc.

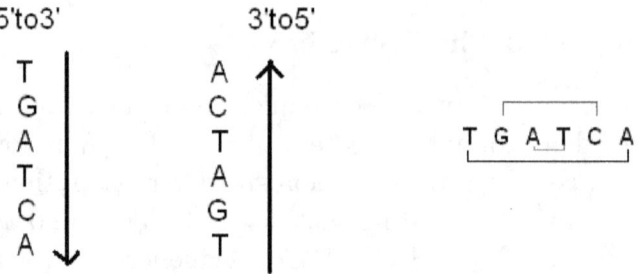

Figure 8.9: Example of palindrome in DNA

Why are palindromes important?

Note: Palindromes are of interest in the work with DNA because the recognition sequence of enzymes that cut double-stranded DNA are usually palindromes. These enzymes are called 'restriction enzymes' or 'restriction endonucleases.' If the sequence of letters that is the target of the restriction enzyme is a palindrome, the enzyme will attach to both strands and will cut the DNA at that point. Those restriction enzymes are commonly called '4-cutters' or '6-cutters' depending on whether their recognition targets have 4 or 6 nucleotides. Some of the well-known restriction enzymes are:

Restriction enzyme	Recognition sequence (5' to 3')
Sau3A	GATC
EcoRI (ekko r 1)	GAATTC
BglII (bagel 2)	AGATCT
HindIII	AAGCTT

Example 8.5: Any 6-letter palindrome

Think of palindromes with 6 letters, as the one in Example 8.4 (Figure 8.9). **Probability of** How many different 6-letter words exist? (See Figure 8.10) Since there are 6 **any 6-letter** positions and each one can be filled in 4 different ways, there are $4^6 = 4096$ **palindrome** different 6-letter words. How many of them are palindromes? There are $4^3 = 64$. This is because only the first 3 cells are free: once the first 3 letters are chosen, the last 3 are automatically determined for the word to be a palindrome. Assuming that the four nucleotides (C,A,G,T) are equally likely and that the nucleotides in different positions are independent, what is the probability that a randomly chosen 6-letter sequence is a palindrome? $64/4096 = 1/64 = 0.015625$. In a sequence of 100719 nucleotides, how many 6-letter sequences can be read? 100714 because in the last 5 positions we cannot start reading a 6-letter sequence. In sequences with 100719 nucleotides, how many 6-letter palindromes would be expected? $100714/64 \sim 1573$ (non-specific) six- letter palindromes.

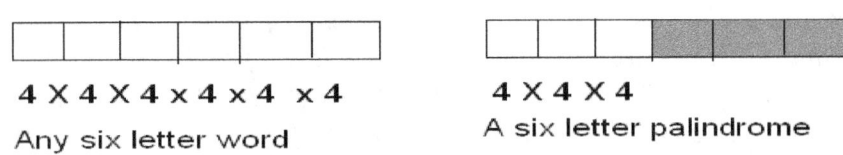

4 X 4 X 4 x 4 x 4 x 4   **4 X 4 X 4**
Any six letter word   A six letter palindrome

Figure 8.10: Possibilities for 6-letter words and 6-letter palindromes

Example 8.6: The BglII ('bagel 2') cutter

What is the probability of finding a specific palindrome, for example the **Probability** palindrome TCTAGA that is the target of BglII?. This amounts to look **of a** for a specific sequence, not for any 6-letter palindrome. There are $4^6 = 4096$ **specific** six-letter words and only one is the sequence of interest. Assume that the nu- **palindrome** cleotides C, A, T and G are equally likely and that the nucleotides in the different positions are independent. What is the probability that a 6-letter word, selected at random, is the palindrome TCTAGA? 1/4096=0.0002441406. Another way of calculating that probability, under independence, is to multiply the probabilities of each letter: P(TCTAGA)=$0.25 \times 0.25 \times 0.25 \times 0.25 \times 0.25 \times 0.25 = 0.25^6 = 0.0002441406$.

In sequences of 100719 nucleotides, how many times would you expect to find TCTAGA (the target of BglII)? In a sequence of 100719 nucleotides, one can start reading 6-letter sequences anywhere except for the last 5 positions, thus there are 100714 sequences of six letters that can be read. Thus, 100714/4096=24.58838~ 25 or 100714 $\times$ 0.0002441406=24.58838~ 25. The conditions of equal probability for each letter (C, A, T, G) and independence between successive letters do not necessary hold; they are assumed to simplify the problem.

Now, drop the assumption that C, A, T and G are equally likely and consider P(A)=0.2, P(T)=0.2, P(G)=0.3, P(C)=0.3. Assuming independence, what is the probability of TCTAGA? P(TCTAGA)=$0.2 \times 0.3 \times 0.2 \times 0.2 \times 0.3 \times 0.2$= 0.000144. Remember that in a sequence of 100719 nucleotides, 100714 six-letter sequences can be read. In a sequence of 100719 nucleotides with GC content of 60%, how many times would you expect to find the palindrome TCTAGA? 100714 $\times$ 0.000144=14.50282~ 15.

Note: There is free software in the internet, such as MOLKIT, in which a sequence of nucleotides can be entered and the software will show where restriction sites are located. Figure 8.11, produced using MOLKIT, shows the location of the palindrome TCTAGA, i.e., the 16 places where the enzyme BglII would cut the sequence of the BCRA1 gene (gene associated with breast cancer) that has 100719 nucleotides. The locations of that specific palindrome in the sequence are 737, 12684, 12961, 13630, 23530, 27042, 31771, 37372, 50250, 50413, 52280, 56988, 69719 ,88267, 88489 ,92469. Notice that the segments are not of the same length.

Actual location of the palindrome TCTAGA in a sequence

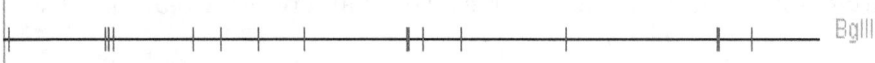

Figure 8.11: Location of the palindrome TCTAGA in the BCRA1 gene

8.2.5   Comparing sequences

One important activity in bioinformatics is to compare two sequences of nucleotides. It is important to realize that coincidences might happen just by chance, especially if short sequences are compared.

Example 8.7 A 30 letter sequence

Consider the following 30 letter sequence taken from the BRCA1 gene:
GCCCCAGAGCATGGGAGAGGAGAGGGACTG
Using a roulette with the 4 letters C,A,T and G, such as the one in Figure 4.13, a random sequence of 30 letters was generated. The random sequence was written below the real sequence in Figure 8.12. Whenever the nucleotides matched (coincidence), a 1 was written below (meaning 'success') and where there was no match (disagreement), a 0 was written (meaning 'failure').

Probability of matches happening just by chance when comparing two DNA sequences

```
GCCCCAGAGCATGGGAGAGGAGAGGGACTG
T CTGCCACACAGTA GAGAGTTT TAAA GGTT
0 1 0 01 00 00 1 100 0 1 1 1 11000 0 0 0 0 0 0 10
```

Figure 8.12: Comparing two short DNA sequencies

There are 10 coincidences out of 30 letters; those coincidences were just due to chance because the second sequence was randomly generated. What is the probability of 10 coincidences happening just by chance in 30 comparisons? This is a job for the binomial distribution. In each position, the probability of a match by chance alone is 1/4 because there are only 4 letters from which to choose. Each position can be considered like a realization of a random experiment and there are 30 replicates of the random experiment. Let us define the variable X: *'number of coincidences in 30 comparisons'*:

$$P(X = 10) = \binom{30}{10}(1/4)^{10}(3/4)^{20}=0.09086524$$

The answer can be calculated with Minitab (CALC > Probability distributions > Binomial > probability) or with **R** by typing **dbinom(10,30,0.25)**. To calculate the probability of **10 or more** coincidences in a sequence of 30 nucleotides, one would need to calculate $P(X \geq 10)$=P(10)+P(11)+.....+P(30) or $1 - P(X \leq 9)$. The latter expression can be calculated with Minitab (CALC>Probability distributions>Binomial> cumulative probability) or with **R** by typing **1- pbinom(9,30,0.25)** to get $P(X \geq 10) = 0.1965934$ (**pbinom** calculates the cumulative probability). Thus, when sequences of 30 nucleotides are being compared, to find 10 coincidences or more is not a rare event at all.

Note: Local similarities (similarities in short sequences within a genome) are quite likely to happen just by chance. That is why comparisons should be done only in longer sequences of at least 100 nucleotides. In bioinformatics, sequences are said to be 'homologous' if at least 70% of the nucleotides are equal.

The longer the sequences to be compared, the less likely they look homologous just by chance

To examine the effect of the length of the sequence on the probability of finding coincidences by chance alone, let us calculate the probability of getting 70% or more of coincidences just by chance when comparing two sequences:

For sequences that are only 10 nucleotides long:

$P(X \geq 7) = 1 - P(X \leq 6) = 1\text{-pbinom}(6,10,0.25) = 0.003505707$.

For sequences that are 100 nucleotides long:

$P(X \geq 70) = 1 - P(X \leq 69) = 1\text{-pbinom}(69,100,0.25) \sim 0.$

Note: Free bioinformatics software such as MOLKIT or CLUSTALW2 are available in the web to compare two sequences of nucleotides. The program BLAST compares one sequence against a data base of sequences. Part of the reason to compare sequences is that if two sequences are homologous, they might have a common ancestor or a similar biological function. Phylogenetic trees are constructed based on the similarities between sequences. The BLAST program and nucleotide sequences are available from the NCBI (National Center for Biotechnology Information) website.

8.3   Exercises

8.3.1   Review questions

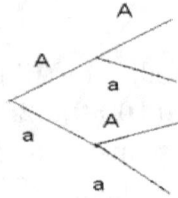

| M | F | Genotype | Phenotype | Probability |

1. This tree is for the offspring of heterozygous parents (M and F). The gene has two alleles A and a, A being the dominant one. The phenotypes are 'type A' and 'no type A.' Write the probability of each branch; write the genotypes and phenotypes and the probability for each one.

2. What is the difference and the relationship between genotype and phenotype?

3. Assume a gene has two alleles A and a. Write the genotype of a heterozygous person. Write the genotype of a homozygous person for the dominant allele. Write the genotype of a homozygous person for the non-dominant allele.

4. Use the probability tree in question 1 to answer the following questions: **Review questions for Chapter 8**

   (a) What is the probability that if both parents are heterozygous, the offspring will be heterozygous?

   (b) What is the probability that if both parents are heterozygous, the offspring will show the dominant trait (Type A)?

   (c) What is the probability that if both parents are heterozygous, the offspring will show the recessive trait (not Type A)?

5. Use your answers to question 1 to calculate the following probabilities:

   (a) If heterozygous parents have 3 children, what is the probability that none of the children will be heterozygous?

   (b) If heterozygous parents have 3 children, what is the probability that at least one of them will be heterozygous?

   (c) If heterozygous parents have 3 children, what is the probability that none of them will show the recessive trait ?

6. Build a probability tree to calculate probabilities for the genotype and phenotype of the offspring for a gene with two alleles, A and a, assuming that one parent (M) is heterozygous and the other (F) is homozygous for the dominant allele.

7. Assuming that each of the letters C, G, A, and T are equally frequent, what is the probability that a 6-letter word picked at random is CCAGGG?

8. Assume you are working with a section of DNA for which the GC content is 60%, P(G)=P(C)=0.3, and P(A)=P(T)=0.2. What is the probability that a 6-letter word picked at random is CCAGGG?

9. What is a palindrome in the DNA context? Why are palindromes important in the analysis of DNA?

10. Is AGGT a palindrome? Is AAGGCCTT a palindrome ?

11. How many 4 nucleotide different sequences (or 'words') do exist? How many of them are palindromes?

12. GATC is a palindrome target of the restriction enzyme Sau3A. If a 4-letter sequence is selected at random from a genome, what is the probability that it is GATC? Assume independence and that the four nucleotides are equally likely. How many 4-letter sequences can be read in a sequence of 100003 nucleotides? How many GATC would be expected in a sequence of 100003 nucleotides?

Review questions for Chapter 8

13. Use binomial tables to answer the following questions.

    (a) Assume you are comparing two DNA sequences of 6 nucleotides each one. What is the probability of 5 or more coincidences by chance alone?

    (b) Assume you are comparing two DNA sequences of 12 nucleotides each one. What is the probability of getting 10 or more coincidences just by chance?

14. When are two sequences of nucleotides said to be homologous? At least how long should the sequences be in order to compare them to find out if they are homologous? Why is it not acceptable to compare shorter sequences?

8.3.2   Exercises for discussion or homework

1. **Blood types.** The gene ABO determines the presence or absence of antigens A and B in the surface of red blood cells. The gene comes in 3 alleles: A (antigen A is present), B (antigen B is present,) and O (neither A nor B are present). In determining the phenotype of a person, A and B are dominant with respect to O, and A and B are 'co-dominant' with respect to each other. The phenotype of a person is popularly called 'blood type.' A woman XX is told that she has blood type A.

    (a) What are the two possible genotypes for XX if the phenotype of her parents is unknown?

    (b) If XX's mother has blood type O, what is the only possible genotype for XX?

(c) Suppose XX marries a man YY who has phenotype O. What are the options of genotype and phenotype for their children? (given the information in b)?

(d) What is the probability of blood type A for the offspring of XX and YY?

(e) Assume XX and YY plan to have 3 children. Define the variable X: *number of children with blood type A.* Build a probability table for X (*Hint: Consider 'A' as 'success.' What is n? What is p?*).

2. **Rh- and Rh+.** The Rh factor was discovered around 1940. The blood type of a person can be Rh+ or Rh-. If an antiserum agglutinates the red cells, the person is Rh+; if not, he/she is Rh-. The possible phenotypes for a person are DD, Dd, dd ; D indicates the dominant allele. People with genotype DD, Dd are Rh+, people with genotype dd are Rh- , meaning they do not have Rh antigen on the surface of the red blood cells.

Exercises for discussion or homework in Chapter 8

(a) Assume two people, XX and YY, both Rh+, have a Rh- child. What is the genotype of XX and YY?

(b) Consider two people Rh+ who are heterozygous with respect to the Rh+ factor. What are the possible genotypes and phenotypes for their offspring?

(c) Consider two people Rh+ who are heterozygous with respect to the Rh+ factor. Assume they plan to have 3 children. Define the variable X: number of children with blood type Rh-. Prepare a probability table for X. (*Hint: Consider Rh- as 'success.' What is n? What is p?*)

3. **Palindromes.** Check if the following sequences are palindromes or not: AGGT, AGCT, GATC, GGATCC, GAATTC, and AAGGCCTT.

4. **Probability of a given palindrome.** Assume equal frequency for A,C,G,T and independence between consecutive nucleotides. What is the probability of finding the palindrome GAATTC? That palindrome is the target of the EcoRI enzyme extracted from the *E coli* (*Escherichia coli*) bacteria. In sequences that are 100005 nucleotides long, how many GAATTC are expected?

5. **Probability of coincidences.** Assume that two sequences that are 50 nucleotides long are compared. Assume independence and that the

4 nucleotides are equally frequent. What is the probability of finding 35 coincidences just by chance? What is the probability of finding 35 coincidences or more by chance alone? Now, assume that the sequences are 100 nucleotides long. What is the probability of finding 70 coincidences just by chance? What is the probability of finding 70 coincidences or more just by chance?

6. **Traits associated with recessive alleles.** There are diseases such as cystic fibrosis, thalassemia, and sickle cell anemia that are associated with recessive alleles. Here is a list of probability questions about traits associated with recessive alleles: Assume that H and h represent the alleles at a particular locus. H is dominant over h. If a person has genotype 'hh,' the person gets the disease at some point in their life. People with genotypes Hh or HH do not get the disease.

Exercises for discussion or homework in Chapter 8

(a) Assume heterozygous parents (Hh). Prepare a probability tree with the possible genotypes and phenotypes for the offspring. Calculate the probabilities for the genotypes (HH,Hh,hh) and phenotypes (gets or does not get the disease).

father mother   genotype phenotype   p

(b) Assume there are heterozygous parents who plan to have 3 children. Prepare a table to display the mass or probability function for the variable X: 'Number of children with the disease.' Plot the mass function. Calculate and interpret $E(X)$.

x	0	1	2	3
p(x)				

(c) If heterozygous parents plan to have 3 children, what is the probability that at least one will have the disease?

(d) If heterozygous parents plan to have 2 children, what is the probability that at least one of them will have the disease?

(e) If they plan to have 4 children, what is the probability that at least one of them will have the disease?

(f) Compare the answers to questions d), c), and e). Do you see any pattern? If you see a pattern, write a sentence describing it.

(g) Now assume that a heterozygous person (Hh) marries a homozygous person (HH): Is it possible for them to have children who get the disease?

(h) Assume a Hh and a HH person have children together. Can the allele h be passed to the next generation (the grandchildren of the couple) even if the children of the couple do not have the disease?

(i) Imagine a person with genotype hh who marries a person who is heterozygous Hh. What is the probability that at least one of the children will get the disease?

(j) Imagine a person with genotype hh that marries a person that is also homozygous HH. What are the possible genotypes for their children? Is it possible for them to have children who get the disease? Can the allele h that gives the disease be passed to the next generation (the grandchildren of the couple)?

Exercises for discussion or homework in Chapter 8

7. **Traits associated with dominant alleles.** Diseases such as Huntington's chorea and porphyria variegate are diseases associated with rare but dominant alleles. The onset of the disease associated with dominant alleles can happen later in life, after people have had children already. Solve a list of exercises about dominant alleles. Assume that the alleles at that particular locus are represented by D and d. D is dominant over d; if a person has genotype Dd or DD, the person gets the disease at some point of life. People with genotype dd do not get the disease.

   (a) Assume heterozygous parents (Dd). What are the possible genotypes for the offspring? What are the corresponding phenotypes? What are the probabilities for each one of those genotypes and phenotypes?

   (b) If heterozygous parents (Dd) plan to have 3 children, define the variable X: *Number of children who get the disease*. Prepare a table for the mass or probability function of X. Calculate and interpret $E(X)$.

   (c) If heterozygous parents (Dd) plan to have 3 children, what is the probability that at least one of them will get the disease?

(d) If a heterozygous (Dd) person marries a homozygous person (dd), what are the possible genotypes and phenotypes for the offspring? What are the probabilities for each one of those genotypes and phenotypes? (Hint: prepare a probability tree)

(e) If a heterozygous (Dd) person marries a homozygous person (dd), and they plan to have 3 children, what is the probability that at least one of them will get the disease?

(f) A heterozygous (Dd) person marries a homozygous person (dd) and plans to have 2 children. Prepare a table with the mass function for the variable X : *number of children that get the disease.* Calculate and interpret $E(X)$.

(g) Assume that a person has Huntington's chorea. Is it possible that both his parents were homozygous (dd)? Is it possible that neither of his parents had the disease?

(h) Assume that a person has Huntington's chorea. Is it possible that none of his/her children will get the disease but that one of his/her grandchildren will get it?

References

Claverie, J.M. and Notredame, C. (2007) Bioinformatics for dummies. Second edition. Indianapolis: Wiley.

CLUSTALW2 http://www.ebi.ac.uk/Tools/msa/clustalw2/ (Last accessed: June 19, 2011)

Ewens, W.J. and Grant G.R. (2005) Statistical Methods in Bioinformatics-An Introduction, Second Edition. New York: Springer.

MOLKIT Molecular Toolkit http://www.vivo.colostate.edu/molkit/index.html (Last accessed: June 19, 2011).

NCBI website http://www.ncbi.nlm.nih.gov/ (Last accessed: June 19, 2011).

Zvelebil, M. and Baum, J.O. (2008) *Understanding Bioinformatics* New York: Garland Science.

Chapter 9

Continuous distributions

Several probability distributions (Bernoulli, binomial, Poisson, and geometric) have been introduced in Chapters 4 and 7, but all of them are for discrete random variables. Continuous random variables (variables that can take not only integer values) are also important. The uniform, normal, chi-square and exponential distributions will be studied in this chapter. The normal distribution is used as a probability model for many variables in the real world and it is intensively utilized in statistical inference, as will be seen in Chapter 11. The chi-square distribution is used to calculate p-values when testing if a model is an appropriate choice for a given situation, as will be seen in Chapter 10. The exponential distribution is used to describe the behavior of the variable 'interval of time or space in between occurrences of a given phenomenon.' Some of the applications found in the literature for the exponential distribution are: distance from natal nest (Biology), survival time (Medicine, Public Health), and distance between genes (Bioinformatics).

9.1   Density function

Consider the two histograms in Figure 2.9 in Chapter 2. The first histogram shows that the pulse rate (number of heartbeats per minute) of 210 students has a fairly symmetric distribution, where the central values (70 to 80 heartbeats) are more frequent and the very low or very high number of heartbeats are much less frequent. The second histogram displays the age of 503 respondents to a survey conducted by phone (the survey targeted people 18 years and older). The histogram shows a moderately skewed distribution with a longer right tail. The most frequent age interval is $30 - 40$ years and the frequency of the interval decreases as age increases because there are fewer

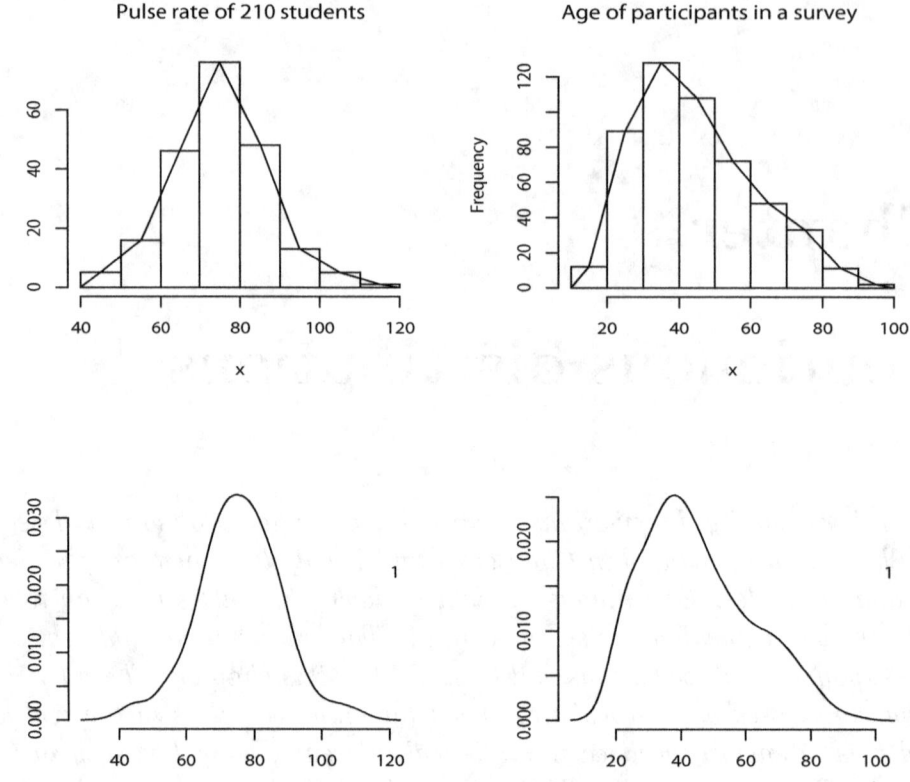

Figure 9.1: Histograms, polygons and densities

older people in the population from where the sample was selected.

The first row of Figure 9.1 displays the same histograms with added lines that connect the mid points of the intervals; the plot formed by those segments is called a **frequency polygon**. Smoothed versions of the frequency polygons are shown in the second row of Figure 9.1. Notice also the change of scale in the vertical axis. Instead of showing frequencies, the scale has been changed so that the total area under the curve is 1 (or 100%). Based on observations, curves have been produced that summarize the behavior of variables.

The **density curves** in the lower row of Figure 9.1 are ad-hoc curves obtained from the data. One might want to go a step further and select an even smoother curve, for the pulse rate example, from a set of available models. The density curve for heartbeats in Figure 9.1 resembles the density in the central plot of Figure 9.4. It is desirable to have curves that summarize the behavior of random variables. Those probability models can be used to calculate the probability that the random variable takes its value in an interval that is of special interest for some reason. The density curves are the graphical representation of density functions that can be used to calculate probabilities. Examples of density functions are in expressions (9.1), (9.3) and (9.6) in this chapter. A density function $f(x)$ is an expression in terms of x such that:

1. Only one value $f(x)$ is assigned to each value x of the variable.

2. It only takes non-negative values. ($f(x) \geq 0$)

3. The area under the curve is 1.

The **distribution** of a continuous random variable X describes the behavior of the variable by specifying in which interval the variable can take values and the corresponding density function. Probability distributions are also called **probability models**. When working with random variables one might want to assume a probability model and calculate probabilities using that model. Contrary to the case of discrete random variables, we do not calculate the probability of a single value x, rather the probability of intervals of values. The density function $f(x)$ is written in terms of x but it also includes **parameters** (constants whose value we might not know and would need to estimate from data). Those parameters determine the mean of the distribution and how spread out it is. Some distributions have the same skewness, kurtosis, and general shape, regardless of the value of the parameters; this is what happens with the uniform, normal and exponential distributions. For some other distributions, such as the Chi-square, the skewness and kurtosis (and so the shape) depend on the values of the parameters.

Density functions are written in terms of the variable and parameters, never take negative values, and the area under the curve is 1

When dealing with a random variable, such as pulse rate, we can choose a probability model (normal in this case) to represent its behavior based on the shape of the histogram for pulse rate. Which specific normal model (that is, with what mean and what standard deviation) we use depends on the mean and standard deviation observed for pulse rate from real data. It is common practice to choose a distribution and estimate its parameters based on data. Four important continuous distributions will be studied next, but many more exist.

Using statistical software:
*The commands **simple.freqpoly** and **simple.densityplot** from the 'Us-* **R:** *ingR' package Verzani (2004, 2011) were used in the production of Figure* *frequency polygons* *9.1. First install the package UsingR in your computer: from the R menu,* *and* *select 'Packages > Install package,' select a location near you, and select the* *density plots*

package 'Using R'. Once the package is installed, load it by selecting from the R menu: 'Packages > load package.'

The following commands read the data file (assuming it is in the 'e:' drive) and obtain the first column of plots in Figure 9.1.

```
x<-scan('e:pulserate.dat')
simple.freqpoly(x)
simple.densityplot(x)
```

9.2   Uniform distribution

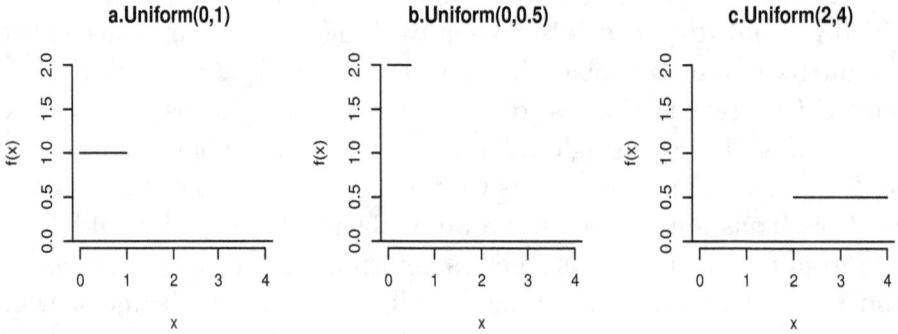

Figure 9.2: Density function of 3 uniform distributions

Assume there is a random variable X that can take values in a given interval $[a, b]$. The values a and b are the parameters of the distribution. The parameters are the values that identify the specific uniform distribution. The length of the interval is $b - a$ and the total area under the density function needs to be 1. Thus, the density function of the continuous uniform distribution is:

$$f(x) = \frac{1}{b - a} \qquad\qquad a \le x \le b \qquad\qquad (9.1)$$

Figure 9.2 displays the density function for various uniform distributions with different values for the **parameters** a and b. The notation $U(a, b)$ represents the uniform distribution defined over the interval $[a, b]$. We write $X \sim U(a, b)$ to mean that a random variable X has uniform distribution over the interval $[a, b]$. The uniform distribution is symmetric and does not have a peak or tails, it is flat. The expected value (or mean) and the median have the same value, they are at the center of the distribution. The mean and variance are:

$$E(X) = \frac{a+b}{2} \qquad\qquad Var(X) = \frac{(b-a)^2}{12} \qquad (9.2)$$

As mentioned before, for continuous variables, the probability is not calculated for a single value of the variable; rather the probability is calculated for an interval of values of the variable. For example, if $X \sim U(0,1)$, as in Figure 9.2a, $P(0.8 \leq X \leq 1)$ is the area between 0.8 and 1. That is the area of a rectangle with base 0.2 and height 1, thus: $P(0.8 \leq X \leq 1) = 0.2$.

9.3 Normal distribution

Revisiting some examples from Chapter 2

Figure 9.3 displays frequency distributions for four data sets. Figure 9.3.a displays the data in Example 2.5.1, the self-reported pulse rate (number of heart beats per minute) for 210 college students while sitting in our introductory statistics class. Pulse rate is usually reported as an integer value, but as mentioned in Chapter 2, when a variable has a long range of possible values, its values are displayed using the same statistical graphs used for continuous variables. Figure 9.3.c displays the same data set as Figure 2.21a, the length (mm) of 270 cells from the apical region of onion roots.

Example 9.1 Bone mineral density in the lumbar region

The bone mineral density (g/cm^2) is usually measured in four locations of the lumbar region in osteoporosis studies. Figure 9.3.b displays the BMD, in one of those locations, for 108 women over 50 years of age.   BMD in spine

Example 9.2 Hemoglobin

The hemoglobin (g/dl) of 380 adult women was measured. Those values are displayed in Figure 9.3.d. The last distribution in Figure 9.4 could be used as a model to represent the behavior of the variable hemoglobin.   Hemoglobin

The four frequency distributions in Figure 9.3 show some similarity in the shape, even when their center and spread are different. The distributions are fairly symmetric. The values of the variable at the center of the distribution are more frequent, while the very low and very high values are equally rare. This is a pattern frequently shown by other variables in nature, especially

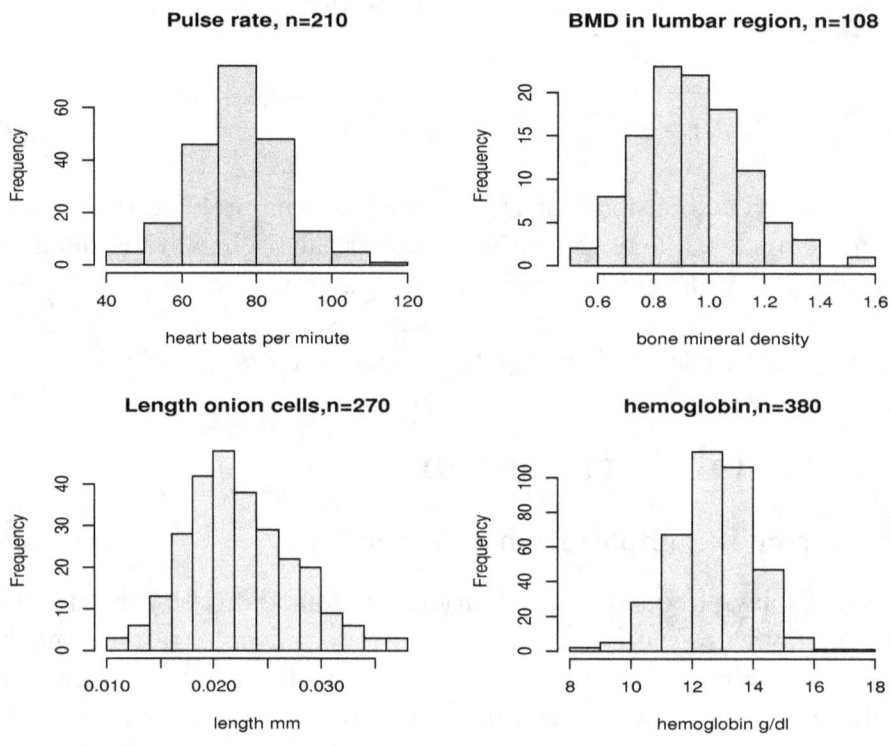

Figure 9.3: Frequency distributions

when observations for a large number of individuals are available. It is a pattern that indicates individual variability around a population mean. A probability distribution that has a symmetric shape, with high concentration of mass in the center and relatively thin tails, is used as a model for many variables in nature. This distribution is the normal, or Gaussian, distribution; the mathematical expression for it was first defined by Abraham de Moivre in 1733.

Normal distribution: symmetric, $E(X) = \mu$, $Var(X) = \sigma^2$, **values at the center are more frequent, 99.7% of the area is between** $\mu - 3\sigma$ **and** $\mu + 3\sigma$

Normal distributions are symmetric and all have the same kurtosis. The **parameters** μ and σ determine the location and spread of the distribution, they are the expected value (mean) and standard deviation respectively. The expression $X \sim N(\mu, \sigma^2)$ indicates that a variable X has a normal distribution with mean μ and variance σ^2. The normal distribution with a mean of 0 and variance of 1, $N(0, 1)$, is called the **standard normal distribution**. In theory, a variable X with normal distribution can take any value, it does not have bounds, i.e., $-\infty < X < \infty$. However, many variables, whose distribution are described using the normal model, have a minimum and a maximum

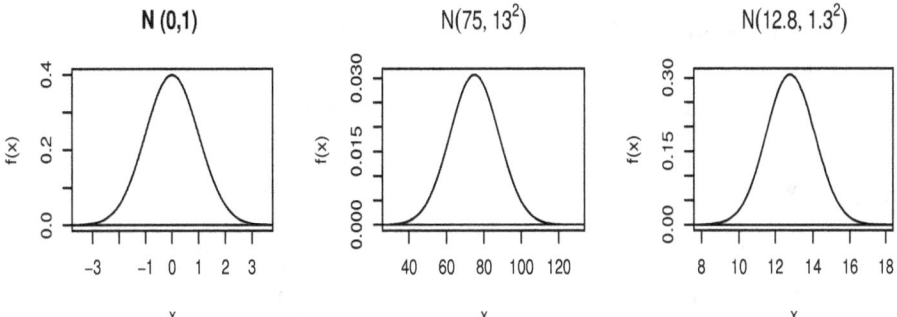

Figure 9.4: Three normal distributions with different parameters

value in real life. Figure 9.4 shows the density function of the standard normal and two other normal distributions. Notice that while the shapes are similar, the center of the distribution and the spread can vary (look at the horizontal and vertical scales). Notice that the mean μ of the distribution is always at the center and that most (99.7%) of the area under the curve is comprised between $\mu - 3\sigma$ and $\mu + 3\sigma$, meaning that it is extremely likely that the value of a variable with normal distribution will fall in the interval $(\mu - 3\sigma, \mu + 3\sigma)$. This interval is 6 standard deviations wide. A feature of the normal distribution is the fact that:

- in the interval $\mu \pm \sigma$ is 68% of the total area

- in the interval $\mu \pm 2\sigma$ is 95% of the total area

- in the interval $\mu \pm 3\sigma$ is 99.7% of the total area

The 68%, 95%, 99.7% rule

The density function of the normal distribution, $N(\mu, \sigma^2)$, which describes the shape of the curves in Figure 9.4, is:

$$f(x) = \frac{1}{\sigma\sqrt{2\pi}} \exp -\frac{(x - \mu)^2}{2\sigma^2} \qquad -\infty < x < \infty \qquad (9.3)$$

You do not need to work with this expression, all the calculations can be done using statistical software or scientific calculators.

Standardization

There is a relationship between the distribution $N(\mu, \sigma^2)$ and the standard normal distribution $N(0, 1)$. Indeed,

$$If \quad X \sim N(\mu, \sigma^2) \quad then \quad Z = \frac{(X - \mu)}{\sigma} \sim N(0, 1). \qquad (9.4)$$

Standardization: subtract the mean from each observation and divide by the standard deviation

The process of subtracting the mean and dividing by the standard deviation is called **standardization**. If a variable has a normal distribution with a mean μ and standard deviation σ, its standardized version has a standard normal distribution (mean equal to 0 and standard deviation equal to 1).

Example 9.3 BMD in hip

BMD in hip

The bone mineral density is expressed in g/cm^2. Assume that the bone mineral density in the hip for women in a given age group has an approximately normal distribution with a mean of $0.8\ g/cm^2$ and a standard deviation of $0.16\ g/cm^2$. Notice that the mean and the standard deviation are expressed in the same units as the variable X, while the standardized variable Z is free of units. If the bone mineral density in the hip of a woman in that age group is $0.64\ g/cm^2$, the standardized version is:

$$\frac{(0.64g/cm^2 - 0.8g/cm^2)}{0.16g/cm^2} = \frac{-0.16g/cm^2}{0.16g/cm^2} = -1.$$

The mean of the standardized values is 0 and their standard deviation is 1

The value -1 indicates that the BMD of the woman is 1 standard deviation below the mean of the population to which she belongs. Also notice that we can write the relationship between the values of X and Z in two ways:

$$z = \frac{(x - \mu)}{\sigma} \qquad or \qquad x = \mu + z\sigma. \qquad (9.5)$$

We can go from the value of the variable to the standardized value or vice versa.

Finding areas under the normal curve

What is the probability that a randomly selected women in this age group has BMD equal to $0.64\ g/cm^2$ or lower than $0.64\ g/cm^2$? Figure 9.5a shows the area under the normal density function (with mean 0.8 and standard deviation 0.16) to the left of the value $X = 0.64$. The standardized value for $X{=}0.64$ is $Z = -1$. Figure 9.5b shows the area under the standard normal distribution to the left of the value $Z = -1$. The area to the left of 0.64 under the $N(0.8, 0.16^2)$ curve is equal to the area to the left of -1 under the standard normal curve. It is not easy to calculate the area in Figures 9.5a and 9.5b by hand; tools from integral calculus would be needed. However, there are normal tables available from the web and most statistics textbooks. Also scientific calculators and statistical software can be used to calculate the areas under the curve.

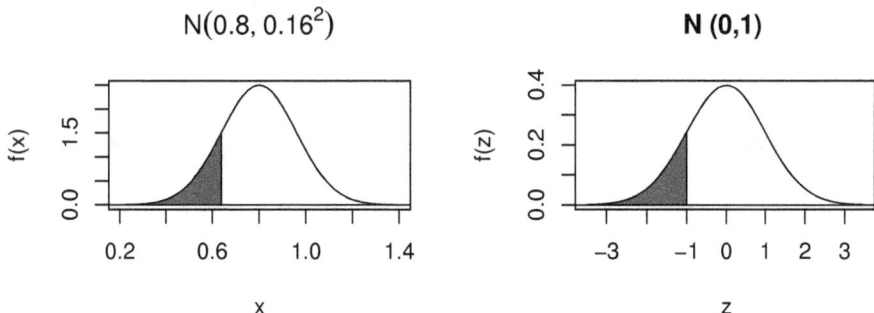

Figure 9.5: Area under a normal and a standard normal density function

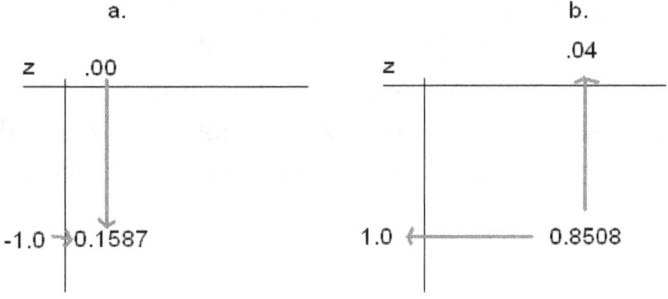

Figure 9.6: Finding probabilities (a) and values of the variable (b) in a normal table

Using normal tables

To work with a table, first calculate the standardized value

$$z = \frac{x - mean}{standard\ deviation}$$

because the only normal distribution for which tables are available is the standard normal. Remember that the area to the left of x in $N(\mu, \sigma^2)$ is equivalent to the area to the left of z in the standard normal distribution. Once the standardized value z is calculated, locate the row corresponding to the integer part and first decimal of z (-1.0 in the example). Then, advance in that row until finding the column corresponding to the number in the second decimal location, in this case 0 (see Figure 9.6a). Tables and statistical software calculate areas to the left of the value of z; that area represents $P(Z \leq z)$. In order to calculate areas to the right of a given value of z, or $P(Z \geq z)$, remember two important facts: the total area under the curve is 1, and the distribution is symmetric. In Figure 9.7, using the normal table

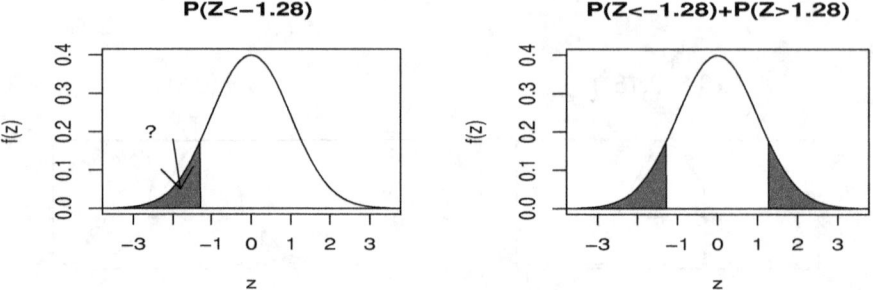

Figure 9.7: Given a value of the variable, find the area

or software $P(Z \leq -1.28) = 0.1002726$. Since the distribution is symmetric, the area to the right of $z=1.28$ also is 0.1002726. The area of the two tails together is just $2 \times 0.1002726 = 0.2005452$. Those types of calculations are useful to find p-values in the context of hypothesis testing.

When using statistical software you do not need to standardize the values, you can enter the mean and standard deviation of the distribution as will be seen next.

Using statistical software:
In R, to calculate the area under the curve in Figure 9.5a from the extreme left up to the point of interest $x = 0.64$, write:

$$\text{pnorm}(0.64, 0.8, 0.16)$$

The output of the command is **0.1586553***. Since the total area under the curve is 1, the area to the right of 0.64 will be*

$$\text{1-pnorm}(0.64, 0.8, 0.16) = \textbf{0.8413447}.$$

In general, to calculate the area under the curve from the extreme left up to x, replace the value of x, as well as the mean and the standard deviation of

MINITAB & R: *the distribution, in:*
normal
probabilities
$$\text{pnorm}(\text{x}, \text{mean}, \text{standard deviation})$$

MINITAB also calculates areas under the normal curve. From the Menu select **CALC<Probability distributions<Normal***, select cumulative probability, enter the values of the mean and the standard deviation, and for 'input constant' type the value 0.64.*

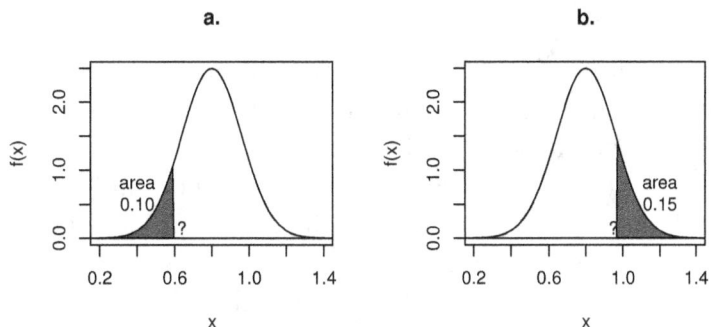

Figure 9.8: Given an area, find the value of the variable

Given an area, find a value

Continuing with Example 9.3, consider a research project that will involve women who are in the lower 10% of bone mineral density. What is the maximum BMD possible in order to qualify for the study? The question is in the opposite direction of the previous section where given a value of z we had to find an area. Given an area (0.1), the value of z that separates the lower 10% of the total area from the upper 90%, as in Figure 9.8a, is the quantile(0.1) of the normal distribution with parameters 0.8 and 0.16. Using software, it is quite easy to find that quantile. In **R**, type: **qnorm(0.1,0.8,0.16)**. The answer is 0.5949517, meaning that the BMD of a woman has to be below 0.5949517 in order to qualify for the study. Notice the three values written as input for the command **qnorm(p,μ,σ)**: $p = 0.1$ is the area to the left of the number, $\mu = 0.8$ is the mean, and $\sigma = 0.16$ is the standard deviation of the population.

A different study is looking for women in the upper 15% of BMD in order to analyze their lifestyle with the hope of finding clues about what makes bones strong. How high of a BMD should a woman have in order to qualify for the study? The value that separates the upper 15% from the lower 85%, as in Figure 9.8b, is the quantile(0.85) of the normal distribution with parameters 0.8 and 0.16. In **R**, type:

R & MINITAB: finding normal quantiles

$$\text{qnorm}(0.85,0.8,0.16)$$

The answer is 0.9658293, so the BMD of a woman has to be above that value to qualify for the study. Notice that the value of p in the command **qnorm(p,μ,σ)** is 0.85, because if the area to the right of the value is 0.15, the area to the left is 0.85.

Working with MINITAB, the same answers are found using CALC>Probability distributions>Normal and selecting 'inverse cumulative probability.'
Working with a normal table, remember that the areas are in the body of the table. Thus, for the example being discussed, look for the area that is the closest to 0.85 and then look at the borders of the table to find the value of z to which that area corresponds (See Figure 9.6b). The table gives the value of z, not of x. Thus, after the value $z = 1.04$ has been identified, x needs to be calculated as $x = \mu + z*\sigma$. For the example, $x = 0.8 + 1.04*0.16 = 0.9664$. The small difference with the value found with **R** is due to the fact that the normal table only works with four decimal places.

Generating normal random values *

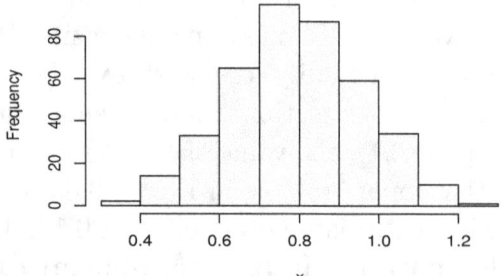

Figure 9.9: Random values generated with $N(0.8, 0.16^2)$

Probability models are used not only to describe, in a somewhat idealized way, the behavior of variables, they are also useful to generate random values that can be used in simulations. Nobody says that simulated values are real data; nevertheless, having simulated values helps sometimes to visualize what can happen under certain conditions. To generate n values with a normal model with mean μ and standard deviation σ, using **R** type the command **rnorm(n,μ,σ)**. The histogram in Figure 9.9 displays 400 values generated with the distribution $N(0.8, 0.16^2)$ using the command $rnorm(400, 0.8, 0.16)$. A scientific journal might report that the bone mineral density in the hip, for a given subpopulation of women, has an approximately normal distribution with a mean of 0.8 and standard deviation of 0.16. Generating values with the distribution and producing a histogram like the one in Figure 9.9 helps us to visualize how the frequency distribution of BMD looks like in the subpopulation of interest.

R: generating random values

The normal distribution is particularly useful for simulating random departures or errors. When simulating data with a more complex model, frequently a purely random component 'ϵ' is added. The values of that random component can be simulated with a standard normal distribution using the command **rnorm(n)**, where n is the number of values to be generated.

9.4 Normal approximation to the binomial *

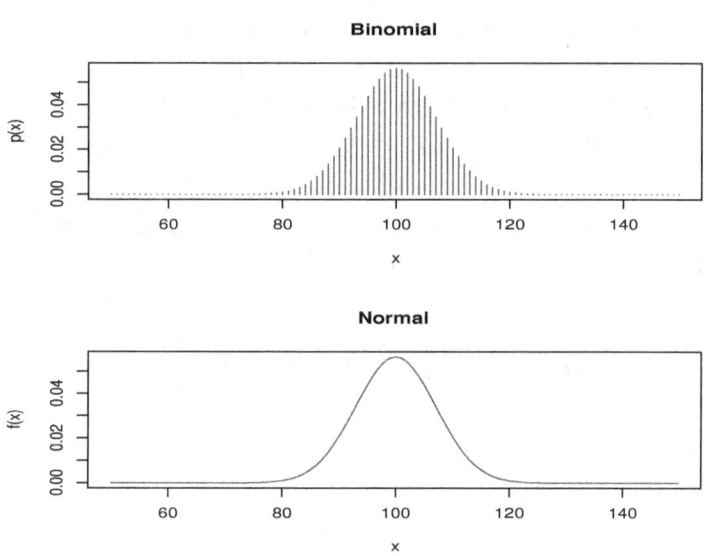

Figure 9.10: Binomial(200,0.5) and N(100,50)

Binomial probabilities can be calculated for practically any value of n and p using appropriate statistical software. Before computing facilities were available, it was very important to find mathematical relationships that helped with the calculations; even if the results were only approximate. One of those important tools originated in the similarity between the normal distribution (a continuous distribution) and the binomial distribution (a discrete distribution). For a large value of n, the binomial distribution with **parameters** n and p looks quite similar to the $N(\mu, \sigma^2)$ distribution where $\mu = np$ and $\sigma^2 = np(1 - p)$. For a given p, the larger the value of n, the better the approximation. The approximation works best when p is neither too close to 0 nor too close to 1, rather in a vicinity of 0.5. Figure 9.10 displays the mass function for Binomial(200,0.5) and $N(100, 50)$.

Assume that we want to calculate $P(X \leq 80) = \sum_{x=0}^{80} \binom{200}{x} 0.5^x (0.5)^{200-x}$. It would be quite laborious to calculate that expression by hand. Using **R**, the command

pbinom(80,200,0.5)

produces the answer **0.002842578**. The same answer is obtained with MINITAB (CALC>Probability distributions>Normal>Cumulative probability).

Before computers were widely available, the normal table was used to calculate the probability in the example above, relying on the approximation of the normal to the binomial distribution. In the case of the example, $\mu = np = 200(0.5) = 100$ and $\sigma^2 = np(1-p) = 200(0.5)(0.5) = 50$, thus $\sigma = 7.071068$. The normal distribution is $N(100, 7.071068^2)$. Standardizing the value 80, $z = \frac{(x-\mu)}{\sigma} = \frac{(80-100)}{7.071068} = -2.828427$. Now the probability will be calculated using the normal distribution: $P(X \leq 80) = P(Z \leq -2.828427) = $ **0.002338868**. The answer is similar to the one obtained with the exact distribution (binomial) up to the third decimal.

9.5   Chi-square distribution *

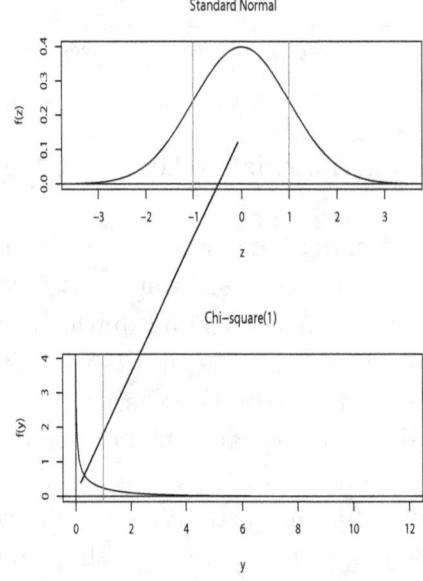

Figure 9.11: The density function for a standard normal variable and its square

The normal distribution has a very important role in the theory of statistics, and several other distributions are derived from it. One of them is the Chi-square or χ^2 distribution. Assume that there is a random variable Z that has a standard normal distribution. If the square of Z is calculated, what is the distribution of Z^2? The squares of the negative values of Z are positive, thus the distribution of $Y = Z^2$ will always be non-negative. Remember that there is a big concentration of mass in the center of the normal distribution near $Z = 0$ (68% of the mass is between the values -1 and 1). The square of a number between 0 and 1 is smaller than the original number, for example if $Z = 0.8$, $Y = Z^2 = 0.64$. Hence, in the distribution of Y there will be a higher concentration of mass near $Y = 0$. On the other hand, for the values $Z > 1$, $Y = Z^2 > Z$. For example, if $Z = 2$, $Y = Z^2 = 4$, thus the tail in the normal distribution is stretched out in the Chi-square distribution. Think of the shape of a distribution, where the variable only takes positive values, there is a high concentration of values near $Y = 0$, and a very long right tail.

The distribution of the square of a standard normal variable is Chi-square(1)

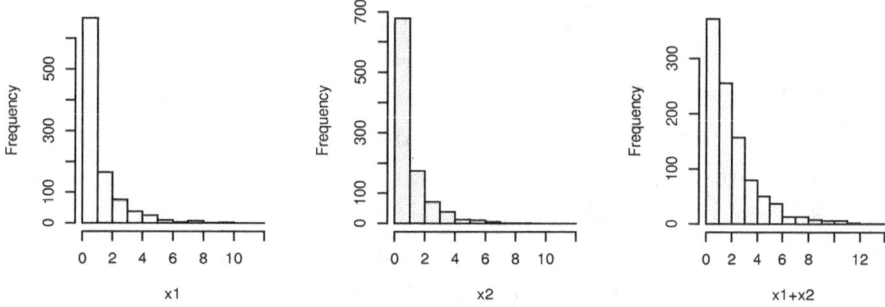

Figure 9.12: Simulated values with two independent χ_1^2 and their sum

Figure 9.11 displays the density functions of the standard normal and the $\chi_{(1)}^2$ distribution; the subindex (1) indicates that only ONE standard normal variable Z was used in the creation of the variable Y. Vertical lines have been added to indicate that the central area $(-1 < z < 1)$ in the normal distribution corresponds to the area for the region $0 < Y < 1$ in the $\chi_{(1)}^2$. Those two areas are equal, although they might not look equal in the figure because of the differences in scale.

When k independent $\chi_{(1)}^2$ variables are added up, the new variable is said to have a $\chi_{(k)}^2$ distribution. The number k is called the **degrees of freedom**. In general, the term **degrees of freedom** in statistics is associated with the number of independent variables that are combined together. Figures 9.12a and 9.12b display the histograms for 1000 randomly generated values with

degrees of freedom

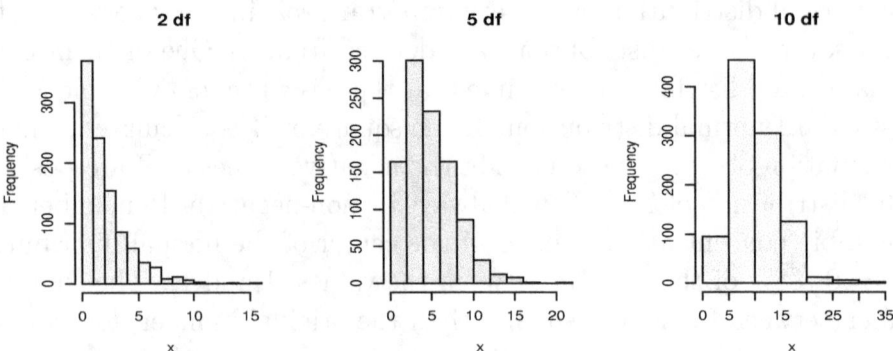

Figure 9.13: Simulated values with three Chi-square distributions with different degrees of freedom

two $\chi^2_{(1)}$ distributions. The two columns, of 1000 values each, were put side by side, and the elements of each pair added. Figure 9.12c shows the histogram of those 1000 sums. Notice that the distribution of the sums is less skewed than each one of the original frequency distributions.

Chi-square distributions with different number of degrees of freedom

Figure 9.13 contains three histograms with data simulated with three different Chi-square distributions: $\chi^2_{(2)}$, $\chi^2_{(5)}$, and $\chi^2_{(10)}$. Notice that as the number of degrees of freedom increases, the shape becomes less skewed. Looking at the values in the horizontal scale one realizes that as the number of degrees of freedom increases, the variable tends to take higher values. The mean or expected value of a variable with χ^2_k distribution is k.

9.6   Exponential distribution *

The Poisson distribution was studied in Chapter 7. The Poisson model is used to represent the behavior of the variable 'number of occurrences of a certain event in an interval of time or length.' The variable **time or space in between occurrences** has an exponential distribution. Some of the variables modeled with the exponential distribution are time between nerve impulses, time between detections by a Geiger counter, or distances in between occurrences, such as distances between salamanders in a transect or distances between genes in the DNA sequence.

The exponential distribution is skewed with a long right tail, as shown in Figure 9.14. Its shape indicates that, even when the lower values are more likely to happen, on occasion the variable can take very large values. One

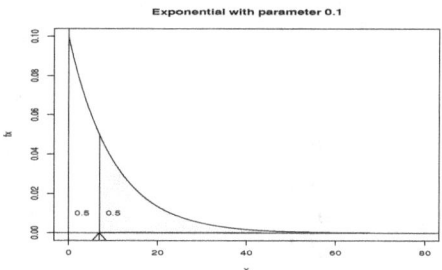

Figure 9.14: Location of the median $0.693147/\lambda$ in an exponential distribution $\lambda=0.1$

way of writing the density function of the exponential distribution is

$$f(x) = \lambda e^{-\lambda x} \qquad x \geq 0. \tag{9.6}$$

The **parameter** that distinguishes one exponential distribution from another is λ. The mean AND the standard deviation of the distribution are equal to $\frac{1}{\lambda}$. An alternative way of writing the density function is in terms of the mean (a symbol such as θ or μ could be used instead of the word 'Mean'): **Exponential distribution**

$$f(x) = \frac{1}{Mean} e^{-\frac{x}{Mean}} \qquad or \qquad f(x) = \frac{1}{\mu} e^{-\frac{x}{\mu}} \qquad x \geq 0. \tag{9.7}$$

One property of the exponential distribution, with density function written as in (9.6), is that the area to the right of the value a is

$$P(x > a) = e^{-\lambda a}. \tag{9.8}$$

In any distribution, the median is the value such that the area under the curve to the left and the right of it is equal to 0.5 (see Figure 9.14). Thus, the median of the exponential distribution is $\frac{0.693147}{\lambda}$ because $e^{-\frac{0.693147\lambda}{\lambda}} = e^{-0.693147} = 0.5$. The median is smaller than the mean $\frac{1}{\lambda}$, something to be expected due to the right skewness of the distribution. Each of the histograms in Figure 9.15 displays 1000 values simulated with exponential models with different values of the parameter λ. Notice that all the distributions are equally skewed with a tail to the right, but the mean value, the median, and the spread, change with the value of the parameter. In some applications, the interpretation of the median is quite important. When the exponential distribution is used to model the decomposition time of the molecules of a toxic chemical, the median is called the **half-life** of the toxic material, i.e., the approximate time in which half of the molecules remain toxic.

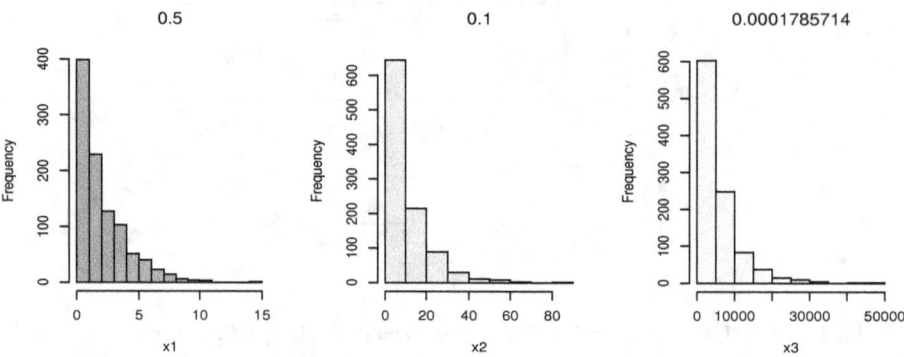

Figure 9.15: Simulated values with the exponential distribution for 3 different values of λ

Example 9.4 Red-shouldered hawks

Distances
traveled
by hawks

Diskstra et al. (2004) use the exponential distribution with parameter $\lambda = 0.048$ to model the distance from their natal nest, at the moment of the encounter, for previously banded red-shouldered hawks from SW Ohio.

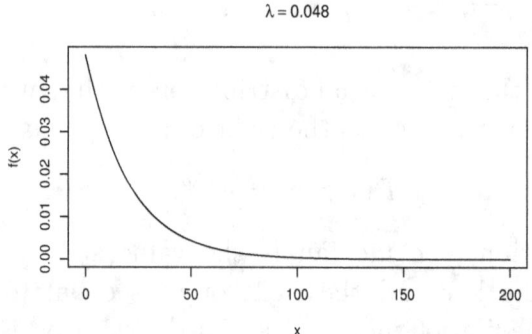

Figure 9.16: Probability model for distances (km) of hawks from their natal nest

The parameter of the distribution indicates that the average distance from their natal nest at which the hawks are encountered is $1/0.048=20.83$ kilometers. The median distance is $\frac{0.693147}{\lambda}= \frac{0.693147}{0.048}= 14.44$ km. Thus, according to the model, half of the birds of that species and region are encountered at less than 14.44 kilometers of their natal nest. The density function, plotted in Figure 9.16 is

$$f(x)=0.048\ e^{-0.048x}$$

where x is the distance in kilometers from the natal nest. The model can be used to answer probability questions about distances or to calculate the percent of hawks that are located beyond (or within) a given distance.

If we are curious to know what percent of hawks might be found at distances larger than 62 kilometers from their natal nest, the area to the right of $x = 62$ needs to be calculated. The area to the right of $x = 62$ is $P(x > 62) = e^{-0.048 \times 62} = exp(-2.976) = 0.05099641 \sim 0.05$. Thus, only 5% of the previously banded hawks are expected to be found at more than 62 kilometers from their natal nest.

Dykstra et al. (2004) also used the exponential distribution as a model for the age at recovery of the banded hawks. Their chosen value of λ is 0.557 when X: *age at encounter*. As an exercise, write the density function for age. Find and interpret the mean and median of the distribution. What proportion of hawks are recovered when they are 2 years old or older than 2 years?

Using statistical software:
In MINITAB, CALC > Probability distributions > Exponential can be used to calculate areas (cumulative probabilities) given a value, and values given an area (inverse cumulative probabilities).
In R, use 'pexp' to calculate areas. To calculate $P(X \leq 62)$ in Example 9.5,   *MINITAB & R:*
type **pexp(62,0.048)**, *and* **1-pexp(62,0.048)** *to calculate $P(X \geq 62)$.*   *Exponential*
   distribution
    To find values, given an area, use the command for quantiles of the exponential distribution 'qexp.' For example **qexp(0.75,0.048)** *will produce the upper quartile. The answer is 28.88113 $\sim$ 29. Thus, the probability that a hawk of that species and region is found at more than 29 kilometers of its natal nest is approximately 0.25. To find the median, use* **qexp(0.5,0.048)**.
    If you type **dexp(62,0.048)** *you will get the value of the density function for $x = 62$, something that is useful only if you plan to do plots such as Figure 9.16. In that case, first you need to create values of the variable X. Figure 9.16 was created with the following commands:*

```
x<-seq(0,200,by=0.1)
fx<-dexp(x,0.048)
plot(x,fx,'l',ylab='f(x)',main=expression(lambda==0.048))
```

The command **rexp(1000,0.048)** *will generate 1000 values with the exponential model with $\lambda = 0.048$. In R, we use the letters* **d**, **p**, **q** *and* **r** *to denote, respectively: the value of the* **d***ensity or probability function, the cumulative* **p***robability, the* **q***uantile and to generate* **r***andom values. The corresponding letter is followed by the name of the distribution.*

9.7   Exercises

9.7.1   Review questions

1. When is a random variable continuous?

2. The density function is used to describe the behavior of a continuous random variable. What are the main properties of a density function?

3. What are the parameters of the uniform distribution? Where can you 'see' them in the plot of the density function ?

4. What are the parameters of the normal distribution? Where can you 'see' them in the plot of the distribution?

5. Describe, in your own words, as many characteristics as you can remember of the normal distribution.

6. Answer the following questions about the normal distribution:

   (a) What is the shape of the normal distribution?

   (b) Theoretically, the variable can take values from _____ to _____

   (c) 99.7% of the area under the normal curve corresponds to values that are within how many standard deviations from the mean?

   (d) The interval where most (99.7%) of the area under the curve is located has a length of ___ standard deviations

   (e) What are the mean and standard deviation of the $N(\mu, \sigma^2)$?

   (f) What are the mean and standard deviation of the standard normal distribution?

   (g) What is at the center of the normal distribution? The mean, the median, both or neither?

   (h) What is the formula to get the standardized version of a value x?

   (i) What is the formula that allows us to go from a z value to a x?

7. Use a normal table to find the area to the right of the point $z= -0.36$ in a standard normal distribution. Check your answer using software.

8. Use the normal table to find the value z in the standard normal distribution such that the area under the curve, to the right of z, is 0.1. Check your answer using software.

9. Assume the pulse rate among people in the age group 18-30 has an approximately normal distribution with a mean of 76 and a standard deviation of 12. What percent of the population in that age group has pulse rate of at most 60 heartbeats per minute? What percent of the population in that age group has more than 80 heartbeats per minute? Solve these questions using a normal table, and then check your answers using software.

10. In one of the exercises in Chapter 2, the following questions were asked about the pulse rate data: How many times is the standard deviation contained in the range of the distribution? What percent of the observations are within one standard deviation of the mean? What percent of the observations are within two standard deviations of the mean? Compare your answers in Chapter 2 to the answers you would expect to get if the distribution of pulse rate was normal.

11. Assume that the pulse rate among people in the age group 18-30 has an approximately normal distribution with a mean of 76 and a standard deviation of 12. We want to select people who have very low pulse rates in order to study them. We would like to select people whose pulse rate is in the lower 20% of the population. What would be the maximum pulse rate in order to qualify for this study? Solve this exercise using a normal table and check your answer using software.

12. Assume that the pulse rate among people in the age group 18-30 has an approximately normal distribution with mean 76 and standard deviation 12. We want to select people who have very fast pulse rates in order to study them. We would like to select people whose pulse rate is in the upper 20% of the population. What would be the minimum pulse rate in order to qualify for this study? Solve this exercise using a normal table and check your answer using software.

13. What is the parameter of the Chi-square distribution? What is the mean or expected value of a variable that has a Chi-square distribution with k degrees of freedom?

14. Questions about the Chi-square distribution.

    (a) If $z \sim N(0, 1)$, what is the distribution of z^2?

    (b) What is the shape of the Chi-square(1)?

    (c) What is the parameter of the Chi-square distribution χ_k^2?

(d) As the number of degrees of freedom k increases, does the Chi-square(k) distribution become more or less skewed?

(e) The expected value or mean of a Chi-square(k) distribution is ___

(f) Can a variable that has a χ_k^2 distribution take negative values?

References

1. Dykstra, C.R., Hays, J.L., Simon, M.M., Holt, J.B. and Austing, G.R. (2004) Dispersal and mortality of red-shouldered hawks banded in Ohio. *Journal of Raptor Research* **38**: 304-311

2. Ewens, W.J. and Grant G.R. (2005) Statistical Methods in Bioinformatics-An Introduction. Second Edition. New York: Springer.

3. Verzani, J. (2004) *Using R for Introductory Statistics*. Chapman and Hall/CRC, Boca Raton.

4. Verzani, J. (2011) *UsingR package* http://cran.r-project.org/web/packages/UsingR/UsingR.pdf. Last accessed June 20, 2011.

Chapter 10

Checking models and assumptions

Several probability models have been studied in previous chapters. Probability models can be used to describe the behavior of random variables. Given a variable of interest, a certain probability model can be assumed to describe its behavior. It is necessary to check if such an assumption is wrong. Data are collected and compared with the predictions made by the model. If the observed values are too far from the expected values, the model is considered inadequate.

Another situation in which it is necessary to contrast expected values with real data is when two variables are assumed to behave independently. We need to check if the assumption of independence is reasonable or not.

Four cases are considered in this chapter:

- *Goodness of fit test*

- *Test of independence*

- *Test of homogeneity*

- *Test of normality*

Goodness of fit tests are used in the study of genetics. A section on the famous data set from Mendel's experiments is included.

10.1 Goodness of fit test

In a general sense, 'goodness of fit' is related to the following question: *How well does a given probability model describe the behavior of a particular ran-*

What is goodness of fit?

dom variable? The **Chi-square goodness of fit test** is applied in situations in which there is a variable with a given number, k, of possible categories, classes, or intervals. The model assigns a probability to each one of the categories. The Chi-square goodness of fit test was created by Karl Pearson circa 1900. This test is frequently used in biology courses, particularly in Mendelian genetics. In the first part of this section we will explain the ideas behind the goodness of fit test in an intuitive way. The reader can go directly to Section 10.1.2, if so preferred, where we explain how to apply the goodness of fit test to real data.

10.1.1   Understanding the goodness of fit test *

A model

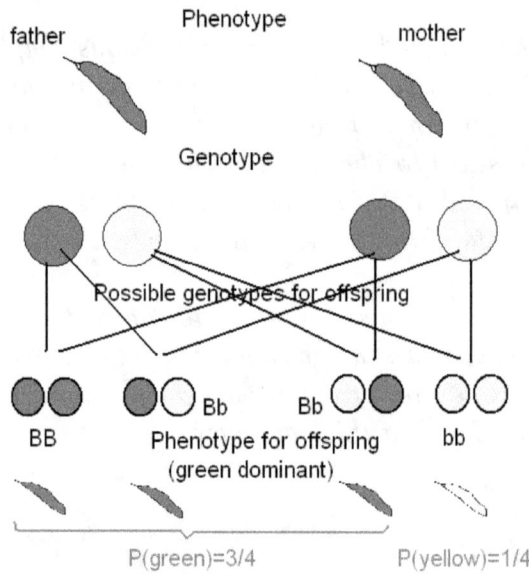

Figure 10.1: Probabilities for the phenotype of the offspring when parents are heterozygous (only two generations are considered)

The phenotype (the way the individual looks with respect to a given characteristic) depends on the genotype of the individual (which alleles the individual has for the gene associated with the characteristic), and which allele is dominant. A parent has two alleles for a particular gene; the one given to the offspring is determined at random, like flipping a coin. Although the genotype of the offspring is randomly determined, in the case of heterozygous offspring the phenotype depends on which allele is dominant. In the example of the pea pod color, the alleles are green and yellow (with green being dominant). In the case of heterozygous parents (parents who have the two alleles: green and yellow), the options for genotype and phenotype, with their corresponding probabilities, are displayed in Figure 10.1. The model says that, in the case of heterozygous parents, P(dominant trait)=3/4 and P(recessive trait)=1/4. Thus, if there are n individuals, offspring of heterozygous parents, the expected frequencies are $3n/4$ individuals with the dominant trait and $n/4$ with the recessive trait.

In simulations done with a model that has a random component, outcomes are NOT necessarily equal to the expected values.

Experimental variability

Even when a model is valid, the observed frequencies do not always coincide exactly with the expected frequencies under the model. A simulation will be used to illustrate this fact. Using a coin or a two-color plastic chip, generate the color of 12 hypothetical pea-pods. According to the model used to generate the data, the expected frequencies are 9 green and 3 yellow pea pods. How many of each color are obtained in the simulation? It should be emphasized that we are NOT yet comparing real data with the predictions done by a model. The intention of this exercise is just to observe that even when the model is valid, variability in experimental results can happen by chance.

Hands-on activity: simulating the genotype of the offspring using coins

Take a couple of fair coins (quarters) or two 2-color plastic chips and mark one as 'father' and one as 'mother.' Toss the coins 12 times and keep track of the results, with 'heads' being the green and 'tails' being the yellow alleles. This generates the genotype of the hypothetical offspring. Assuming green is dominant, write the phenotype (green or yellow) for each one of the 12 hypothetical offspring. Use the table in Figure 10.2 to record the results. How many green and how many yellow pea pods were generated?

The result of our own simulation is shown in Figure 10.3. Later, the experiment of generating 12 simulated offspring was repeated 100 times. The number of green pods in each one of the 100 simulations are listed below. The values are tabulated in Table 10.1.

9, 11, 10, 6, 9, 7, 7, 10, 8, 10, 10, 7, 9, 11, 8, 12, 11, 9, 9, 10, 7, 10, 8, 6, 10, 9, 8, 10, 8, 10, 11, 8, 6, 11, 8, 9, 9, 10, 10, 5, 10, 10, 11, 11, 10, 7, 10, 9, 11, 8, 9, 11, 9, 10, 11, 9, 11, 10, 8, 9, 10, 8, 9, 7, 8, 9, 7, 7, 10, 10, 9, 8, 8, 8, 10, 8, 11, 10, 9, 11, 7, 10, 12, 9, 9, 9, 10, 8, 8, 10, 8, 9, 12, 10, 10, 7, 5, 9, 12, 7

Notice that when the data are generated with the coin model, the observed frequencies are NOT necessarily identical to the expected frequen-

Toss #	Father	Mother	From coins to colors	Genotype of offspring	Phenotype of offspring
1	H	T	●○	Bb	╲
2	H	T	●○	Bb	╲
3	T	H	○●	bB	╲
4	H	T	●○	Bb	╲
5			○○		╲
6			○○		╲
7			○○		╲
8			○○		╲
9			○○		╲
10			○○		╲
11			○○		╲
12			○○		╲

Figure 10.2: Complete your simulation with the coin model

Table 10.1: Number of green pea pods out of 12 in 100 simulations

green pea pods	5	6	7	8	9	10	11	12
frequency	2	3	11	18	22	27	13	4

cies; simulated data show variability. Thus, it sounds reasonable that, when working with real data, having observed frequencies different from expected frequencies does not imply necessarily that the model should be rejected.

How far are the observed from the expected frequencies?

In the simulation shown in Figure 10.3, 10 pea pods were green and 2 pea pods were yellow. How far are those counts from the expected counts of 9 and 3? It would be reasonable to compare the frequencies (observed and expected) for each color. In order to prevent the negative and positive differences from canceling each other, we square the differences: $(10-9)^2 + (2-3)^2$. However, a difference of 1 between the expected and the observed value might not look the same in all circumstances. If the expected value is 9, as in the example, a difference of 1 does not look as small as if the expected value was 239 and the observed value was 238. Thus, it would make sense to

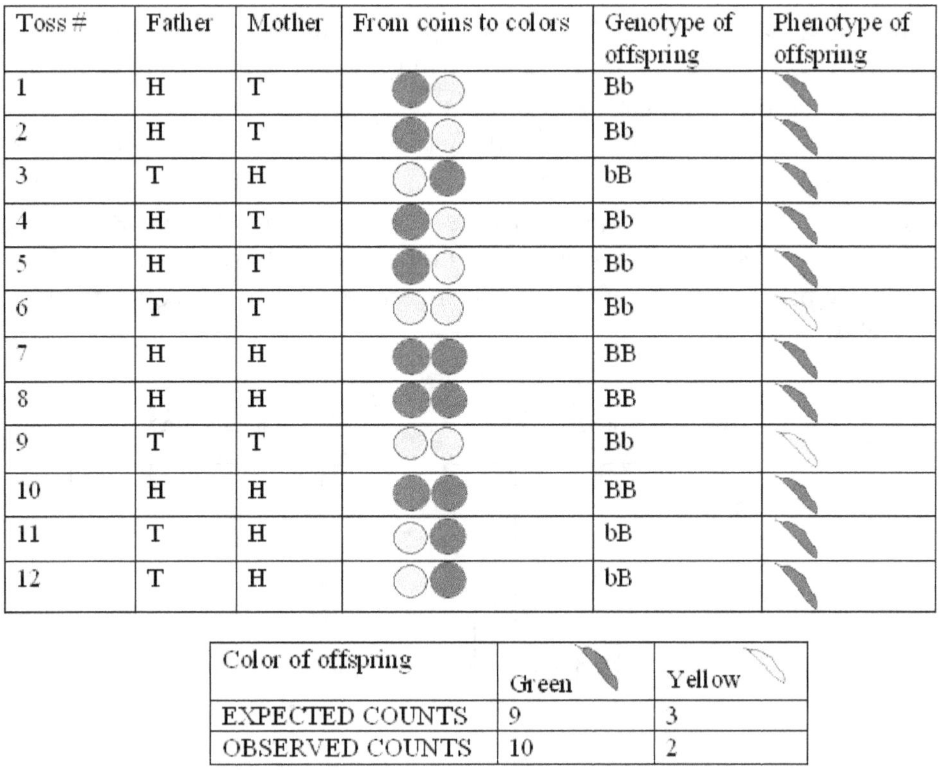

Toss #	Father	Mother	From coins to colors	Genotype of offspring	Phenotype of offspring
1	H	T		Bb	
2	H	T		Bb	
3	T	H		bB	
4	H	T		Bb	
5	H	T		Bb	
6	T	T		Bb	
7	H	H		BB	
8	H	H		BB	
9	T	T		Bb	
10	H	H		BB	
11	T	H		bB	
12	T	H		bB	

Color of offspring		Green	Yellow
EXPECTED COUNTS		9	3
OBSERVED COUNTS		10	2

Figure 10.3: Example of one simulation with the coin model

standardize the differences by comparing them with the expected values: $\frac{(10-9)^2}{9} + \frac{(2-3)^2}{3} = \frac{4}{9}$. Considering that in other examples there could be k classes instead of 2 (green and yellow), the general expression for the statistic that compares observed and expected frequencies is:

$$\sum_{i=1}^{k} \frac{(O_i - E_i)^2}{E_i} \qquad (10.1)$$

Comparing observed with expected counts

Distribution of the statistic

How does the value of the statistic defined in equation 10.1 vary just by chance when the model is valid? The reason for this question is that, as discussed in previous chapters, **the p-value is the probability of getting the value of the statistic actually obtained, or a more extreme one, when the null statistical hypothesis is true.** We need to know the distribution of the statistic, when the null hypothesis is true, in order to calculate the p-value. In the case of pea pods, since the only possible outcomes are *green* (with probability 3/4) and *yellow* (with probability 1/4), the binomial distribution can be of help. The probability for each possible number x of green pea pods, out of a total of n pea pods, is $P(x \ green \ peapods) = \binom{n}{x}0.75^x(0.25)^{(n-x)}$. Those probabilities are displayed in Figures 10.4a and c for two different numbers of pea pods, $n=12$ and $n = 120$. Both distributions look centered at the expected value (9 and 90 respectively). When the model is valid, the number of green and yellow pea pods are NOT necessarily equal to the expected values; there is variability.

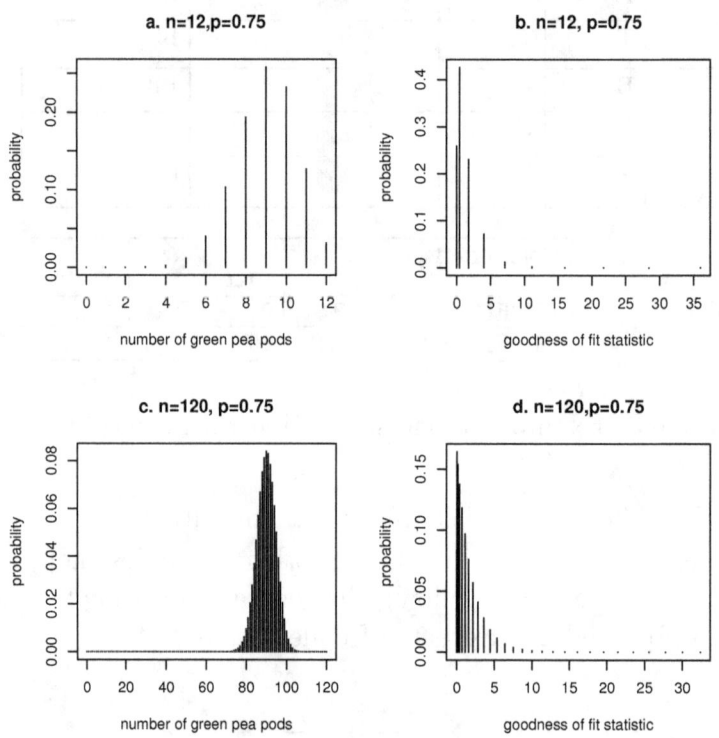

Figure 10.4: Probability distribution of the number of green peas and the goodness of fit statistic $\chi = \sum_{i=1}^{2} \frac{(O_i - E_i)^2}{E_i}$

We used the probability distributions in Figures 10.4 a) and c) to calculate the probability of each value that the statistic in equation 10.1 can take. That is 'the sampling distribution of the statistic' defined in equation 10.1, when the model (P(green)=3/4, P(yellow)=1/4) is valid. Figures 10.4 b) and d) display the sampling distribution of the statistic in (10.1) when $n = 12$ and $n = 120$ for this particular example. The distributions in Figures 10.4 b) and

c) are skewed with a long right tail and a high concentration of values near 0. Notice that the discrete distributions in Figures 10.4 b) and d) suggest the shape of the continuous χ_1^2 distribution in Figure 9.11. The Chi-square distribution with k-1 (k-1=1 in the pea pods example) is the approximated distribution assumed for the statistic defined in equation 10.1. By comparing Figures 10.4 b) and d) with Figure 9.11, it is evident that approximating the distribution of the statistic in (10.1) by the chi-square distribution might work well when the sample size n is large, but not when it is small. The statistic defined in equation (10.1) is called the **chi-square statistic** χ^2.

Getting data

In the previous subsections, data has been generated using the model in order to get some insight into experimental variability and the distribution of the goodness of fit statistic when the proposed model is valid. This was done just to understand why the Chi-square distribution is used to find the p-values and the reason for preferring large samples. However, in order to perform a test to decide if the assumed model should be rejected, it is necessary to conduct a survey or an experiment to obtain REAL data. The observed frequencies in the data are contrasted with the expected frequencies produced by the model. We will use data from Mendel (1865). Section 10.2 (optional) describes Mendel's experiments in more detail.

Real data is needed to contrast observed counts with the frequencies predicted by the model

10.1.2 Applying the goodness of fit test

The section 'Understanding the goodness of fit test' was included to intuitively explain the rationale behind the test statistic and its approximated distribution. This section focuses on the mechanics of the application of the 'Chi-square goodness of fit test' using data.

Consider a categorical, or categorized, variable with k classes or categories $A_1, A_2, ..., A_k$ and a model (based on a theory or hypothesis) that assigns theoretical probabilities $p_1, p_2, ..., p_k$ to those categories. In the case of quantitative variables, consider that the variable can take k values or intervals of values, with probabilities $p_1, p_2, ..., p_k$. The null statistical hypothesis is that the model is valid and it can be written as:

H_o: The probabilities for classes $A_1, A_2, ..., A_k$ are $p_1, p_2, ..., p_k$

In order to test H_o, it is necessary to:

- calculate the expected frequencies under the assumed model.

- collect data through experimentation or observation.

- compare the observed frequencies or counts with the expected frequencies using the statistic defined in equation 10.1.

- use the chi-square distribution with $k - 1$ degrees of freedom to calculate the p-value (i.e., the probability that such a discrepancy between observed and expected counts, or a more extreme one, can happen by chance alone when the model is valid).

Steps in the goodness of fit test

- reject the model if the p-value is very small.

Example 10.1: Pea pod color

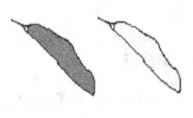

For a trait associated with a single gene with two alleles, one of which is dominant, consider the hypothesis that if the parents are heterozygous the probability of the offspring getting the dominant phenotype is 0.75. The model can be tested for the case of pea pod color, assuming the two alleles are green and yellow, with green being dominant.

$$H_o: \qquad \text{P(green)}=0.75 \qquad \text{P(yellow)}=0.25$$

A sample of 580 pea pods is collected. What are the expected number of green and yellow pods? The expected values are $580 \times 0.75 = 435$ and $580 \times 0.25 = 145$. Mendel got 428 green pods and 152 yellow pods in his experiment. Are Mendel's data evidence against the proposed model?

Goodness of fit test example

counts	green	yellow
expected	435	145
observed	428	152

The chi-square statistic defined in (10.1) is calculated with the observed data:

$$\sum_{i=1}^{k} \frac{(O_i - E_i)^2}{E_i} = \frac{(428 - 435)^2}{435} + \frac{(152 - 145)^2}{145} = 0.4505747 \qquad (10.2)$$

Since in the example there are only two ($k = 2$) categories (green and yellow) the chi-square distribution with $k - 1 = 1$ degree of freedom is used to find the p-value. The p-value is the area to the right of the value (0.4505747) obtained for the statistic in (10.2). The p-value or area can be calculated using software. In **R**, type: **pchisq(0.4505747,1)**; the output is 0.4979378. However, the output given by the software is the area to the left of 0.4505747.

Thus, the p-value is $1 - 0.4979378 = 0.5020622$. Figure 10.5a displays the Chi-square (with one degree of freedom) density function. The p-value is the area under the curve to the right of the value of the statistic (marked with an arrow). The p-value is a probability, therefore it can only take values in the interval $(0,1)$. In the p-value world, 0.5020622 is considered a large p-value. Mendel's data with respect to pea pod color DOES NOT contradict the hypothesis that says that, if parents are heterozygous, the probability that the offspring gets the dominant phenotype is 0.75.

Note: This is no surprise because Mendel used the data from his experiments to establish the ratio 3:1 for the dominant vs. the recessive trait.

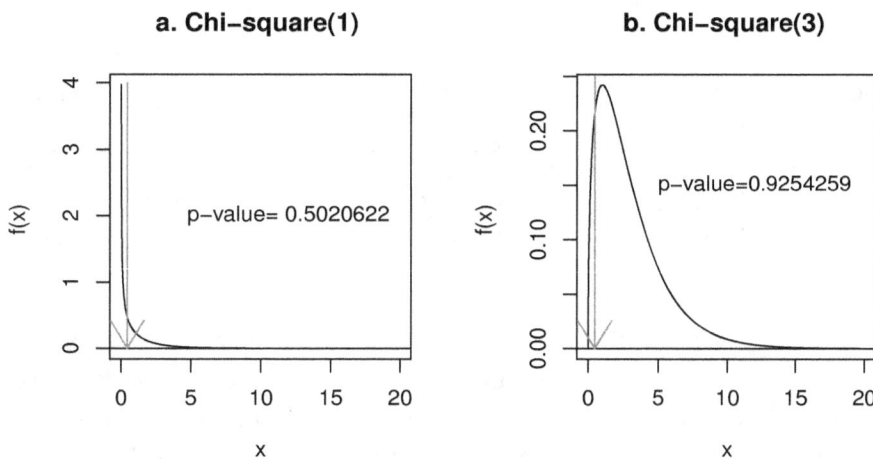

Figure 10.5: Calculation of p-values for Examples 1 and 2

R & MINITAB: Chi-square goodness of fit test

Using statistical software:
The chi-square goodness of fit test can be performed with a simple command. In R, it is enough to enter the observed frequencies for the categories, the probabilities indicated by the model, and to apply the test with the command **chisq.test***:*

```
x <- c(428,152)  ;   p <- c(0.75,0.25)
chisq.test(x,p=p)
```

The output is:

```
Chi-squared test for given probabilities
data:  x     X-squared = 0.4506, df = 1, p-value = 0.5021
```

In MINITAB, the data must be written in two columns, one column for the observed frequencies and the other for the probabilities specified by the model. From the menu, use Stat> Tables> Chi-square goodness of fit test, enter the name of the column where the observed counts are, choose the option specific proportions, and enter the name of the column where those proportions are located.

Counts, NOT percentages, are compared

It is important to notice that in the chi-square goodness of fit test, COUNTS or FREQUENCIES (expected and observed) are compared, NEVER the proportions or percentages. The sample size affects the strength of the evidence. To better understand this, consider Mendel's data and two hypothetical cases in which the observed percentages of green and yellow pods are the same (73.8% and 26.2% respectively), but the sample sizes are different:

Case	n	expected counts green	yellow	observed counts green	yellow	χ^2	p-value
1	136	102	34	100	36	0.15686	0.69206
2	580	435	145	428	152	0.45057	0.50206
3	5800	4350	1450	4280	1520	4.50575	0.03378

When the sample size is very large, small differences become significant

Even when the observed proportions of green and yellow pea pods are the same in the three cases, evidence against the null hypothesis looks stronger (p-value goes down) as the sample size increases. Even small discrepancies become statistically significant when the sample size is VERY LARGE.

Example 10.2: Color and shape of pea seeds

The 9:3:3:1 ratio

Consider two characteristics of seeds: form and color. The form can be round with an smooth surface (S, dominant) or angular (wrinkled) (s). The color of the seed can be yellow (Y, dominant) and green (y).

If the form and the color of pea seeds are independent and the parents are heterozygous, the probability for each one of the combinations of form and color for the offspring can be calculated using the probability tree in Figure 10.6. The probabilities are 9/16 for the combination of the two dominant traits, 3/16 for each of the two combinations of the dominant allele in one gene with the recessive allele in the other gene, and 1/16 for the combination of the two recessive traits. In biology this is known as the 9:3:3:1 ratio. Mendel actually conducted an experiment involving seed form and color, in which he obtained 556 seeds. The expected frequencies, or counts, for the 4

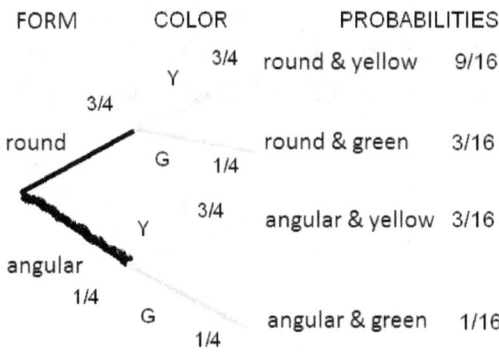

Figure 10.6: Probability tree for form and color of pea seeds

combinations of form and color, according to the model, are:

$$556 \times (9/16) = 312.75$$
$$556 \times (3/16) = 104.25$$
$$556 \times (3/16) = 104.25$$
$$556 \times (1/16) = 34.75$$

It is not a problem to have expected counts that are not integers because the notion of 'expected' value is related to the idea of 'long term average'. Those expected counts, together with the actual frequencies obtained by Mendel in his experiment, are displayed below.

Expected counts do not necessarily take integer values

counts	round& yellow	round& green	angular& yellow	angular& green
expected	312.75	104.25	104.25	34.75
observed	315	108	101	32

The value of the Chi-square statistic is :

$$\chi^2 = \frac{(315-312.75)^2}{312.75} + \frac{(108-104.25)^2}{104.25} + \frac{(101-104.25)^2}{104.25} + \frac{(32-34.75)^2}{34.75} = 0.470024.$$

In this example there are 4 categories. Thus, the Chi-square distribution with 3 degrees of freedom is used to calculate the p-value (see Figure 10.5b). To calculate the area under the curve to the right of 0.470024, type (in **R**):

1-pchisq(0.470024,3)

The output is 0.9254259, a very large p-value indeed, indicating that Mendel's data does not contradict the hypothesis of the 9:3:3:1 ratio. Again, no surprise, since Mendel established the 9:3:3:1 ratio based on his experimental data.

Statistical software can be used to perform the test with a single command. In R, using the names 'obs' for the observations and 'prob' for the probabilities given by the model, type :

R:
Goodness
of fit test

```
obs <- c(315,108,101,32)  ## read observed frequencies
prob <- c(9/16,3/16,3/16,1/16) ## model probabilities
chisq.test(obs,p=prob)
```

The output is: X-squared = 0.47, df = 3, p-value = 0.9254

10.2   Mendel's experiments *

Gregor Mendel (1822-1884) grew up on a farm in a region that is now in the Czech Republic. He became a priest, studied at the University of Vienna, taught physics and performed many experiments in the garden of his monastery. Based on his results he was able to deduce important genetic processes. Between 1853 and 1863 Mendel experimented with garden peas (*Pisum savitum*), a species of the family Leguminosae. He published his results in 1865 but his work was ignored until the 1900's when other scientists arrived at similar conclusions and became aware of his seminal work. An English translation, by Druery & Bates, of Mendel's 'Experiments in Plant Hybridization' is available at the website *http://www.mendelweb.org/Mendel.plain.html*.

Mendel focused on the following characteristics of the pea plant:

**Mendel's
experiments
with peas**

1. form of ripe seeds: round or angular (wrinkled)

2. color of the seed albumen : yellow or green

3. seed coat: non-white (grey-brown) or white

4. form of ripe pods: inflated or constricted

5. color of unripe pods: green or yellow

6. position of flowers: axial or terminal

7. length of stem: tall or short

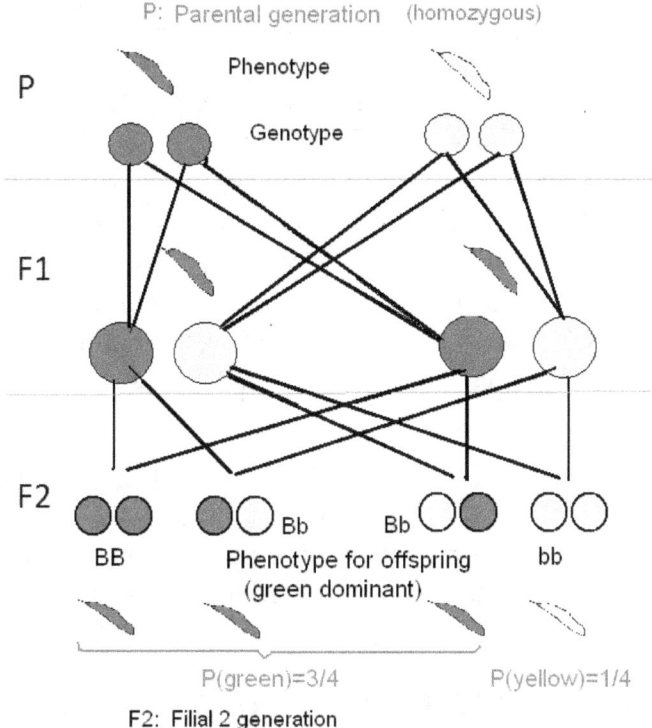

Figure 10.7: Three generations (P,F1,F2) involved in an experiment

The cells of pea plants have seven pairs of chromosomes in their nuclei. Each of the genes associated with the seven characteristics listed above are located on a different chromosome, and those seven characteristics behave independently one from the other. Table 10.2 summarizes the results from Mendel's experiments.

In Figure 10.1, the male and the female are considered to be heterozygous, but where did those heterozygous plants come from? Experiments, such as those that Mendel performed, start with homozygous plants (also called 'true breeders'). A **BB** crossed with a **bb** yields only heterozygous **Bb** offspring. Those heterozygous plants will be the parents of the following generation. One generation is added at the beginning of Figure 10.1 to obtain Figure 10.7. The heterozygous plants, obtained from the crossing of the homozygous plants, are crossed amongst themselves producing a third generation in which some are bb, some BB, and the rest Bb. In the genetic literature, the first generation is called the parental generation (P). The next cross is the F1 (first filial) generation in which all are heterozygous. The offspring of the F1

Table 10.2: Mendel's experiments with *Pisum savitum* (peas)

Charac-	n	Dominant phenotype			Recessive phenotype		
teristic			Count	%		Count	%
Form (seed)	7324	round	5474	74.7	angular	1850	25.3
Color (seed)	8023	yellow	6022	75.1	green	2001	24.9
Coat(seed)	929	non-white	705	75.9	white	224	24.1
Form (pod)	1181	inflated	882	74.7	constricted	229	25.3
Color (pod)	580	green	428	73.8	yellow	152	26.2
Flower	858	axial	651	75.9	terminal	207	24.1
Stem	1064	long	787	74.0	short	277	26.0

generation constitute the F2 (second filial) generation. In Figure 10.1, only the two filial generations are considered, but Figure 10.7 includes the three generations. Based on the outcomes of his experiments, Mendel arrived at the conclusions now known in biology as Mendel's laws.

10.3   Chi-square test of independence

Assume that there are two characteristics A and B studied for the same population. The null statistical hypothesis is:

$$H_o: A \text{ and } B \text{ are independent}$$

Example 10.3 Independence of color and form

Table 10.3: Mendel's data on form and color of pea seeds

form	color		total
	yellow	green	
round	315	108	423
angular	101	32	133
total	416	140	556

Peas can be round with a smooth surface or angular (wrinkled), and they can be yellow or green. Is being angular or round independent from color? Are yellow peas more (or less) prone to be wrinkled than green peas? Gregor Mendel conducted extensive experiments with peas; some of those experiments involved color and form (see Table 10.3). In Example 10.2 it was assumed that color (yellow or green) and form (round or angular) are independent characteristics of the pea seeds. How can this claim be checked?

The Chi-square test of independence compares the observed frequencies with the frequencies expected under independence. The way to calculate the expected counts under the assumption of independence is based on the definition of independent events studied in Chapters 4 and 6. Events A and B are said to be independent if $P(A \text{ and } B) = P(A) \times P(B)$ (*Note*: $P(A \text{ and } B)$ can be written as $P(A \cap B)$). When color and form are independent:

$$P(yellow \text{ and } round) = P(yellow) \times P(round).$$

Consider the data from Mendel's experiment involving color and form in Table 10.3. Notice that before preparing the two-way table, only the total (556 in the example) is known; the row and column totals are not pre-fixed. If a pea seed is selected at random from the set of 556 described in Table 10.3, the probability of being round is 423/556, and the probability of being yellow is 416/556. If being yellow is independent of being round, P(round and yellow)=P(round)$\times$P(yellow)=$\frac{423}{556} \times \frac{416}{556}$. There is a total of 556 pea seeds. Thus, if being yellow is independent of being round, the expected number of round yellow pea seeds, under the assumption of independence, is $556 \times \frac{423}{556} \times \frac{416}{556} = \frac{423 \times 416}{556} = 316.4892$. The value 423 is the total for the row 'round' in Table 10.3, 426 is the total for the column 'yellow' and 556 is the total number of peas. Under the hypothesis of independence, the expected value for a given cell of the two-way table is calculated as:

Calculation of expected values in the test of independence

expected value= (total of the row * total of the column)/total

The same rule is written using mathematical notation in (10.4).

After calculating the expected values for each one of the cells, the expected and observed counts are compared using the same Chi-square statistic employed in the goodness of fit test. However, to write expression (10.3), a double subindex is used for E and O to account for the fact that the counts correspond to cells that have a location with respect to columns AND rows. In expressions (10.3) and (10.4), i represents the row number, and j represents the column number. There are r rows and c columns.

Test statistic

$$\sum_{i=1}^{r} \sum_{j=1}^{c} \frac{(O_{ij} - E_{ij})^2}{E_{ij}} \tag{10.3}$$

Expected values in test of independence

$$E_{ij} = \frac{n_{i.} \times n_{.j}}{n} \tag{10.4}$$

The number of degrees of freedom for the statistic in expression (10.3) is $(r-1) \times (c-1)$, where r and c represent the number of rows and columns respectively. Table 10.4 is the generic version of a 2×2 two-way table using symbols instead of numbers for the counts. The symbol n represents the

Table 10.4: Generic 2×2 two-way table

Variable 1	Variable B		total
	Category B_1	Category B_2	
Category A_1	n_{11}	n_{12}	$n_{1.}$
Category A_2	n_{21}	n_{22}	$n_{2.}$
total	$n_{.1}$	$n_{.2}$	n

total number of observations, $n_{i.}$ represents the total of the ith row, and $n_{.j}$ represents the total of the jth column. The frequencies or counts in the individual cells are represented with the symbol n_{ij}, indicating that the cell is in the ith row and the jth column. A dot is commonly used in mathematics to indicate *sum* with respect to a subindex; for instance, $n_{i.}$ represents the sum of all the cells that are in the ith row. For Example 10.3, the number of degrees of freedom is $(2-1) \times (2-1) = 1$. Tables 10.3 and 10.4 are 'two-way' tables because they display data for two variables. Two-way tables can have more rows and columns. Tables such as 10.3 and 10.4 are also called 2×2 tables because they only have two rows and two columns. The expected values, calculated with the formula in (10.4), appear (in italics and inside parentheses) next to the observed counts in the table below:

Test of independence example

form	color				total
	yellow		green		
round	315	*(316.49)*	108	*(99.51)*	423
angular	101	*(106.51)*	32	*(33.41)*	133
total	416		140		556

Once the expected values are obtained, the test statistic is calculated using equation (10.3):

$$\sum_{i=1}^{r} \sum_{j=1}^{c} \frac{(O_{ij}-E_{ij})^2}{E_{ij}} =$$

$$\frac{(315-316.49)^2}{316.49} + \frac{(108-106.51)^2}{106.51} + \frac{(101-99.5)^2}{99.5} + \frac{(32-33.41)^2}{33.41} = 0.1163.$$

The area to the right of the value 0.1163, under the density function of the χ_1^2 distribution, is 0.733. The p-value is large, thus the hypothesis of independence is not rejected.

R & MINITAB:
Test of
independence

Using statistical software:
The expected values, the value of the test statistic, and the p-value can be
obtained all at once. For the data in table 10.3, type:

```
x<-matrix(c(315,101,108,32),nc=2)
chisq.test(x, correct=FALSE)
```

The output is

```
        Pearson's Chi-squared test
data:  x
X-squared = 0.1163, df = 1, p-value = 0.733
```

The option correct = FALSE indicates that we are not applying the conti-
nuity correction to improve the approximation of the Chi-square distribution.
The correction consists of adding 0.5 to each difference between the observed
and the expected counts. The option 'nc=2' indicates that the table has two
columns.

In MINITAB, the test for independence is performed by selecting STAT >
Tables, and then 'Cross tabulation and Chi-square test' if using raw data
(observations for each individual). If the data are already tabulated, type in
the frequencies and use 'Chi-square test.'

Not all the two-way tables are 2×2; each one of the variables can have
more than two categories. However, the test statistic (10.3) is the same. The
only thing that varies with the number of rows and columns is the number
of degrees of freedom of the Chi-square distribution used to calculate the
p-value, df=(r-1)(c-1).

When the null hypothesis of independence is rejected, it is important to
remember that association or lack of independence does not necessarily imply
a cause-effect relationship, both variables might be associated with a third
one.

Lack of independence does NOT imply a cause-effect relationship

10.4   Test of homogeneity

Assume that there are two or more groups or populations to be compared
in terms of a categorical variable that has k categories or classes. Using the
symbol p_i for the proportion of individuals in the ith category, $i = 1, 2, ..., k$,
the null statistical hypothesis is:

H_o: the proportions p_i are similar in all the groups

The null hypothesis says that there is no difference among populations with regard to the behavior of the categorical variable. The variable does not need to be intrinsically categorical, we can also apply the homogeneity test when the variable of interest is quantitative but it has been categorized in the sense that we work with intervals of values.

The test of homogeneity compares two or more groups in terms of a categorical or categorized variable

Example 10.4 Vegetation & habitat

Does a certain type of bird prefer to live in places with more understory vegetation? Or is there no difference with regard to understory vegetation cover between randomly selected places and places known to be habitat for this bird? (Understory vegetation is the area of the forest that grows in the shade of the forest canopy). A researcher collected information about understory vegetation cover, both in places known to be habitat for this species and some random places. The data are in Table 10.5 (Clark-Schubert, 2009). What test can we apply to answer the research question? From the computational

Table 10.5: 561 sites classified by type and % of understory vegetation cover

Site	0-20%	20-40%	40-60%	60-80%	80-100%	Total
Random	77	49	45	39	45	255
	30.20%	19.22%	17.65%	15.29%	17.65%	100%
Habitat	36	43	101	75	51	306
	11.76%	14.05%	33.01%	24.51%	16.67%	100.00%

point of view, the test of homogeneity is similar to the test of independence. Expressions (10.4) and (10.3) are used to calculate the expected values and the test statistic, respectively. However, the context is a little different. The test of homogeneity is used to compare two or more groups with regard to the distribution of a categorical (or categorized) variable. The test of homogeneity can be applied to data from experiments to compare treatment vs. placebo, like in the polio experiment described in Table 2.12. Before conducting the experiment, the number of individuals in each group is known.

R:
Test of homogeneity

Using statistical software:
The test of homogeneity and the test of independence are performed using the same commands. For the data in Table 10.5, type:

```
x <- matrix(c(77,36,49,43,45,101,39,75,45,51), nc=5)
chisq.test(x, correct=FALSE)
```

The output is:

```
        Pearson's Chi-squared test
data:  x , X-squared = 44.2194, df = 4, p-value = 5.777e-09
```

In order to check that the data were entered correctly, and to calculate the expected values, one can type:

```
chisq.test(x)$observed ; chisq.text(x)$expected
```

The output is:

```
> chisq.test(x)$observed
     [,1] [,2] [,3] [,4] [,5]
[1,]   77   49   45   39   45
[2,]   36   43  101   75   51
> chisq.test(x)$expected
         [,1]     [,2]     [,3]     [,4]     [,5]
[1,] 51.36364 41.81818 66.36364 51.81818 43.63636
[2,] 61.63636 50.18182 79.63636 62.18182 52.36364
```

The p-value is very small, thus the null hypothesis that the distribution for the variable *%cover* is the same for both types of places, random and habitat, is rejected. The bar-charts in Figure 2.35 clearly show that places that are habitat for this species of birds tend to have a larger % of understory cover.

10.5 Test of normality

Given a random variable X, the null hypothesis in a test of normality is:

$$H_o: X \text{ has a normal distribution}$$

Although randomization methods are becoming popular, statistical inference methods that were developed under the assumption that the variable has a normal distribution are still the most extensively used in applications. Some of those methods will be presented in Chapter 11. Developing methods to test for normality is a research area that has received considerable attention. A wide variety of tests for normality exist, new tests are always being developed.

The histogram in Figure 10.8 was prepared with a large data set and seems to suggest that the variable has a normal distribution. When the sample size is small, the histogram might not clearly suggest a normal distribution even if the data actually come from a normal distribution. The histogram in Figure 10.9a is not totally convincing with regard to the normality of the variable, even when those 20 values were artificially generated using a normal model. When the sample size is small, it is quite possible not to 'see' the normal shape in the frequency distribution, even if the values come from an actual normal distribution; that is why normality tests are used. First we will give a brief overview of different types of tests of normality. The reader can go directly, if so preferred, to Section 10.5.2 where the Shapiro-Wilk test of normality is explained.

In a normality test, the null hypothesis is that the variable has a normal distribution in the population from where the sample was selected

Example 10.5 Does pulse rate have a normal distribution?

Can the variable 'pulse rate' for college students be considered to have a normal distribution? Figure 10.8 displays the number of heart beats per minute for a sample of 210 college students. Does this data set present evidence against the assumption of normality? What test can be applied in order to answer this question?

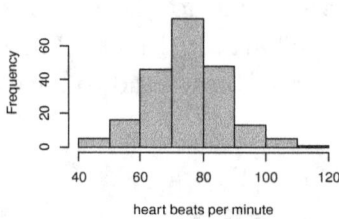

Figure 10.8: Number of heart beats per minute for a sample of 210 students

10.5.1   Different types of tests of normality *

There are several types of tests for normality, and each statistical software might include more than one test. One of the first tests for normality to be applied was the Chi-square goodness of fit test studied in Section 10.1. For that test, the range of possible values for the variable is divided in intervals and the number of observations that fall in each interval is counted. The observed frequencies are compared with the frequencies expected under normality using the statistic in expression (10.1). However, many other tests for normality exist nowadays, and some of them are more 'powerful' than the Chi-square test in the sense that they have a higher probability of finding out if the data does not come from a normal distribution. Some tests are more powerful against a specific type of non-normality (skewness or high kurtosis). Researchers who define new tests of normality usually focus on one or more characteristics of the normal distribution (studied in Section 9.2) and develop ways to see if the data present those characteristics. The null hypothesis is that the variable has a normal distribution in the population from where the sample was selected. Three main types of test of normality are:

- **Skewness and Kurtosis tests.** These tests compare the values of skewness and kurtosis statistics for the sample with the theoretical values for the skewness and kurtosis of the normal distribution. The notions of skewness and kurtosis were discussed in Chapter 2. The normal distribution is symmetric and has skewness 0. These tests, first summarize the data (with regard to skewness and kurtosis) and then compare those values with the skewness and kurtosis values for the normal distribution.

- **Empirical distribution function tests.** These tests compare the cumulative frequencies (studied in Chapter 2) with the cumulative distribution function (CDF) of the normal distribution studied in Chapter 9 (Figure 10.9b). The two best known tests of this type are the Anderson-Darling and the Kolmogorov-Smirnov tests. The Kolmogorov-Smirnov test focuses on the maximum difference between the empirical and the theoretical cumulative distribution functions. The Anderson-Darling test looks at the average difference between the empirical and theoretical CDFs. The Anderson-Darling test tends to be more powerful than the Kolmogorov-Smirnov. These two tests are performed by many statistical software.

- **Regression type tests.** In some way, the values of the observations are compared with the values they would have had if the distribution was normal. The Shapiro-Wilk test belongs to this type of normality tests and will be explained in the next section. Generally, regression-type tests are accompanied by a normal quantile plot or a normal probability plot to visualize the comparison. These plots are scatterplots that compare the ordered observations with the quantiles of a normal distribution; dots that suggest a straight line indicate normality.

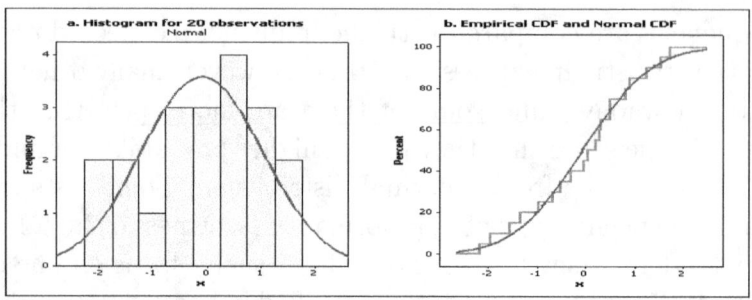

Figure 10.9: Comparing a frequency distribution with the normal distribution

10.5.2   The Shapiro-Wilk test

The Shapiro-Wilk test is quite powerful against a wide variety of non-normal distributions. It is a very popular test, and it is available in statistical software.

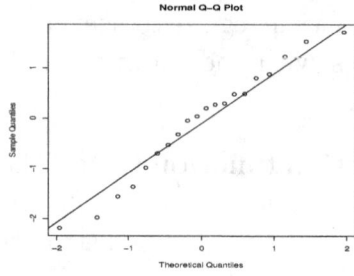

Figure 10.10: Normal quantile plot for 20 values generated by a normal model

quantile-quantile plot to explore if the data seem to come from a normal distribution

The special type of scatterplot in Figure 10.10, is called 'normal quantile-quantile' plot and is used to have a visual representation of the test for normality. The 20 observations in the histogram of Figure 10.9a are plotted against 20 quantiles of the normal distribution (the 20 values that would slice the normal distribution in 21 parts of equal area). As mentioned in Chapter 9, between any variable X with normal distribution, with mean μ and standard deviation σ, and the standard normal variable Z, there is a linear relationship $X = \sigma Z + \mu$. Thus, if the observations came from a normal distribution it would be expected to see points close to the straight line.

The normal quantile-quantile plot can be easily obtained by typing the command **qqnorm** in **R** while the command **qqline** adds the straight line to the plot. The closer the dots are to the straight line, the less likely that the hypothesis of normality will be rejected.

R & MINITAB:
quantile-quantile
plots

Using statistical software:
To obtain the plot in Figure 10.10, the following commands were used:

```
x<-c(-2.18751,-1.97832,-1.56082,-1.36684,-0.98292,-0.69815,-0.52746
,-0.31729,-0.04333, 0.03984, 0.20021, 0.27260, 0.29497, 0.48123
, 0.48703, 0.80128, 0.87959, 1.22938, 1.52928, 1.71519)
qqnorm(x) ; qqline(x)
```

A similar plot can be obtained with MINITAB by selecting the option GRAPH
> Probability plot.

The Shapiro-Wilk test statistic, as originally defined in 1965, is:

$$W = \frac{(\sum_i^n a_i x_{(i)})^2}{\sum_i^n (x_i - \bar{x})^2} \tag{10.5}$$

where $x_{(i)}$ are the ordered observations or 'order statistics,' and the a_i are numbers calculated based on the expected values of the order statistics under normality. The original paper provided the values a_i and the critical values for the test statistic when the sample size was at most 50 (Shapiro and Wilk, 1965). For each value of n, a different set of $a_1, ..., a_n$ constants and a critical **Shapiro-Wilk** value for the statistic were needed. Other authors have worked on extensions **test of normality** of the Shapiro-Wilk test for larger samples and on approximations that do not require special tables. Royston's version (1992) of the Shapiro-Wilk test is implemented in **R**. Since we already typed the data for the quantile-quantile plot, now it is enough to type **shapiro.test(x)**. For the 20 observations in Figures 10.9a and 10.10, the output is:

```
        Shapiro-Wilk normality test
data:  x
W = 0.9687, p-value = 0.7277
```

The null hypothesis 'H_o: **X** has a normal distribution,' is NOT rejected because the p-value is large.

Continuing with Example 10.5

The histogram in Figures 10.8 and 10.11a displays the values of the pulse rate of 210 students. Assuming that the 210 students are a random sample from a large population of college students, the null statistical hypothesis says that pulse rate in that large population has a normal distribution. Strictly speaking, the normal distribution is continuous, and pulse rate is discrete (number of heart beats per minute); however, the range of the variable pulse rate is quite large. Figure 10.11b displays the quantile-quantile plot for the pulse rate data.

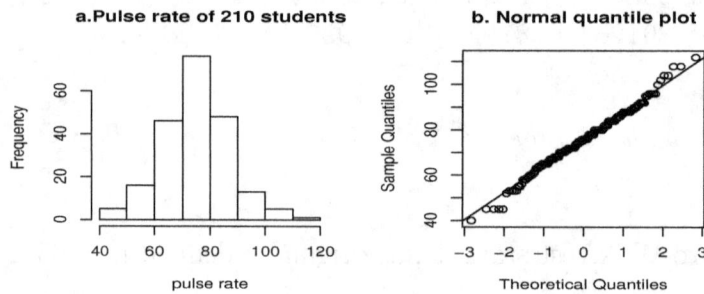

Figure 10.11: Normal quantile plot for pulse rate of 210 students

The output of the Shapiro-Wilk test is:

```
        Shapiro-Wilk normality test
data:  x  ,   W = 0.9911, p-value = 0.223
```

The hypothesis of normality is not rejected because the p-value is NOT small.

Using statistical software:
The pulse rate data set is in the data file pulserate.dat located, for us, in the folder 'DATdatafiles' in drive e:. You will need to indicate in the scan command the location of the data file in your computer. The following commands were used to read the data, produce the two plots in Figure 10.11, and apply the Shapiro-Wilk test:

R & MINITAB:
Testing for normality

```
x<-scan('e:/DATdatafiles/pulserate.dat')
par(mfcol=c(1,2)) # two plots in one figure
hist(x, xlab="pulse rate",main="a.Pulse rate of 210 students")
qqnorm(x, main="b. Normal quantile plot" )
qqline(x)  ; shapiro.test(x)
```

The base package of R calculates five different tests for normality, among them: Anderson-Darling, Kolmogorov-Smirnov, Pearson Chi-square test, and Shapiro-Wilk test for normality. Other packages written in R apply additional tests for normality.

10.6 Beyond biology

Tests of independence and homogeneity are widely applied in the analysis of surveys in social studies, opinion polls, market, and educational research, among other fields. The answers to two questions of a survey are cross-tabulated, and a test of independence is performed to see if the two characteristics or opinions of the individuals are independent or not. As an example, the answers to two questions from the General Social Survey of 2006 are tabulated in Table 10.6. One question is gender and the other is about how the respondent sees his/her own political views (extremely liberal, liberal, slightly liberal, moderate, slightly conservative, conservative or extremely conservative). The p-value for the Chi-square test is 0.008, reject-

Table 10.6: Political views by gender

Gender	extr. lib.	lib.	sligh. lib.	mod.	sligh. cons.	cons.	extr. cons.	Total
Male	57	213	241	714	286	343	77	1931
Female	82	311	276	969	332	342	90	2402
Total	129	524	517	1683	618	685	167	4333

ing the null hypothesis that self-perceived political views are independent of gender. When a null hypothesis of independence is rejected, it is recommended to further explore the association between the variables. The odds ratio to compare the odds of being conservative or extremely conservative vs. being liberal or extremely liberal, for males and females in the USA, is: $(420/270)/(432/393)=1.415123$. The odds of seeing oneself as conservative or extremely conservative, as opposed to liberal or extremely liberal, are 41% higher for males than for females.

Tests of normality are also commonly applied in many disciplines, psychology and education among them, before applying classical inference methods (to be studied in Chapter 11). The normal distribution is assumed as the distribution of many variables inside and outside biology. Tests of normality are applied in many fields of study.

10.7   Exercises

10.7.1   Review questions

1. Flowers of the pea plant can be located on the axial or the terminal (at the top of the plant) position. According to Mendel, the axial position is 'dominant' and when crossing heterozygous plants, 75% of the offspring will have flowers in the axial position, and 25% of the plants will have flowers in the terminal position. Among 858 plants resulting from the cross of heterozygous plants, 651 had flowers in the axial position and 207 in the terminal position. Prepare a table of expected and observed values. Calculate the Chi-square statistic. Find the p-value and arrive at a conclusion. Is this data set evidence against Mendel's statements?

2. What is the difference between a test of independence of characteristics and a test of homogeneity? Which of them would you apply in each one of the following cases? For each case, write the null hypothesis in plain English.

   (a) Each kernel of a corn cob was classified according to two characteristics, shape (smooth or wrinkled) and color (yellow and purple).

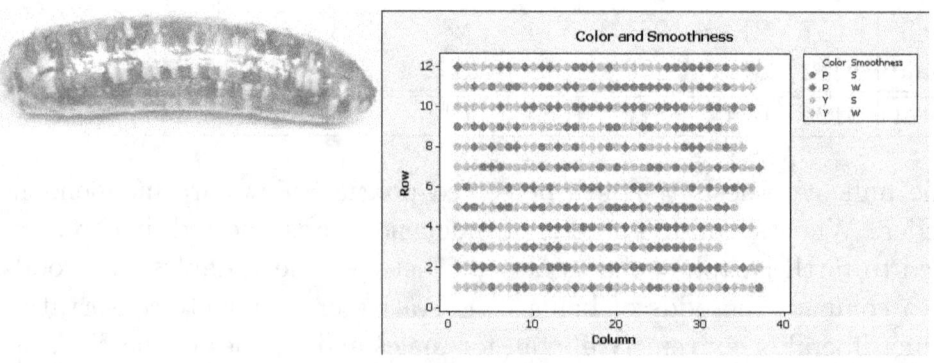

	Smooth	Wrinkled	Total
Purple	102	112	214
Yellow	115	98	213
All	217	210	427

   (b) In an experiment, 16,608 post-menopausical women were randomly assigned to one of two groups. One group received hormone replacement therapy and the other group received a placebo. They were followed for some time and the number of women who had venous thromboembolism was recorded.

Treatment	Venous Thromboembolism Yes	No	Total
Estrogen + progestin	167	8339	8506
Placebo	76	8026	8102
All	243	16365	16608

3. Calculate the expected values and the Chi-square statistic for the corn example in question 2a).

Review questions in Chapter 10

4. Calculate the odds ratio for being smooth to compare purple and yellow kernels. We learned about odds ratio in Chapter 2.

5. Assume that, for the data in question 2b), the Chi-square statistic is 30.255 and the p-value is less than 0.0001. What is your conclusion?

6. Calculate the relative risk of getting venous thromboembolism for women who had hormone replacement therapy as compared with those who did not.

7. Just like in Figure 10.7, in this example there are 3 generations (Parental, Filial 1 and Filial 2), except that they are three generations of mice, not peas. The focus is on the genotype instead of the phenotype. In an article on mapping mouse traits (Bahlo & Speed , 2006), the authors report data for 120 female mice, with respect to a marker on chromosome 7. The alleles are **a** (albino) and **b** (black). It is reported that in the parental generation, the mice were homozygous either of the albino-strain (a/a) or the black-strain (b/b). Thus, it is understood that all the mice in the F1 generation were heterozygous. The mice for which the data are reported are from the F2 (second filial) generation. The observed frequencies for the genotype are:

Genotype	a/a	b/b	a/b
Frequency	24	29	67

(a) What are the probabilities for each genotype (a/a, b/b,a/b) according to the model in Figure 10.7?

(b) What is the expected frequency for each genotype according to the probability model? (The total number of observations is 120).

(c) Which test, from the four studied in this chapter, would you apply to see if the data are evidence against the model?

(d) What is your conclusion if the p-value is 0.359?

8. When is the Shapiro-Wilk test applied ? What is the null hypothesis in that case?

9. The following commands in **R** were used to analyze the the length of the petiole of a random sample of 20 sugar maple leaves. Look at the output. What is your conclusion?

```
petiole<-c( 6.2,3.6,5.3,5.6,5.2,6.0,4.2,6.1,5.3,
6.9,4.3,4.8,5.5,5.0,5.2,4.9,5.5,4.6,5.9,6.4)
shapiro.test(petiole)
qqnorm(petiole)
qqline(petiole)
hist(petiole)
```

The output of those commands is:

```
Shapiro-Wilk normality test
data:  petiole
W = 0.9924, p-value = 0.9997
```

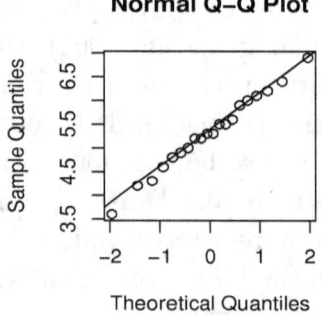
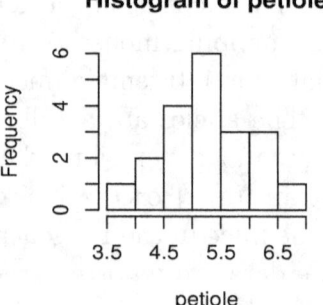

10.7.2   Lab for Chapter 10

Use MINITAB, **R** or the statistical software of your choice to solve Exercises 1, 4, 7, and 9 in Section 10.7.3 'Exercises for discussion or homework'. For each question, write a short paragraph with your answers and insert the necessary computer output.

10.7.3 Exercises for discussion or homework

1. Select one of the characteristics that Mendel studied, look at the corresponding frequencies in Table 10.2, and use them to test the hypothesis that when the parents are heterozygous, the probability that the offspring will show the dominant trait is 0.75.

2. Assume that you examine 500 individuals for the presence of antigens A and B in the surface of their red blood cells. You summarize the results in the two-way table below. What is appropriate to apply here, a test for independence or a test of homogeneity? Why? Apply the test and interpret the results. If you reject the hypothesis of independence, calculate the odds ratio and interpret it.

	B present	B not present	TOTAL
A present	20	210	230
A not present	50	220	270
Total	70	430	500

3. Snee (1974) reports data collected from 592 students about hair and eye color. Here is a reduced version of the table, where only two categories are considered for hair color 'B: black or brown' and 'R: red or blond'. Only two categories of eye color are considered: 'blue' and 'not blue'. Perform a statistical test to answer the question **Is having blue eyes independent of hair color?**

Hair color	Eye color		TOTAL
	blue	non-blue	
red/blond	111	87	198
black /brown	104	290	394
total	215	377	592

Note: For the complete table with 4 colors of hair and 4 eye color see Snee (1974) This data set is one of the R data sets. Type HairEyeColor in the R commands window and two tables will appear, one for males and one for females.

4. During 1980-2000, post-menopausal women who received hormone replacement therapy (HRT) were commonly told that the treatment had the added benefit of protecting their heart. That changed after a clinical trial (1994-2002), called the Women's Health Initiative (WHI), was

completed. Prentice et al. (2005) provide information about the incidence of coronary heart disease among women in the treatment (Estrogen + progestin) and the placebo groups. Perform a test where the null hypothesis is that both groups have the same incidence of coronary disease. Is the null hypothesis rejected at the $\alpha = 0.05$ level? Sometimes a p-value is not small enough (somewhere between 0.05 and 0.10), as to reject a null hypothesis at the 0.05 significance level, but small enough as to 'suggest' a trend. What is the p-value for this example? Is this value considered 'suggestive' of a trend? In what direction would the association go? That is to say, would HRT seem to increase or decrease the risk of CHD?

	Coronary heart disease		
Treatment	Yes	No	Total
Estrogen + progestin	188	8318	8506
Placebo	147	7955	8102
All	335	16273	16608

Another disease studied in the WHI was stroke; the data appears in the next table. Perform a homogeneity test, and if the null hypothesis is rejected, calculate the relative risk (studied in Section 2.15.1) of having a stroke for women using estrogen + progestin as compared with women who do not receive this treatment.

	Stroke		
Treatment	YES	NO	Total
Estrogen + progestin	151	8355	8506
Placebo	107	7995	8102
All	258	16350	16608

Question 2b) of the review questions section of this chapter includes data from the same study about the incidence of thromboembolism in both the treatment and placebo group. Perform a test of homogeneity with that information. **Write a short paragraph summarizing and interpreting the results of the analysis you did for the three tables with data from the WHI.**

Note: The WHI was an interesting study because the clinical trial apparently contradicted the observational study with regard to the effect of hormone replacement therapy in cardio-vascular disease. For a discussion beyond the scope of this book see Prentice et al. (2005) and Freedman & Petitti (2005).

5. Look at the histograms in Figure 9.3. In Section 10.5 (Figure 10.11) the Shapiro-Wilk test was applied to the pulse rate data set. Take any one of the other examples in Figure 9.3 and test the hypothesis that the variable has a normal distribution. The data files are *BMD*, *lumbar108*, and *HEMOGLOBIN380*. Obtain a quantile plot and apply the Shapiro-Wilk test using software.

6. Table 6.1 displays information for 4304 individuals who were asked if they had ever been smokers and if they had ever used marijuana. Use Table 6.1 to test the null statistical hypothesis that having used marijuana is independent of having been a smoker.

7. The data file *drugsurvwg* contains information from an illegal drug and alcohol survey for a sample of 503 individuals . It contains observations for the variables: gender, tobacco use, marijuana use and age group, among others. Prepare two-way tables and answer the following questions:

   (a) Is having used marijuana independent of gender? If your answer is 'no,' calculate the odds ratio comparing the odds of using marijuana for men and women.

   (b) Is having used marijuana independent of age group? If your answer is 'no' calculate the odds ratio comparing the odds of using marijuana for the 'under 25' age group vs. the '40-64' age group.

   (c) Is having used marijuana independent of having been a smoker? If your answer is 'no,' calculate the odds ratio to compare the odds of having used marijuana for smokers vs. non-smokers.

8. A group of students measured several variables for a sample of 10 adult crickets, among them weight (grams), distance jumped (mm), length of thorax (mm), and length of hind legs (mm). Does the data below provide enough evidence against the assumption that these variables have a normal distribution in the population of crickets from where the sample was drawn?

```
weight<-c(0.291,0.183,0.359,0.313,0.327,0.232,0.193,0.38
,0.219,0.314)
```

```
thorax<-c(16,14,14,14,14,13,13,12,12,14)
hindlegs<-c(10,16,17,19,19,18,17,19,18,18)
distance<-c(160,190,180,310,210,350,240,170,360,300)
```

Is the hypothesis of normality rejected for any of these variables? If your answer is 'yes,' get a histogram for that variable and explain why the hypothesis of normality was rejected.

9. Consider the sample of 10 rhododendron leaves in Figure 3.8a. The lengths of the leaves were: 9.70, 10.40, 8.90, 10.70, 9.85, 10.30, 9.00, 10.00, 10.45, and 8.65. Test the hypothesis that the length of the leaves of that particular rhododendron has a normal distribution. Produce also a quantile normal plot.

10. The data file *sugarmaple* contains observations for twenty randomly selected sugar maple leaves. The variables considered are: length, width, length of petiole (all in centimeters), and mass of the leaf (grams). Test the null hypothesis of normality for the distribution of each one of these variables in the population formed by all the leaves of that tree.

Exercises for discussion or homework in Chapter 10

11. DNA is a sequence of nucleotides, which can be A, G, C or T. If we read two subsequent nucleotides (di-nucleotides), there are 16 possible di-nucleotides (AA, AG,...). In the coding region, a sequence of three nucleotides, or tri-nucleotide, codes for an aminoacid. Since there are four possibilities (A, G, C, T) for each position of a tri-nucleotide, there are $4 \times 4 \times 4 = 64$ different trinucleotides. However, there are only 16 different aminoacids, so two different tri-nucleotides might code for the same aminoacid (for example AAA and AAC, both code for Lys). A few trinuclotides are a stop (coding) signal instead of coding for an aminoacid (for example, in mitochondria, AGA and AGG are stop signals). Tables like the one below can be obtained with bioinformatics software (such as Genomatix) asking for the dinucleotides. One enters a sequence and the software produces a two-way table with the frequencies of the 16 different dinucleotides. The table below was prepared with the sequence of 100719 nucleotides of the BRCA1 (BX255925) gene associated with breast cancer. The frequency (4048) indicates that when the 100718 2-nucleotide sequences are read, the pair AA is found 4048 times. The question is: Are subsequent letters independent one from the other? Perform a test of independence, and say if you reject the hypothesis of independence or not. Do you think the result of the independence test makes sense? Why?

First nucleotide	Second nucleotide				Total
	A	G	C	T	
A	4048	7706	4557	2856	19167
G	5901	11388	8935	5020	31244
C	7098	4154	10615	7868	29735
T	2120	7997	5627	4828	20572
Total					100718

10.7.4 Small projects

1. (Experiment) Think of a categorical variable (or categorized version of a quantitative variable). Pose a research question on whether a given factor can have an impact on that variable. Think of the two levels of that factor: 'treatment' and 'control.' Design an experiment, implement it and produce data in order to prepare a two-way table (with one row for treatment and one for control). Perform a homogeneity test, where the null hypothesis is that the treatment does not really affect the response variable as compared with the absence of treatment ('control'). Write a short report describing the study and interpreting the results.

2. (Observational study) Think of a question related to the opinion that individuals have about an issue. The options could be 'yes/no,' or 'agree/neutral/disagree.' Think of two populations that could be compared with respect to the opinion they have about the issue. For example men and women, old and young, members and non-members of a given profession, etc. Select a random sample from each one of those populations. Perform a homogeneity test, where the null hypothesis is homogeneity, i.e., that the distribution along the different categories of the variable is the same for both populations. Write a short report describing the study and interpreting the results.

3. (Observational study) Think of a population (not necessarily human) and two categorical variables, which may or may not be independent, (it could be something as simple as wondering if having the ring finger longer than the index finger is independent of gender). Select a random sample from that population and prepare a two-way table to display the data. Perform a test of independence. Write a short report describing the study and interpreting the results.

4. (Observational study) Think of a quantitative variable that can be easily measured in individuals such as trees or animals (from one single species) or humans (either adults over 18 or children of a similar age). Select a random sample of individuals, do the measurements, and test the hypothesis that the variable has a normal distribution in the population from where you took the sample.

References

1. Bahlo, M. & Speed, T. (2005) How many genes? Mapping Mouse Traits- *Statistics a guide to the unknown, 4th edition.* (Peck et al. Editors) Pacific Grove: Duxbury Press.

2. Clark-Schubert, N.D. (2009) *Fall Migration Ecology of the Sora (Porzana carolina) at Four Rivers Conservation Area in Missouri.* Master Thesis, University of Arkansas.

3. Freedman, D.A., and Petitti, D.B. (2005) Comments on the Women's Health Initiative. *Biometrics, Vol. 61*: 918-920.

4. Mendel, G (1865) Experiments in Plant Hybridization.
http://www.mendelweb.org/Mendel.plain.html
Last accessed: June 22, 2011.

5. Royston, P. (1992) Approximating the Shapiro-Wilk W-test for non-normality. *Statistics and Computing* **2**: 117-119

6. Shapiro, S. and Wilk, M.B. (1965) An analysis of variance test for normality. Biometrika **52**: 591-611

7. Snee, R. D. (1974) Graphical display of two-way contingency tables. *The American Statistician* **28**: 9-12.

8. Prentice, R.L. et al. for the Women's Health Initiative Investigators (2005) Combined Postmenopausal Hormone Therapy and Cardiovascular Disease: Toward Resolving the Discrepancy between Observational Studies and the Women's Health Initiative Clinical Trial. *American Journal of Epidemiology* **162:** 404-414.

Chapter 11

Inference with the normal and t

Statistical inference is a process in which statements are made about populations based on observations from randomly selected samples. Conclusions can also be drawn about treatments, based on carefully designed experiments. Statistical inference involves ESTIMATION and HYPOTHESIS TESTING. The statements made about populations or treatments generally mention a certain degree of uncertainty. Confidence is never 100% for confidence intervals, and α (probability of Type I error) is never 0.

In confidence interval estimation, an interval is calculated in which it is thought, with a given confidence, that the unknown value of the parameter is contained. In Chapter 3, confidence intervals are calculated using bootstrapping. In this chapter, confidence intervals will be calculated based on the knowledge about 'sampling distributions.'

In hypothesis testing, 'p-value' is defined as the probability that the statistic takes the value actually obtained based on the survey or experimental data, or a more extreme one, when the null hypothesis is true. If the p-value is very small, then the null statistical hypothesis is rejected. In order to calculate the p-value it is necessary to have a distribution that describes the behavior of the statistic when the null hypothesis is true. In Chapter 3, an empirical distribution obtained by randomization is used to calculate the approximated p-value. In Chapter 5, when testing for the value of a population proportion, the binomial distribution is used to find the exact p-value. In Chapter 10, when testing for the independence of characteristics, the Chi-square distribution is the distribution of choice. The knowledge of a distribution, which describes what can happen when the null hypothesis is true, allows us to judge if the values from the sample or the experiment are unusual enough as to be considered evidence against the null hypothesis.

In this chapter, p-values will be calculated using the normal distribution and a distribution associated with the normal distribution (t-Student). The cases that will be studied are:

- *t-test for one mean*

- *t-test for the mean difference in paired data*

- *two-independent sample t-test*

- *test for one population proportion (when the sample size is large)*

- *test for two proportions*

The necessary sample size will be discussed in the context of both estimation and hypothesis testing. The basis for the inferential methods to be studied in this chapter is the notion of sampling distribution to be introduced in Section 11.1

11.1   Sampling distributions

11.1.1   What is a sampling distribution?

Each time a random sample of size n is selected from a large population, different individuals may be in the sample. Thus, the value of the sample mean (or of any other statistic of interest) can be different for each sample; this is called **sampling variability**.

sampling variability

As an example, consider a large population of college students and the variable *pulse rate*. A sample of 20 students is selected from that population, and the mean pulse rate of the 20 students happens to be 70 heartbeats per minute. If a different sample is selected, the mean of the 20 observations could be 72, or maybe 73.5, because other students might be in the sample.

To draw a random sample is a random experiment

Drawing a random sample can be considered a random experiment. The sample mean $\bar{x}$ is considered a random variable, its value depending on the outcome of the random experiment (i.e., who is in the sample). Some questions come to mind: Is there a pattern? If the mean of a population is μ, what can be said about the possible values of the means $\bar{x}$ of random samples taken from that population? Are the values of the sample means $\bar{x}$ close to μ? How close? Are the values of $\bar{x}$ closer to μ more likely to happen than values distant from μ? All those questions can be answered when the 'sampling distribution' of the statistic $\bar{x}$ is known.

The sample mean can be considered a random variable

The **sampling distribution** of a statistic, when the sample size is n, is formed by the values that the statistic can take when considering ALL the possible samples of size n that can be selected from the population. The sampling distribution describes the behavior of the statistic considering all the possibilities with regard to the n individuals who could be in the sample. This section focuses on the sampling distribution of the sample mean ($\bar{x}$). **sampling distribution**

Example 11.1 Pine trees per square mile

Think of a small forest with an area of only 3 square miles. The forest is divided in 3 smaller areas of 1 square mile each. The 'individuals' in this case are the 3 one-square-mile regions. Assume that the variable of interest is X: *number of pine trees per square mile*. Assume that the number of pine trees in those one square mile regions are 0, 5 and 10 respectively (Figure 11.1a). The mean of the population is $\mu = (0 + 5 + 10)/3 = 5$ and the standard deviation is $\sigma = \sqrt{((0-5)^2 + (5-5)^2 + (10-5)^2)/3} = 4.08248$. All the

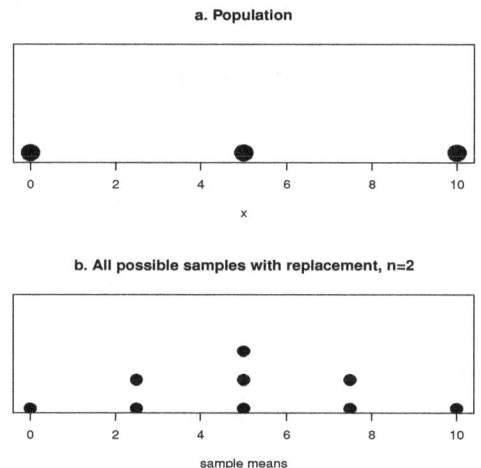

Figure 11.1: A small population (N=3) and the means of all the possible samples (n=2) we can draw from it

possible samples of size 2 that could be obtained from that population, if sampling with replacement, are displayed in Table 11.1. The last column of the table contains the sample means. The means of each one of those possible samples are displayed in Figure 11.1.b. Notice that there is sampling variability: the sample means are different, depending on which elements of the population are in the sample. The center of the distribution of the sample means coincides with the center of the population. Notice also that in the distribution of the sample means, the values in the center or close to the center have a higher probability of occurrence than the values far from the center. This is not something that happens just by pure chance in this particular example; there is indeed a known pattern for the behavior of the sample means.

Table 11.1: All possible samples of size 2 from a population of size 3

sample	regions in sample	values in sample		$\bar{x}$
1	1 and 1	0	0	0.0
2	1 and 2	0	5	2.5
3	1 and 3	0	10	5.0
4	2 and 1	5	0	2.5
5	2 and 2	5	5	5.0
6	2 and 3	5	10	7.5
7	3 and 1	10	0	5.0
8	3 and 2	10	5	7.5
9	3 and 3	10	10	10.0

Mean of sample means:

$\mu_{\bar{x}}$

What is the mean of all the sample means?

$$\mu_{\bar{x}} = (0+2.5+5+2.5+5+7.5+5+7.5+10)/9 = 5$$

What is the standard deviation of the sample means?

$$\sigma_{\bar{x}} = \sqrt{((0-5)^2 + 2 \times (2.5-5)^2 + 3 \times (5-5)^2 + 2 \times (7.5-5)^2 + (10-5)^2)/9}$$
$$= 2.8867$$

The standard deviation of the population was $\sigma = 4.08248$. Notice that $\frac{4.08248}{\sqrt{2}} = 2.8867$, where 2 is the sample size.

Standard deviation of sample means:

$\sigma_{\bar{x}}$

11.1.2   Sampling distribution of the sample mean

The mean and standard deviation of the sample means

When samples of size n are randomly selected from a population with mean μ and variance σ^2, the mean, $E(\bar{x})$ or $\mu_{\bar{x}}$, of the sample means of all the possible samples of size n, is equal to the population mean. The standard deviation of the sample means is equal to the standard deviation in the population divided by the square root of the sample size:

$$\mu_{\bar{x}} = \mu \qquad and \qquad \sigma_{\bar{x}} = \sigma/\sqrt{n}. \qquad (11.1)$$

Thus, the larger the sample size, the smaller the variability in sample mean from sample to sample.

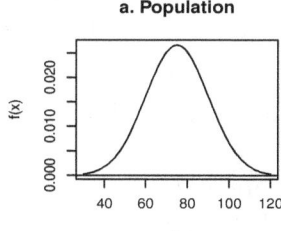 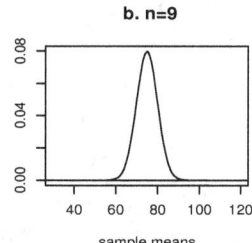 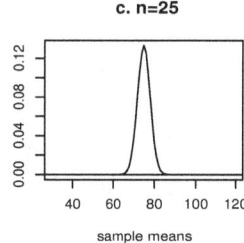

Figure 11.2: A population with normal distribution and the sampling distribution of the sample mean

The shape of the sampling distribution when the variable has a normal distribution

If the variable X has a normal distribution, $X \sim N(\mu, \sigma)$, then the sample mean has a normal distribution regardless of the sample size, $\bar{x} \sim N(\mu, \frac{\sigma^2}{n})$. Thus, if the variable $\bar{x}$ is standardized, we will have:

If $X \sim Normal$, then $\bar{x} \sim Normal$, for any sample size

$$Z = \frac{\bar{x} - \mu}{\sigma/\sqrt{n}} \sim N(0,1) \qquad (11.2)$$

Example 11.2 Sample means for samples from $N(75, 15^2)$

Figure 11.2a displays a normal distribution with a mean of 75 and a standard deviation of 15. In the normal distribution the variable has no bounds $(-\infty < x < \infty)$; however, the distribution is commonly used to represent the behavior of variables that are bounded, such as pulse rate, using the expression 'X has an approximated normal distribution.' Figures 11.2b and 11.2c display the distribution of the sample mean for samples of size 9 and 25, randomly selected from that population. Notice that all are centered in the same value (75), but the spread decreases as sample size increases. The sample means show less variability from sample to sample when there are more individuals in each sample.

The shape of the sampling distribution when the variable does not necessarily have a normal distribution

If the sample size is large, the distribution of the sample mean is assumed to be approximately normal. This is related to a mathematical result known as the *Central Limit Theorem*. The Central Limit Theorem assumes random sampling from an infinite (not necessarily normal) population with finite variance. As the sample size increases and n tends toward ∞, the distribution of the sample mean tends toward normality. In real life, we do not work with samples of nearly infinite size. What sample size is needed for the assumption of approximate normality of the distribution of the sample mean to be considered reasonable? The answer to this question depends on how skewed the distribution of the variable X is. The more skewed the distribution of X, the larger n needs to be for $\bar{x}$ to have an approximately normal distribution.

For large sample sizes, $\bar{x}$ has an approximately normal distribution even if X does not

Example 11.3 Distribution of the sample mean when the variable has a skewed distribution

Assume that the lifetime of a given type of organism follows the clearly skewed distribution in Figure 11.3a (exponential distribution with $\lambda=0.1$).

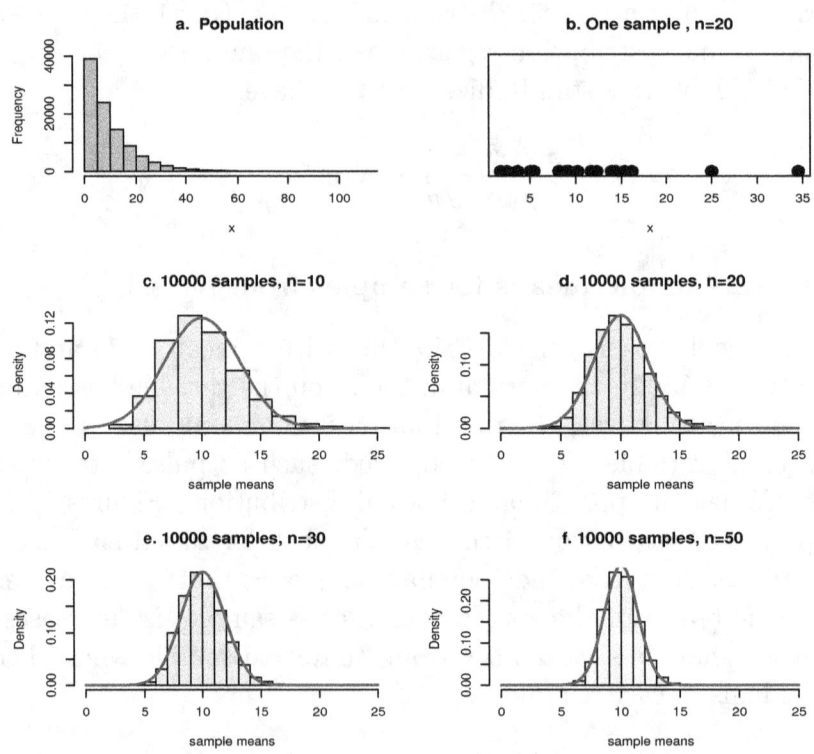

Figure 11.3: A population, a sample and the sample means of 10,000 samples of sizes $n=10,20,30,50$

A single random sample of size 20 was drawn from that population, the length of the life of each one of those 20 individuals is displayed in Figure

11.3.b. Using a computer, 10,000 random samples of sizes 10, 20, 30 and 50 were generated with the same distribution. The means of those samples are shown in Figures 11.3 c-f. These are not the true sampling distributions because not all the possible samples of a given size are being considered, just 10,000 of them. However, the simulation gives a pretty good idea of the shape of the distribution of the sample means. It is clear that the distribution of the sample means is much less skewed than the distribution of the variable in the population. Notice that the distribution looks more symmetric and close to the normal as the sample size increases. Notice also that the variability of the sample means becomes smaller as the sample size increases. It seems that for a skewed distribution such as the one in Figure 11.3.a, a sample size of 50, or even 30, would be enough to feel comfortable with the assumption of normality for the sample mean. Less skewed distributions require smaller sample sizes, and more skewed ones (with distant outliers at one side) will require larger sample sizes in order to assume normality for the sample mean.

11.1.3   The *t*-Student distribution

t-**Student distribution**

Equation (11.2) describes the distribution of the standardized version of the sample mean when the distribution of the variable X is normal, and it clearly involves the standard deviation (σ) of the population. When we do not know the value of the standard deviation of the population - which is usually the case - we replace it with the standard deviation of the sample (s). However, this change has a consequence; the distribution of the statistic is no longer normal. Around 1908, William S. Gosset, under the pen name 'Student,' derived a new distribution now called the '*t*-Student' distribution. He used a pen name because the brewery, where he worked as a scientist, did not allow its employees to publish papers with their names. We quote two of the conclusions in Student (1908): "A curve has been found representing the frequency distribution of values of the means of such samples, when these values are measured from the mean of the population in terms of the standard deviation of the sample. It has been shown that this curve represents the facts fairly well even when the distribution of the population is not strictly normal". R.A. Fisher also worked on the mathematical aspects of the *t*-distribution assuming a normal population.

The variable t defined in expression (11.3) will be used instead of the variable Z in expression (11.2) to work in inference about the population mean when the standard deviation of the population is not known:

$$t = \frac{\bar{x} - \mu}{s/\sqrt{n}} \sim t - Student_{(n-1)}. \tag{11.3}$$

If you know σ, use z in (11.2)

If you don't know σ, replace it by s and use t in (11.3)

The t-distribution is symmetric, and its mean is 0. Its shape, in terms of variability and kurtosis, depends on the parameter called **degrees of freedom**. For the statistic in expression (11.3), the number of degrees of freedom is $n-1$. The number of degrees of freedom depends on the sample size. For low numbers of degrees of freedom (small samples), the t distribution has visibly heavier tails than the normal distribution. However, as the number of degrees of freedom increases (large samples), the t-distribution resembles more and more the normal distribution (see Figure 11.4). Tables for the t-Student are available from the web and most statistical textbooks.

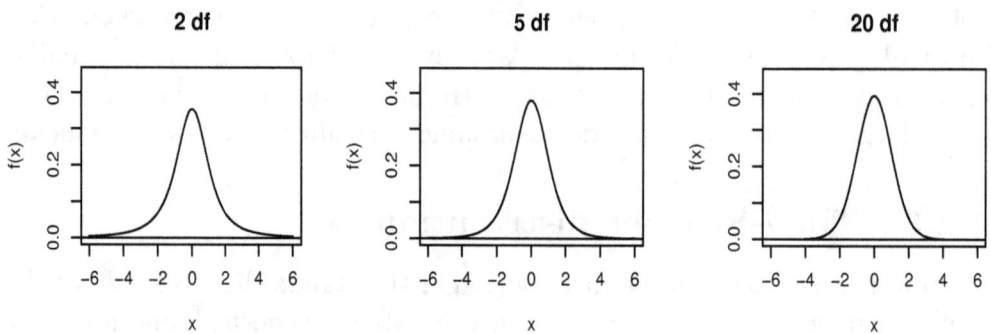

Figure 11.4: t-Student distribution with 2, 5 and 20 degrees of freedom

11.2   Confidence interval for a population mean

In Chapter 3, confidence intervals for the mean are constructed using the empirical distribution of bootstrap replicates obtained by re-sampling. The upper and lower tails of the distribution are located leaving at the center an area similar to the desired confidence, as in Figure 3.10. The quantiles that separate the upper and lower tails are then taken as the endpoints of the confidence interval. In this chapter, a confidence interval for the population mean is obtained based on the knowledge of the sampling distribution of the sample mean described in expression (11.2) when σ is known and expression (11.3) when σ is not known. Distributions such as those in Figures 11.2 and 11.4 will be used instead of the empirical bootstrap distribution in Figure 3.10.

11.2.1 Case 1: The z-confidence interval

Expression (11.2) says that $Z = (\bar{x} - \mu)/(\sigma/\sqrt{n})$ has a standard normal distribution. The normal distribution is symmetric. Thus, two values $-z_{\alpha/2}$ and $z_{\alpha/2}$ (also called $-z^*$ and z^*) can be found, in a normal table or using software, such that

$$P(-z_{\alpha/2} \leq Z \leq z_{\alpha/2}) = 1 - \alpha \qquad (11.4)$$

as in Figure 11.5a.

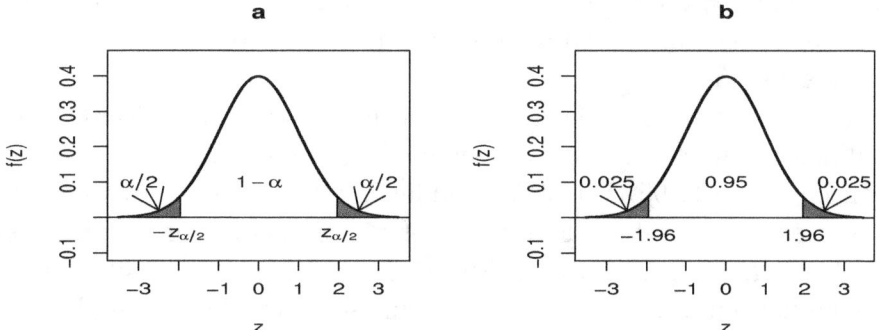

Figure 11.5: Central area and tails in the standard normal distribution

By replacing Z in expression (11.4) with its value in expression (11.2), we obtain:

$$P(-z_{\alpha/2} \leq \frac{\bar{x} - \mu}{\sigma/\sqrt{n}} \leq z_{\alpha/2}) = 1 - \alpha. \qquad (11.5)$$

Doing some algebra so that only μ remains in the center, it follows that

$$P(\bar{x} - z_{\alpha/2} \times \frac{\sigma}{\sqrt{n}} \leq \mu \leq \bar{x} + z_{\alpha/2} \times \frac{\sigma}{\sqrt{n}}) = 1 - \alpha. \qquad (11.6)$$

Expression (11.6) is telling us that the probability that the population mean is between $\bar{x} - z_{\alpha/2}\frac{\sigma}{\sqrt{n}}$ and $\bar{x} + z_{\alpha/2}\frac{\sigma}{\sqrt{n}}$ is $1 - \alpha$. The interval with endpoints $\bar{x} - z_{\alpha/2}\frac{\sigma}{\sqrt{n}}$ and $\bar{x} + z_{\alpha/2}\frac{\sigma}{\sqrt{n}}$ is called the $(1 - \alpha) \times 100\%$ confidence interval. When writing mathematical expressions it is not necessary to write the multiplication sign. It is enough to write the two quantities one next to the other to indicate multiplication. Thus, the expression for the confidence interval is written as:

$$\bar{x} \pm z_{\alpha/2}\frac{\sigma}{\sqrt{n}}, \qquad (11.7)$$

where $z_{\alpha/2}$ is the name given to the value of the standard normal variable that separates the upper tail of area $\alpha/2$, as depicted in Figure 11.5a. Sometimes the value $z_{\alpha/2}$ is called 'critical value.' A simpler notation can be used by replacing $z_{\alpha/2}$ with the symbol z^*. The expression for the confidence interval is then written as:

$$\bar{x} \pm z^* \frac{\sigma}{\sqrt{n}}. \tag{11.8}$$

The area under the normal density function between the values -1.96 and 1.96 is 0.95, as shown in Figure 11.5b. When $z_{\alpha/2}$ is replaced by 1.96, the interval with endpoints $\bar{x} - 1.96\sigma/\sqrt{n}$ and $\bar{x} + 1.96\sigma/\sqrt{n}$ is called the 95% confidence interval. To have 95% confidence means that, if the process of selecting samples to calculate confidence intervals was to be repeated many times, in the long run, approximately 95% of random samples will produce confidence intervals (calculated in this way) that will contain the true value of the population mean.

The meaning of confidence

The practitioner calculating the confidence interval should decide the desired confidence, which can be any percentage. However, it is common to work with either 90%, 95% or 99% confidence intervals just because tables for such values are available. The values of z^* corresponding to some confidence levels are given in Table 11.2. More values of z^* can be easily obtained either from a normal probability table or using software. We should be aware that if the area in the center is $1 - \alpha$, each tail has the area $\alpha/2$. Thus, in order to find the value of z^* that separates an area of size 0.95 in the center, each tail must be of size 0.025 and we should look for the 0.025 or the 0.975 quantile. In **R**, by typing qnorm(0.975), the quantile or critical value z^* $=1.959964 \sim 1.96$ is obtained. Typing qnorm(0.025) gives -1.959964 ~ -1.96 because the normal distribution is symmetric.

Table 11.2: Values of $z_{\alpha/2}$ or z^* for different values of confidence

Confidence	80%	90%	95%	98%	99%
$z_{\alpha/2}$ or z^*	1.28	1.645	1.96	2.33	2.576

95% confidence interval for the population mean

The expression for the 95% confidence interval becomes:

$$\bar{x} \pm 1.96 \frac{\sigma}{\sqrt{n}}. \tag{11.9}$$

Only seldom is the value of σ (the standard deviation of the population) known. When selecting a random sample in order to estimate with confidence the mean of a population, the t-confidence interval (section 11.2.3) is more commonly used. However, the z-confidence interval in expression (11.8) has some very important uses. Among them are designing quality control charts for industry (where the value of σ is known when the process is under control), and determining the necessary sample size to achieve a certain confidence and precision when estimating the population mean (see Section 11.2.2).

Margin of error is the name given to the quantity $z^* \frac{\sigma}{\sqrt{n}}$ in expression (11.8). The 'point estimator' of the population mean is the sample mean $\bar{x}$, in the sense that it only gives one estimated value for the population mean, as opposed to the confidence interval that gives a whole interval of values where, with a certain confidence, the population mean is thought to be. Expressions (11.7)-(11.9) can be understood as:

$$point\ \ estimator \pm\ \ margin\ \ of\ \ error \qquad (11.10)$$

General form of a confidence interval

11.2.2   Necessary sample size to estimate the mean

The margin of error in expression (11.8) is:

margin of error

$$m = z^* \frac{\sigma}{\sqrt{n}}. \qquad (11.11)$$

One of the big questions when planning a survey or experiment is: *How large should the sample be?* If the purpose of the study is to estimate a population mean using a confidence interval, then equation (11.11) suggests the answer to the question. Solving for n, equation (11.11) becomes:

$$n = [z^* \frac{\sigma}{m}]^2. \qquad (11.12)$$

Finding the necessary sample size for estimating a population mean

To calculate the sample size, the values of z^*, m, and σ are needed. The first two values are determined by the person designing the study; information is needed for the third one:

The necessary sample size depends on: confidence, precision and variability

1. z^* is determined by the confidence desired for the estimation (values are found in Table 11.2).

2. m is the maximum value that we allow the margin of error to have.

3. σ is the standard deviation of the population. In case it is not known, it either needs to be 'borrowed' from similar studies, or a pre-survey needs to be done in order to estimate it.

In summary, the necessary sample size depends on the **confidence** and **precision** desired, considering the **variability** existing in the population.

Example 11.3 How many leaves?

Assume that a sample of leaves will be taken from a mature rhododendron in order to estimate the mean length of the leaves. Assume also that it is decided that the margin of error should not be larger than 0.5 centimeter and that the estimation should be done with 95% confidence. Some information about the variance of the length of rhododendron leaves is needed. One can look in research journals, books, or the web. Imagine that no information about the variance or standard deviation of the length of rhododendron leaves could be found in the literature. However, in Example 3.3, a random sample of rhododendron leaves was drawn. That sample could be used as auxiliary information in order to have some idea about the variability in length among the leaves of that species. The standard deviation of the 10 values in the sample was 0.72. Just to be on the safe side, the standard deviation in the population will be assumed to be 1 (larger variability in the population requires larger samples). Now, expression (11.12) will be used to calculate the necessary sample size to estimate, with a 95% confidence interval and a margin of error of 0.5 centimeter, the mean length of all the leaves in the rhododendron:

sample size example

$$n = (1.96 \times 1/0.5)^2 = 15.3664 \sim 16 \ .$$

A sample of 16 leaves would be needed, since sample sizes are always rounded up.

11.2.3   Case 2: The t-confidence interval

The confidence interval in expressions (11.7)-(11.9) was developed based on equation (11.2). In a similar way, when the value of the standard deviation of the population is not known, the following confidence interval can be derived from expression (11.3):

$$\bar{x} \pm t^* \frac{s}{\sqrt{n}}. \tag{11.13}$$

The value of t^* is the value, in the t-Student distribution with $n-1$ degrees of freedom, that separates the upper tail of area $\alpha/2$.

Revisiting Example 3.3 on Rhododendron leaves

The lengths (cm) of 10 randomly selected leaves from a rhododendron are:

9.70, 10.40, 8.90, 10.70, 9.85, 10.30, 9.00, 10.00, 10.45, 8.65

What can be said about the mean length for all the leaves in that particular plant? In Chapter 3, a 95% confidence interval for the mean length of the rhododendron leaves was built using bootstrapping; that confidence interval was **(9.33, 10.195)**. Now the confidence interval will be calculated using expression (11.13).

The mean and standard deviation of the lengths of the ten leaves are $\bar{x} = 9.795$ and $s = 0.7201273$. In the t-Student distribution with 9 degrees of freedom, the values -2.262157 and 2.262157 separate the lower and upper tails, each one with 0.025 of area, from the central part of the distribution with area 0.95. Those values can be found in a t-Student table or by typing in **R: qt(0.025,9)** (qt for quantile of the t-distribution, 0.025 for the area of the left tail, and 9 for the degrees of freedom). Thus, the confidence interval for the population mean, based on those 10 observations is:

$$9.795 \pm 2.262157 \frac{0.7201273}{\sqrt{10}},$$

which simplifies to 9.795 ± 0.515148 or **(9.279852, 10.31015)**. Based on the 10 observations, it can be said that we are 95% confident that the mean length of all the leaves in that rhododendron is between **9.28** and **10.31** centimeters.

Checking for normality before applying the t

There are two conditions that have to be met in order to trust the result obtained with the t-Student confidence interval. The first one is that 10 leaves were randomly selected, and they were. Randomness is necessary to ensure that the sample is representative of the whole population. The other is that the length of the rhododendron leaves should have a normal, or if not normal at least a fairly symmetric distribution in the population from where the sample comes. The second condition is important when the sample size is small as in Example 3.3. The Shapiro-Wilk test of normality, learned in Chapter 10, is applied next:

```
x<-c(9.70,10.40,8.90,10.70,9.85,10.30,9.00,10.00,10.45,8.65)
shapiro.test(x)
OUTPUT:  Shapiro-Wilk normality test
data:  x,     W = 0.9144, p-value = 0.3122
```

The p-value of the test for normality is large: thus, the hypothesis of normality for the length of the leaves is not rejected, and the results from the t-confidence interval are trusted. Actually, what is really important is not exact normality but symmetry, in order to feel comfortable with the results of the t-procedures. In the case of small samples with severe skewness and distant outliers at one side, the t-procedures such as the t-confidence intervals should be avoided because their results are not reliable, since the t-distribution might not be a good approximation for the true distribution of the statistic.

Comparing the bootstrap and t-confidence intervals *

In the case of the rhododendron leaves, the 95% confidence t-interval is only slightly different than the one obtained in Chapter 3 by the method of boot-strapping. Rounded to the nearest millimeter, the bootstrap confidence interval is **(9.3, 10.2)**, and the t confidence interval is **(9.3,10.3)**.

standard error of the sample mean

The expression $\frac{s}{\sqrt{n}}$ is known as the **standard error of the sample mean**; it is the estimator of the standard deviation of the sample mean described in (11.1). For the length of rhododendron leaves, $\frac{s}{\sqrt{n}} = \frac{0.7201273}{\sqrt{10}}$ =0.2277242 ~0.228. In Chapter 3, for the same example, the bootstrap standard error, i.e., the standard deviation of the bootstrap replicates of the sample mean, is 0.214. The bootstrap method gives a good approximation to the confidence interval and standard error obtained by the classical method based on the theory of sampling distributions.

Now that the z- & t-confidence intervals have been studied, we can mention another way of calculating bootstrap confidence intervals. Replacing $\frac{s}{\sqrt{n}}$ by SE_{boot} in (11.9) the following expression is obtained for the standard 95% bootstrap confidence interval:

$$\bar{x} \pm 1.96 SE_{boot} \tag{11.14}$$

where SE_{boot} is the standard deviation of the bootstrap replicates of the sample mean. Sometimes, for sake of simplification, the value 1.96 is replaced by 2. In the rhododendron leaves example, expression (11.14) becomes:

$$9.795 \pm 2 \times 0.214$$

Thus, the confidence interval for the mean length of the rhododendron leaves becomes (9.367,10.223), not too different from the results obtained with the other methods.

11.3   Testing hypotheses about means

Three cases of tests of hypotheses about population means will be studied in this section. In the **one population case** there is a single population and the statistical hypotheses is about the mean of that population. In the **paired t-test** each individual has been measured twice and the statistical hypotheses are written in terms of the mean difference between the two measurements. The purpose of the **two-sample t-test** is to compare the means of two populations. The comparison of two groups is treated in Chapter 3 using the randomization test; now it will be studied using a different approach. Commonly, σ, the standard deviation of the population, is not known, that is why we work with t-tests in this section. If the variances of the populations were known, z-tests using the normal distribution would be applied.

W.S. Gosset and R.A. Fisher assumed, in their work with the t distribution, that the variable X had a normal distribution. Gosset (1908) mentions in the conclusions about the now known as the t distribution: "It has been shown that this curve represents the facts fairly well even when the distribution of the population is not strictly normal". Efron (1969) proved that exact normality is not required to apply the t-test, but that the milder assumption of symmetry is enough.

Symmetry, not exact normality, is enough to apply the t-test

11.3.1 One population case

The null statistical hypothesis is

$$H_o : \mu = \mu_o$$

where μ_o is a number. For example, in the case of the rhododendron leaves, somebody might claim that the average length of the leaves is 12 centimeters, a hypothesis that should be tested. The null statistical hypothesis is $H_o :$ $\mu = 12$. The alternative hypothesis can be two-sided or one-sided, depending on the nature of the research question. For example, if somebody argues that 12 is too high a value, the alternative statistical hypothesis would be $H_a :$ $\mu < 12$. In the approach followed here and in most introductory textbooks, the three possibilities for the alternative hypothesis are:

Null statistical hypothesis

Alternative statistical hypotheses

$$H_a : \mu < \mu_o,\ H_a : \mu > \mu_o,\ \text{and}\ H_a : \mu \neq \mu_o$$

Following (11.3), the test statistic can be written as:

$$t = \frac{\bar{x} - \mu_o}{s/\sqrt{n}} \quad or \quad t = \frac{\bar{x} - \mu_o}{SE_{\bar{x}}} \tag{11.15}$$

where $SE_{\bar{x}}$ is called the standard error of the sample mean. Once the value of the test statistic (t in this case) has been calculated, the p-value is the area of the corresponding tail or tails, depending on the nature of the alternative hypothesis as shown in Figure 11.6.

The test statistic

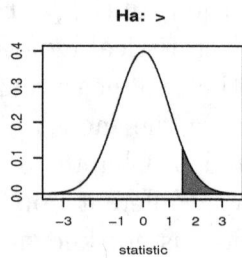 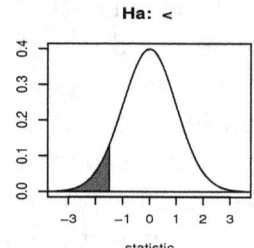 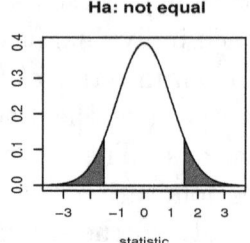

Figure 11.6: p-values according to the type of alternative hypothesis

Revisiting Example 3.3

In the case of the rhododendron example, assume that somebody claims that the mean length of the leaves of a rhododendron bush is 12 centimeters, but we think that they are not so long. The null and alternative statistical hypotheses are:

$$H_o : \mu = 12$$
$$H_a : \mu < 12$$

The mean and the standard deviation in the sample of 10 leaves are $\bar{x} = 9.795$ and $s = 0.7201273$. The test statistic in expression (11.15) becomes:

$$t = \frac{9.795 - 12}{0.7201273/\sqrt{10}} = -9.682763$$

The area to the left of -9.682763, under the density curve of the t-student distribution with 9 degrees of freedom, is very small, $pt(-9.682763, 9) = 2.338245e - 06 = 0.000002338245$ and thus, the null hypothesis is rejected.

R & MINITAB:
One
sample t-test

Using statistical software:
To perform the one sample t-test with R in the previous example, enter the data and use the command 't.test':

```
x<-c(9.70,10.40,8.90,10.70,9.85,10.30,9.00,10.00,10.45,8.65)
t.test(x,alternative = c("less"),mu =12)
```

Similar results can be obtained with MINITAB by selecting STAT>Basic Statistics> one sample t.
The R output is:

```
        One Sample t-test
data:   x
```

```
t = -9.6828, df = 9, p-value = 2.338e-06
alternative hypothesis: true mean is less than 12
95 percent confidence interval:  -Inf 10.21244
sample estimates: mean of x 9.795
```

The output indicates the value of the statistic and the p-value, as well as the nature ('less') of the alternative hypothesis. It also reports the 'one-sided' confidence interval (-Inf , 10.21244). We say with 95% confidence, that the mean length of the leaves in the population (one particular bush in this case) is at most 10.21 centimeters. After rejecting a null hypothesis, a confidence interval should always be calculated. Why? Because people want to know more. If in the rhododendron example you say that the mean length of the leaves is not 12, somebody might wonder: 'if it is not 12, what is it?' In this case, the 95% confidence interval in the computer output indicates that the mean length of all the leaves in the shrub is at most 10.21244 cm. **If $H_o : \mu = \mu_o$ is rejected, calculate a confidence interval for μ**

Two-sided confidence intervals, such as those learned in Section 11.2, can also be used to answer hypothesis testing questions, provided that the confidence matches the value of α (for example, 95% confidence when $\alpha = 0.05$) and that the alternative hypothesis is also two-sided.

11.3.2   Paired-data case

A case of paired observations arises when each individual has been measured twice, such as in 'before/after' or 'pre/post' studies. It can also arise in the study of symmetry such as in measuring the length of the same part of the body of animals in the left and right sides. The research question is whether there is a difference between before and after, pre and post, or left and right. Paired data also happen in the context of a matched-pairs experimental design in which two near-identical individuals each receive one of two treatments. One example of matched-pairs design in agricultural research is an experiment in which each small plot of land is divided in two parts, and each part receives one of two treatments (for example, two different varieties of potatoes are planted and the yield of each half-plot is later measured). In the case of humans, since it is difficult to find nearly identical individuals, each individual might receive the two different treatments (with a 'wash-out' period in between). In all those cases, the analysis focuses on the variable **difference**. The null and alternative hypotheses are written in terms of the mean difference. The null statistical hypothesis, H_o, says that on average there is no difference in the values of the variable between the elements of the pairs. The alternative hypothesis can be one or two-sided: **Paired t-test**

$$H_o : \mu_d = 0,$$
$$H_a : \mu_d < 0, \; H_a : \mu_d > 0, \text{ or } H_a : \mu_d \neq 0 \;.$$

test
statistic
for
paired data

The test statistic is similar to the one in expression (11.15), except that it is applied to the differences:

$$t = \frac{\bar{d}}{s_d/\sqrt{n}} \tag{11.16}$$

Revisiting Example 2.7.1 Cherry juice and arthritis

As part of a larger experiment (Delk, 2009), ten male arthritis patients were given cherry juice daily over a period of four weeks. The pressure needed to produce pain in the knee was measured before (baseline), and after the four weeks of treatment. Assume that the research question is whether cherry juice improves the condition of the patient. That is, if they are able to endure more pressure without feeling pain. Thus, the null and alternative statistical hypotheses are:

$$H_o : \mu_d = 0$$
$$H_o : \mu_d > 0$$

The data set is shown in tables 2.5 and 11.3 The mean and standard devi-

Table 11.3: Pressure needed to produce pain

Patient	1	2	3	4	5	6	7	8	9	10
Before	2.3	2.6	2.5	2.0	2.4	2.4	2.1	2.5	2.0	2.2
After	4.3	4.6	4.9	3.8	4.3	4.2	4.1	4.0	3.9	4.3
Difference	2.0	2.0	2.4	1.8	1.9	1.8	2.0	1.5	1.9	2.1

ation of the differences are: $\bar{x}_d = 1.94$ and $s_d = 0.2319004$. In this case the assumption of normality needs to be checked for the differences, not for the before and after measurements separately, because the variable of interest is $difference = after - before$. The Shapiro-Wilk test does not reject the assumption of normality (p-value = 0.5433). Thus, knowing that the experiment was carefully conducted and that the assumption of normality is not rejected, the t-test can be applied without major concerns:

$$t = 1.94/(0.23190004/\sqrt{10}) = 26.45458$$

The value of the test statistic is very extreme in the context of a *t*-Student distribution with 9 degrees of freedom. The p-value or area to the right of 26.45458 is very small: **1-pt(26.45458,9)**= 3.807402e-10. The data constitute enough evidence against the null statistical hypothesis of no improvement under the treatment with cherry juice.

When working with software, part of the output for the *t*-test is a confidence interval. If the alternative hypothesis is one-sided, then the confidence interval provided in the output is one-sided as well. The confidence interval reported in the output (see below) is interpreted as: 'we are 95% confident that the additional pressure arthritis patients could tolerate on average, after four weeks of drinking cherry juice, is at least 1.8 units.'

R &MINITAB: paired t-test

Using statistical software:
Two alternative ways of performing the t-test for paired data with R will be seen, depending on whether the differences are the input or if the values for 'before' and 'after' are the input. The commands to read the differences 'after − before' and perform the test are:

```
y<-c(2,2,2.4,1.8,1.9,1.8,2,1.5,1.9,2.1)
t.test(y, alternative = c( "greater" ),mu = 0)
```

The output is:

```
        One Sample t-test
data:  y
t = 26.4545, df = 9, p-value = 3.807e-10
alternative hypothesis: true mean is greater than 0
95 percent confidence interval:
 1.805572      Inf
sample estimates: mean of x    1.94
```

If instead of the difference 'after-before,' the individual values for 'before' and 'after' are entered, the paired t-test is applied to the same cherry juice example in the following way:

```
before<-c( 2.3, 2.6, 2.5, 2.0, 2.4, 2.4, 2.1, 2.5, 2.0, 2.2)
after<-c( 4.3, 4.6, 4.9, 3.8, 4.3, 4.2, 4.1, 4.0, 3.9, 4.3)
t.test(before, y = after, alternative = c( "less"), paired = TRUE)
```

The output is:

```
       Paired t-test
data:  before and after
t = -26.4545, df = 9, p-value = 3.807e-10
alternative hypothesis: true difference in means is less than 0
95 percent confidence interval:
      -Inf -1.805572
sample estimates:mean of the differences  -1.94
```

Notice that when the data for both variables ('before' and 'after') are provided, and the test command is written as 't.test(before, y=after...,' R automatically works with the differences 'before-after.' That is the reason for getting negative signs instead of positive in the second option. However, the p-value is exactly the same.

Similar output could be obtained with MINITAB by choosing: STAT > Basic Statistics > Paired t. Make sure to indicate the nature of the alternative hypothesis ('less than') in 'Options.'

Example 11.4 Are flies more active in the second part of the day?

Table 11.4: Activity for first and second 6-hour period of light for 32 *Sarcophaga crasipalpis*

fly	1	2	3	4	5	6	7	8	9	10	11
first	80	341	409	178	317	97	168	222	152	399	159
second	38	788	802	681	479	345	333	296	505	1026	180

fly	12	13	14	15	16	17	18	19	20	21	22
first	86	216	403	115	258	658	324	307	215	303	334
second	84	586	638	262	720	1345	669	935	164	328	493

fly	23	24	25	26	27	28	29	30	31	32	
first	250	467	251	161	512	123	363	81	47	236	
second	162	507	536	56	843	1101	831	125	138	750	

Sarcophaga crasipalpis (flesh flies) are placed in individual plexiglass chambers connected to a device that registers the number of times that each fly interrupts a beam of light. The purpose is to quantify the activity of the fly during a given period of time. Lights are on for a 12-hour period (8 am to 8 pm), and off for the remaining 12 hours of each day (8 pm to 8am of the following day). The researcher hypothesizes that flies are more active during the second part of the 12-hour light period. A random sample of 32 flies is selected and put in the 32 compartments available (data from experiments by Joplin, Moore & Bray). The activity for each fly is recorded as the number of times it interrupts the beam of light. The values are reported in Table 11.4 for each fly and each six-hour period on a given day. Are flies kept in these conditions more active on average during the second part of the day? Why is this a paired-data case?

The null and alternative statistical hypotheses are:

$$H_o : \mu_d = 0$$
$$H_o : \mu_d > 0$$

where the symbol d refers to the difference $d = second - first$.

Fly activity in two parts of light period

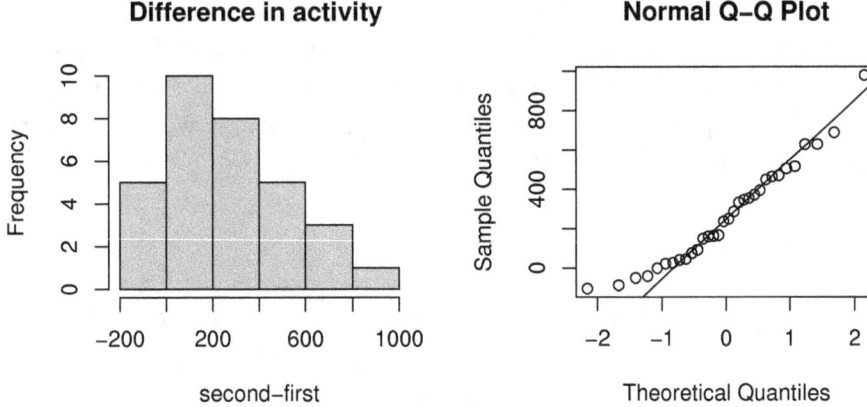

Figure 11.7: Difference in activity between second and first 6-hours periods

The histogram and the normal quantile plot in Figure 11.7 display the differences calculated from Table 11.4. Remember that in the case of paired data, the normality assumption is not about the first or second measurements but about the difference between them. In this case the assumption of normality for the differences is not violated since the p-value for the Shapiro-Wilk test is 0.1767. The histogram in Figure 11.7 shows a mild case of skewness and the sample size is not small. Thus, we feel comfortable with applying the paired t-test. The computer output is:

```
     Paired t-test
data:  second and first
t = 5.8041, df = 31, p-value = 1.073e-06
alternative hypothesis: true difference in means is greater than 0
```

```
95 percent confidence interval:
 188.3390       Inf
sample estimates: mean of the differences 266.0625
```

The null statistical hypothesis, which says that there is no difference in the level of activity between the first and second six-hour lights-on periods, is strongly rejected because the p-value for the paired t-test is 0.000001073. For the flies in the sample, the average difference between the two 6-hour periods was 266. The histogram in Figure 11.7 informs us that only 5 flies displayed more activity during the first 6-hour period, and 27 were more active in the second period. According to the confidence interval provided in the output, it could be said that we are 95% confident that, during the second part of the lights-on period, flies in the population would interrupt the beam of light on average at least 188 more times than during the first 6 hours.

As an exercise apply the paired t-test to Darwin's data in Example 3.2. Discuss the assumption of normality (or at least symmetry) for the differences *cross-self*. Compare the p-value from the paired t test with that of the randomization test in Section 3.1.3.

Using statistical software:
The following R commands were used to read the data in Example 11.4, produce the histogram and the normal probability plot in Figure 11.4, and perform both the normality test and the paired t-test. When typing the t.test command, be careful about the order in which you enter the groups so that the analysis is consistent with the way you want to calculate the differences and the nature of the alternative hypothesis.

R:
Example 11.4

```
first<-c(80,341,409,178,317,97,168,222,152,399,159,86,216,403,115,
258,658,324,307,215,303,334,250,467,251,161,512,123,363,81,47,236)
second<-c(38,788,802,681,479,345,333,296,505,1026,180,84,586,638,262,
720,1345,669,935,164,328,493,162,507,536,56,843,1101,831,125,138,750)
dif<-second-first
par(mfcol=c(1,2))
hist(dif,main='Difference in activity ',
xlab='second-first',col='orange')
qqnorm(dif)
qqline(dif)
shapiro.test(dif)
t.test(second, first, alternative =c('greater'), paired=TRUE)
```

Why we shouldn't apply a test to Example 2.7.2.

Table 2.6 in Chapter 2 displays the typical length of wing (mm) for male and female of 23 species of Australasian robins. In Example 2.7.2, the 'element' or 'individual' is the species, not individual birds. Thus, we know the value of the variable for all the 'individuals' (species) in the population, which is the Petroicidae family. The variable *typical length* is reported for each species. Those 23 species are not a random sample, they are the whole population of species (well, the 23 species out of 30 for which information exists). We would report the statistics calculated in Chapter 2 (see Figure 2.18), stating that they refer to the 23 species of the family but would NOT perform a paired *t*-test. We know from the descriptive analysis done in Chapter 2 that for all the species of the Petroicidae family for which information is available, typically the males have slightly longer wings than the females. The difference 'male-female' in typical length of the wing varies between 0.02 and 9.2 depending on the species, and the mean difference is 3.74 mm. In general, when data for the whole population is available, we have no need for statistical inference. **If data are available for the whole population, we DO NOT apply statistical inference methods**

A more general case of paired *t*-test *

The paired *t*-test can be a little more general. Instead of testing if the mean difference between the elements of the pairs is zero, it can test that the mean difference is a given value δ. The null statistical hypothesis is $H_o : \mu_d = \delta$. One such case could be a variation of Example 11.4. Imagine that the biologist hypothesized that, during the second part of the day, flies interrupt the beam of light more than 100 additional times on average than during the first part of the day. In that case $\delta = 100$, $H_o : \mu_d = 100$, and $H_a : \mu_d > 100$. The test statistic for the general case is: **$H_o : \mu_d = \delta$**

$$t = \frac{\bar{x}_d - \delta}{s_d/\sqrt{n}} \tag{11.17}$$

Using R, it would be enough to add the expression u=100 *inside the* t.test *command to solve the variation to Example 11.4. MINITAB also has a space to input the numerical value of δ when performing the paired t-test.*

11.3.3   The two independent samples *t*-test

The two-sample *t*-test is applied to compare the means of two populations or to compare the mean response of two treatments (or treatment vs. control). Consider the null statistical hypothesis that says that the two population means are equal: **t-test for the means of two populations**

$$H_o : \mu_1 = \mu_2 \qquad \text{or} \qquad H_o : \mu_1 - \mu_2 = 0$$

The alternative statistical hypothesis could be two-sided or one-sided:

$$H_a : \mu_1 \neq \mu_2, \quad H_a : \mu_1 < \mu_2 \quad \text{or} \quad H_a = \mu_1 > \mu_2$$

Consider the difference of the two sample means as a random variable. The mean of that random variable is the difference of the two population means, and the symbol to represent the standard deviation of the difference of means is $\sigma_{\bar{x}_1 - \bar{x}_2}$. That standard deviation needs to be estimated with the data; the symbol used for the estimator is $SE_{(\bar{x}_1 - \bar{x}_2)}$ or standard error of the difference of sample means. Standardizing (subtracting the mean and dividing by the standard deviation) as was done in (11.15), the test statistic becomes:

$$t = \frac{(\bar{x}_1 - \bar{x}_2) - (\mu_1 - \mu_2)}{SE_{(\bar{x}_1 - \bar{x}_2)}} \tag{11.18}$$

test statistic for the 2-sample t-test

If the null statistical hypothesis is $\mu_1 - \mu_2 = 0$, then the test statistic is simplified to:

$$t = \frac{(\bar{x}_1 - \bar{x}_2)}{SE_{(\bar{x}_1 - \bar{x}_2)}} \tag{11.19}$$

The t-Student distribution is used to calculate the p-value. The parameter df (degrees of freedom) will depend on the variances of the two populations being equal or not. If they are equal, $df = n_1 + n_2 - 2$; if they are not equal, df will be estimated using the expression in (11.25).

The assumptions required to apply the two-sample t-test are:

- The two samples have been independently selected from the two populations. In the case of the experiments, the experimental subjects or units have been randomly assigned to one of the two treatments.

- The variable has a normal or at least symmetric distribution in each population.

There are two versions of the test; in one of them, there is an additional assumption:

- The variance of the variable is the same in the two populations

The other version (called Welch's t-test) does not require the assumption of equal variances. From the computational point of view, the difference between the two versions is how the standard error of the difference of means, in the denominator of expression (11.19), is calculated. There is also a difference in the parameter (number of degrees of freedom) of the t-distribution.

The variance of the difference of two random variables is the sum of their variances. If the variance of the variable X is σ_1^2 in the first population and σ_2^2 in the second, the variance of the difference of the sample means is:

$$\sigma_{\bar{x}_1-\bar{x}_2} = \sqrt{\frac{\sigma_1^2}{n_1} + \frac{\sigma_2^2}{n_2}} \qquad (11.20)$$

However, σ_1 and σ_2 are not known and need to be estimated.

Case 1: Variances are assumed to be equal in the two groups

If the variances are equal, then it makes sense to have a single estimated value for both. A *pooled* estimator of the common variance can be obtained based on the data from both samples:

$$s_p = \sqrt{\frac{(n_1 - 1)s_1^2 + (n_2 - 1)s_2^2}{n_1 + n_2 - 2}} \qquad (11.21)$$

The estimator of (11.20) appears as denominator in (11.19) and can now be calculated as:

$$SE_{(\bar{x}_1-\bar{x}_2)} = \sqrt{\frac{s_p^2}{n_1} + \frac{s_p^2}{n_2}} = s_p\sqrt{\frac{1}{n_1} + \frac{1}{n_2}} \qquad (11.22)$$

When the variance of X in both populations is the same, replacing (11.22) in (11.19) results in the following test statistic:

$$t = \frac{(\bar{x}_1 - \bar{x}_2)}{s_p\sqrt{\frac{1}{n_1} + \frac{1}{n_2}}}. \qquad (11.23)$$

test statistic when variances are assumed to be equal

Case 2: Variances are not assumed to be equal

If we cannot assume that the variances are equal, then separate estimates are needed for σ_1 and σ_2, and expression (11.19) becomes:

$$t = \frac{(\bar{x}_1 - \bar{x}_2)}{\sqrt{\frac{s_1^2}{n_1} + \frac{s_2^2}{n_2}}}. \qquad (11.24)$$

test statistic without the assumption of equal variances

The statistic in (11.24) can be assumed to have a *t*-Student distribution, but the number of degrees of freedom is not $n_1 + n_2 - 2$ as the one in (11.23);

instead, the number of degrees of freedom is approximately:

$$df = \frac{\left[\frac{s_1^2}{n_1} + \frac{s_2^2}{n_2}\right]^2}{\frac{1}{n_1-1}\left[\frac{s_1^2}{n_1}\right]^2 + \frac{1}{n_2-1}\left[\frac{s_2^2}{n_2}\right]^2} \qquad (11.25)$$

This is called the Welch-Satterthwaite approximation and dates from 1947. The calculation of expressions 11.18-11.25 seems complicated to do by hand. Fortunately, they are calculated automatically by the statistical software, so we can focus on the interpretation of the results.

Revisiting Example 3.1 Chickens and phytate

Does phytate increase the mean amount of endogenous aminoacids in the excreta of chikens? A diet that includes phytate is being compared with a regular diet for chickens. Eight chickens were randomly assigned to each diet with the response variable being the amount of mucin in the excreta. This data set was already analyzed in Chapter 3 using the randomization test. Now, the two-independent sample t-test will be applied. The null and alternative statistical hypotheses are:

example of the two-sample t-test

$$H_o : \mu_{Phytate} = \mu_{No-phytate}$$
$$H_a : \mu_{Phytate} > \mu_{No-phytate}$$

The data set is presented again in Table 11.5.

Table 11.5: Amount of mucin in the excreta of chickens

Treatment	Amount of mucin (g)								$\bar{x}$	s
Phytate	6.07	7.20	6.61	9.69	9.45	8.95	8.72	6.07	7.845	1.522
Control	2.57	2.39	2.51	2.57	1.80	2.37	6.28	2.91	2.925	1.391

Source: Onyango et al.(2009)

The statistic in (11.19) needs to be calculated. But what version should we use? The one in (11.23) (equal variances version) or the one in (11.24) (unequal variances version)? Does the variable X: *mucin* seem to have a normal or symmetric distribution in the experimental and control groups?
    The standard deviations are:

$$s_{phytate} = 1.521710 \text{ and } s_{control} = 1.390807$$

The p-values for the Shapiro-Wilk test for normality are:

data: treatment, p-value = 0.1504
data: control, p-value = 0.0003382

The standard deviations of the sample are not too different, even if a formal test of equal variances in the populations has not been performed; having unequal variances does not seem a major concern in this example. The hypothesis of normality is rejected for the control group, likely because of an outlier (see Figure 3.1). Having a skewed distribution with outliers at one side, in small samples, is not a good environment for the application of the t-test. The presence of that outlier can compromise how well the t-Student distribution approximates the distribution of the statistic in (11.19). Despite this major concern, the t-test will be applied with illustrative purposes. As a consequence of the lack of symmetry (due to outlier at one side) in one of the groups, the p-value to be obtained will be considered with some reservations.

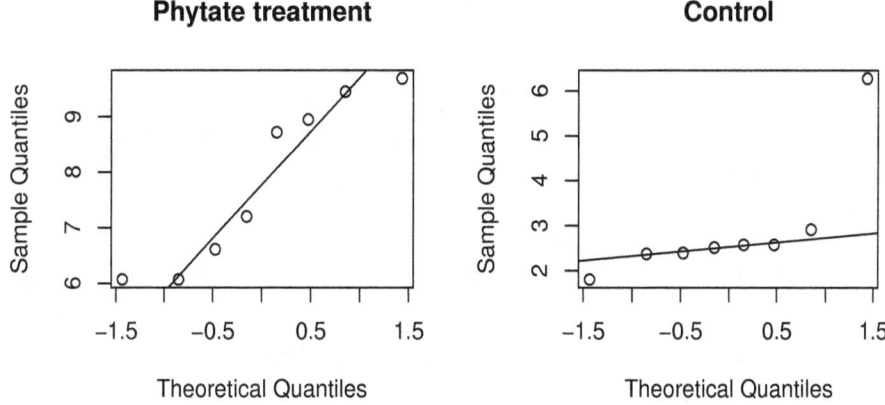

Figure 11.8: Normal quantile plots for the phytate experiment

The *pooled estimated standard deviation* in expression (11.21) becomes: **pooled standard deviation**

$$s_p^2 = \sqrt{\frac{7*1.52171^2 + 7*1.390807^2}{14}} = 1.457729$$

The sample means are $\bar{x}_{treatment} = 7.845$ and $\bar{x}_{control} = 2.925$. Thus, the value of the test statistic in expression (11.23) becomes:

$$t = \frac{(7.845 - 2.925)}{1.457729\sqrt{\frac{1}{8} + \frac{1}{8}}} = 6.750226$$

Since the alternative hypothesis is one-sided ($>$), the p-value will be the area (under the density curve of the t-Student distribution with $n_1 + n_2 - 2 = 14$ degrees of freedom) to the right of 6.750226. This area can be calculated with **R** by typing: **1-pt(6.750226,14)**. The output is 4.657904e-06=0.0000046579, a very small p-value. This p-value needs to be considered with caution because the data from the control group shows skewness and a possible outlier. However, using the randomization test, the null hypothesis was also conclusively rejected because the approximated empirical p-value was around 0.0001.

The t-test can be applied using software instead of going through all the calculations of the previous paragraph by hand (see complete instructions in 'Using statistical sofware' section). The command in **R** is: **t.test(treatment, control, alternative=c("greater"), mu=0, paired=FALSE, var.equal = TRUE)**. The output is:

```
        Two Sample t-test
data:   treatment and control
t = 6.7502, df = 14, p-value = 4.658e-06
alternative hypothesis: true difference in means is greater than 0
95 percent confidence interval:   3.636244      Inf
sample estimates:
mean of x mean of y
   7.845      2.925
```

Observe that when the option **var.equal=FALSE** is chosen, expression (11.24) is used instead of (11.23). The command is now: **t.test(treatment, control, alternative=c("greater"), mu=0, paired= FALSE, var.equal = FALSE)**. The output is:

```
        Welch Two Sample t-test
data:   treatment and control
t = 6.7502, df = 13.888, p-value = 4.854e-06
alternative hypothesis: true difference in means is greater than 0
95 percent confidence interval:
 3.635514      Inf
sample estimates:  mean of x mean of y
                7.845      2.925
```

The number of degrees of freedom (and therefore the p-value) is a little different, 13.88 instead of 14, because the number of degrees of freedom was calculated with expression (11.25). However, the difference in the degrees of freedom between the 'equal variances' option and the 'unequal variances' option is very small because the standard deviations of the sample were quite similar. If the standard deviations of the samples had been very different, then the number of degrees of freedom under the two different options would be quite different too. If in another example, you are unsure whether to apply the 'equal variance' version of the t-test or Welch's test, apply the latter. It is safer to use a more general version.

Using statistical software:
The following commands in R read the data, calculate the standard devia- | *R & MINITAB:*
tions, prepare Figure 11.8, and perform the Shapiro-Wilk test for each group, | *the two-sample*
and perform the two versions (equal and not equal variances) of the two sam- | *t-test*
ple t-test.

```
treatment<-c(6.07,7.20,6.61,9.69, 9.45,8.95,8.72,6.07)
control<-c(2.57,2.39,2.51,2.57,1.80,2.37,6.28,2.91)
sd(treatment)
sd(control)
par(mfcol=c(1,2))
qqnorm(treatment,main='Phytate treatment')
qqline(treatment)
qqnorm(control,main='Control')
qqline(control)
shapiro.test(treatment)
shapiro.test(control)
t.test(treatment, control,alternative = c( "greater"), mu = 0,
paired = FALSE, var.equal = TRUE)
t.test(treatment, control,alternative = c( "greater"), mu = 0,
paired = FALSE, var.equal = FALSE)
```

In MINITAB, the two-sample t-test is found under STAT>Basic Statis-
tics > Two-sample t. It has the two options of equal and unequal vari-
ances, and the type of alternative hypothesis is specified under 'options.'
GRAPH>Probability plots or STAT > normality test can be used to check
for normality.

Revisiting Example 2.9.1 Cherry juice vs. placebo

A two-sample
In some cases, two groups or treatments are being compared, but in each one | **t-test on the**
the individuals have been measured twice. Thus, the two sample t-test is | **after-before**
applied on the individual differences ('after-before,' 'right-left'). Such is the | **differences**

case of the experiment in Example 2.9.1. Ten males were given cherry juice for four weeks and ten males were given a placebo that looked like cherry juice. The pressure needed to produce pain in the knee was measured before (baseline values) and after the four weeks (final values). The differences 'final-baseline' for each group are displayed in Table 2.7 in Chapter 2. The research question is: **Does cherry juice reduce arthritis pain?** Consider the null and alternative hypotheses $H_o : \mu_{cherry} = \mu_{placebo}$ and $H_a : \mu_{cherry} > \mu_{placebo}$, where μ represents the mean of the difference 'final-baseline.' Are the data evidence against the null hypothesis? What is your conclusion? These questions were already answered in Chapter 3 using the randomization test. Now, the two sample t-test will be applied. Looking at Figure 2.24, we realize that there seems to be more variability in the treatment group than in the control group. Thus, we will use the option with non-equal variances or Welch's test. The same commands used for the chickens and phytate example will be used here. We will just change the data, the titles of the figures and will specify that the variances are not equal.

```
treatment<-c(2,2,2.4,1.8,1.9,1.8,2,1.5,1.9,2.1)
control<-c(0.1,0,0.1,0.1,0,0.2,0.2,0.1,0.1,0.1)
sd(treatment)
sd(control)
par(mfcol=c(1,2))
qqnorm(treatment,main='a. Cherry treatment'); qqline(treatment)
qqnorm(control,main='b. Placebo'); qqline(control)
shapiro.test(treatment); shapiro.test(control)
t.test(treatment, control,alternative = c( "greater"), mu = 0,
paired = FALSE, var.equal = FALSE)
```

The output is next and in Figure 11.9

```
sd(treatment)= 0.2319004
sd(control) = 0.06666667
        Shapiro-Wilk normality test
data:  treatment
W = 0.9391, p-value = 0.5433
        Shapiro-Wilk normality test
data:  control
W = 0.8148, p-value = 0.02195
        Welch Two Sample t-test
data:  treatment and control
t = 24.1142, df = 10.478, p-value = 8.038e-11
```

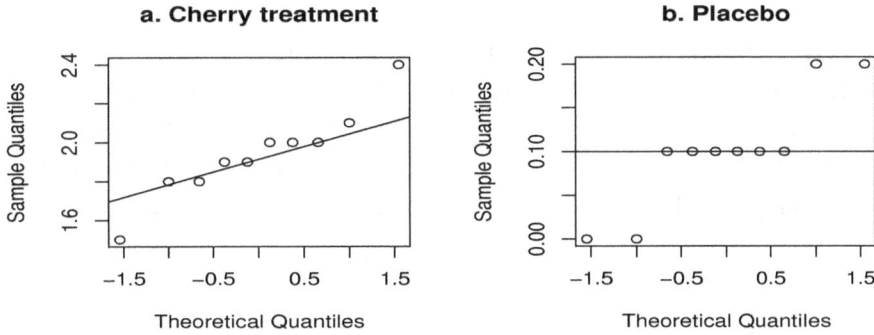

Figure 11.9: Normal quantile plots for the cherry juice experiment

```
alternative hypothesis: true difference in means is greater than 0
95 percent confidence interval:  1.702340        Inf
sample estimates:   mean of x mean of y
                      1.94       0.10
```

The standard deviations are different (0.232 for the treatment group and 0.067 for the control group); we did well choosing the Welch test (non-equal variance version of the two-sample t-test). There are some issues with normality in the placebo group (p-value=0.02195 in the Shapiro-Wilk test). However, we should remember that exact normality is not required, symmetry is enough. Figures 11.9b and 2.24b indicate that we do not have to worry about symmetry in the placebo group. The p-value of the Welch test is 0.00000000008038, a very small value. Thus, the null hypothesis $H_o : \mu_{cherry} = \mu_{placebo}$ is enthusiastically rejected. The results are 'statistically significant.' The issue of 'practical significance' needs to be discussed with the information given by the confidence interval. The one-sided confidence interval is indicating that if arthritis patients drink cherry juice for four weeks, they are able to endure on average at least 1.7 more units of pressure on the knee before feeling pain than if they take just a placebo that looks like cherry juice. It is up to medical doctors or nurses to decide if that is a significant enough gain as to prescribe cherry juice to their patients.

Statistical significance and practical significance

We discuss only the treatment group in Example 2.7.1 and revisit it in the bootstrap section of Chapter 3. Doing an experiment, comparing treatment and placebo, allows the researcher to rule out the placebo effect. The comparison of placebo and treatment groups is discussed in the context of the randomization test at the end of Section 3.1.3.

H_o: $\mu_1 - \mu_2 = \Delta$

A more general version of the two-sample t-test *

There is a more general version of the test in which the null hypothesis is that the difference between the two means is a number Δ not necessarily equal to 0, $H_o : \mu_1 - \mu_2 = \Delta$. The alternative hypothesis states that the difference between the two means is either less, more or just different from Δ. In that case, based on expression (11.18), the test statistic becomes:

$$t = \frac{(\bar{x}_1 - \bar{x}_2) - \Delta}{SE_{(\bar{x}_1 - \bar{x}_2)}} \qquad (11.26)$$

As an example of this more general case of the two-independent sample test, consider the following hypothetical variation of the phytate story. Assume that the research question is whether phytate increases by more than 4 units the mean amount of mucin in the excreta of chickens. The null and alternative hypotheses would then be:

$$H_o : \mu_1 - \mu_2 = 4$$
$$H_o : \mu_1 - \mu_2 > 4$$

The value of the test statistic becomes

$$t = \frac{(7.845 - 2.925) - 4}{1.457729\sqrt{\frac{1}{8} + \frac{1}{8}}} = 1.262237$$

In this case, the value of the t-statistic is 1.262237, a value not extreme enough so as to reject the null hypothesis. In the t-Student distribution with 14 degrees of freedom, the area to the right of 1.262237 is 0.1137429. The only change in the **R** commands is to include the expression $\mu = 4$. Now the command would read: **t.test(treatment, control, alternative=c("greater"), mu=4, paired=FALSE, var.equal=TRUE)**

Note: Frequently the researcher needs to compare more than two groups or treatments, such as in Example 2.8.3 (Figure 2.23) where there are four subtypes of tumors to compare with regard to the intensity in the expression of a particular gene. The method called ANOVA (Analysis of Variance), not included in this book, is used in those cases. ANOVA is a test for the equality of means equivalent to the t-test, but used when more than two groups are to be compared. In Example 2.8.3 a 'one-way' ANOVA would be applied, because there is only one 'factor' (the subtype of tumor) being considered as affecting the expression of that particular gene.

ANOVA is useful also when we need to analyze the effect of more than one 'factor' on the response variable. The experiment with cherry juice (Delk,

2009) is mentioned several times in this book. Example 2.71 mentions only 10 males, for whom a variable was measured before and after the treatment, as an example of paired observations. Example 2.9.1 brings in 10 more males, who receive a placebo, to compare with the 10 males who receive the treatment, in order to rule out the 'placebo effect.' The researcher wants to know if the improvement is due to the treatment or just to the attention the patients receive and autosuggestion. A randomization test or a two-sample t-test compares the two groups and answers the research question. However, the real experiment also includes 20 females, 10 in the treatment and 10 in the placebo. The researcher wants to know if gender has something to do with the way patients respond to the treatment. In the complete version of the experiment there are 2 'factors': treatment (cherry juice or placebo) and gender (male, female). A two-way ANOVA will allow us to analyze the simultaneous effect of the **A** *two factors on the response variable X: change in the tolerated pressure on* **N** *the knee. In the complete experiment there were also four groups (as in the* **O** *tumor subtype example), but those groups are defined by the combinations of* **V** *two factors (treatment and gender).* **A**

In summary, the t-test and the randomization test are very useful tools when you are comparing two treatments or populations, but if you want to compare several groups, or if the groups are defined by the combination of two or more factors, look for ANOVA in the statistical literature and software.

11.4 Large sample inference for proportions

11.4.1 Sampling distribution of the sample proportion

The approximated sampling distribution of the sample proportion, when the sample size is large, is specified in expression (11.29). The reader can go directly there, if so desired. However, we first try to explain the reasons behind that expression.

The exact sampling distribution for the sample proportion is determined by the binomial distribution as seen in Chapter 5. The sample proportion is '(number of successes)/(sample size)' and the variable X: *number of successes* has a binomial distribution. Think of selecting an individual as a random experiment. The probability that a randomly selected individual has a given trait or opinion (in the case of opinion polls) is p. The variable Y: *the individual has the trait or not* is a Bernoulli variable with probability of success p. It should be remembered that $E(Y) = p$ and $Var(Y) = p(1 - p)$. Now, think of repeating that random experiment, i.e., selecting several individuals. If sampling is done with replacement or from an infinite (or extremely large) population, the probability (p) of 'having the trait' is the same in each selection.

Once a sample is drawn, the sample proportion ($\hat{p}$) of individuals with the trait is $\hat{p} = x/n$ where n is the sample size and x is the number of individuals with the trait. As an example, Table 11.6 and Figure 11.10a display the sampling distribution of $\hat{p}$ when $n = 10$ and the proportion of individuals with the trait in the population is $p = 0.3$. Before the sample is drawn, X: *the number of individuals with the trait in the sample* is a random variable and $\hat{p}$ is also a random variable whose value will depend on who will be in the sample, $\hat{p} = X/n$. The variance of X:*number of successes* is $np(1 - p)$; thus the standard deviation of X is $\sqrt{np(1 - p)}$. However, $\hat{p} = X/n$, so its standard deviation is the standard deviation of X divided by n:

$$\sigma_{\hat{p}} = \frac{\sqrt{np(1 - p)}}{n} = \sqrt{\frac{p(1 - p)}{n}}. \tag{11.27}$$

When $X \sim Binomial(n, p)$, $E(X) = np$ and since $\hat{p} = X/n$, $E(\hat{p}) = p$ where p is the proportion in the population. The normal distribution provides a good approximation to the binomial when n is very large and p is neither close to 0 nor to 1.

Table 11.6: Distribution of the sample proportion when $p = 0.3$ and $n = 10$

$\hat{p}$	X	probability
0.0	0	0.0282475249
0.1	1	0.1210608210
0.2	2	0.2334744405
0.3	3	0.2668279320
0.4	4	0.2001209490
0.5	5	0.1029193452
0.6	6	0.0367569090
0.7	7	0.0090016920
0.8	8	0.0014467005
0.9	9	0.0001377810
1.0	10	0.0000059049

Another way of looking at the sampling distribution of the sample proportion is to think of the sample proportion $\hat{p}$ as a sample mean of a Bernoulli variable Y: 'the individual has the trait or not.' The Bernoulli variable Y

takes only the values 0 and 1, thus to add the values Y takes in a sample is equivalent to counting the number of individuals with the trait (or number of successes) in the sample. For example, if a random sample of 10 individuals is drawn, and the observations for the individuals are:

Yes , No, No, Yes, Yes, No, No, No, No, Yes

the corresponding values of the random variable Y are:

1,0,0,1,1,0,0,0,0,1

It is the same to count the number of successes (4) and calculate the sample proportion $\hat{p} = 4/10$, as it is to calculate the mean of Y:

$$\bar{y} = (1 + 0 + 0 + 1 + 1 + 0 + 0 + 0 + 0 + 1)/10 = 4/10$$

Since the sample proportion is a sample mean, it could be claimed that for VERY LARGE sample sizes, taking into consideration the Central Limit Theorem, the sample proportion would have an approximately normal distribution. The expected value would be that of $E(\bar{y}) = E(Y) = p$, and the standard deviation would be that of $\bar{y}$, and it is known from Section 11.1 that $\sigma_{\bar{y}} = \frac{\sigma_Y}{\sqrt{n}} = \frac{\sqrt{p(1-p)}}{\sqrt{n}}$, the same as in expression (11.29).

It can be understood that for $\hat{p} = \bar{y}$ to have an approximately normal distribution, the sample size has to be really large. It is known that the more skewed the distribution of Y is, the larger the value of n that is needed. In situations in which there are too few successes (p close to 0) or very few failures (p close to 1), using the normal approximation for the distribution of $\hat{p}$ might not be a good idea. Remember that Y only takes two values : 0 and 1 with probabilities p and $(1-p)$, so the closer p is either to 0 or 1, the more skewed the distribution of Y is.

Figure 11.10b displays the exact sampling distribution (based on the binomial distribution) for $\hat{p}$, when the sample size is $n = 100$ and $p = 0.3$. Figure 11.10c displays the normal distribution with mean $p = 0.3$ and standard deviation (calculated with expression 11.29) $\sqrt{\frac{pq}{n}} = \sqrt{\frac{0.3*0.7}{100}} = 0.04582576$. Comparing Figures 11.10a and 11.10b, it can be appreciated that when the sample size increases, the standard deviation of the sample proportion decreases. Comparing figures 11.10 b and c, it can be noted that when the sample size is large, it sounds reasonable to approximate the sampling distribution of $\hat{p}$ by a normal distribution. Actually, the normal approximation should be applied only when n is very, very large (in the hundreds) and the trait or event that is under study is neither too abundant or frequent (p close

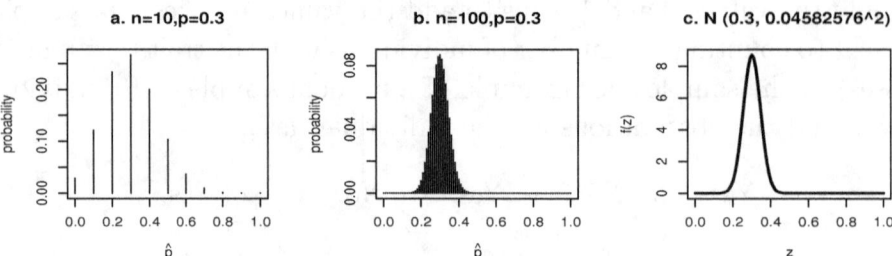

Figure 11.10: Sampling distribution (exact and approximated) of sample proportion $\hat{p}$

to 1) or too rare (p close to 0). Under those conditions, it can be assumed that

$$\hat{p} \sim_{approximately} N(p, (\sqrt{\frac{p(1-p)}{n}})^2).$$ (11.28)

Standardizing (subtracting the mean and dividing by the standard deviation), the following expression is obtained:

$$\frac{\hat{p} - p}{\sqrt{\frac{p(1-p)}{n}}} \sim_{approximately} N(0, 1).$$ (11.29)

Now that statistical software is available, it is possible to work with the exact distribution (based on the Binomial) instead of with this approximated distribution. However, expression (11.29) is useful to calculate confidence intervals in an easy way, and to calculate the necessary sample size when planning a study.

11.4.2 Confidence intervals for a proportion

Applying simple algebra (as in Section 11.2), the following expression for the confidence interval for the population proportion is obtained from (11.29):

$$\hat{p} \pm z^* \sqrt{\frac{p(1-p)}{n}}.$$ (11.30)

However, since the value of p is not known, it is replaced by its sample estimator and (11.30) becomes:

$$\hat{p} \pm z^* \sqrt{\frac{\hat{p}(1-\hat{p})}{n}}$$ (11.31)

There is an improved version of the confidence interval in (11.31) that works with $\tilde{p} = \frac{successes+2}{n+4}$ instead of $\hat{p}$.

Example 11.5 Purple kernels

A random sample of 427 kernels is selected from a large population of corn kernels (from corn cobs similar to the one in Section 10.7.1) and 214 of them happen to be purple. Applying the formula in expression (11.31), the 95% confidence interval for the proportion of purple kernels in the population is:

$$\frac{214}{427} \pm 1.96 \sqrt{\frac{\frac{214}{427}(1-(\frac{214}{427})}{427}}.$$

Thus, the endpoints of the confidence interval are: $0.501171 - 0.04742539$ and $0.501171 + 0.04742539$. The 95% confidence for the proportion of purple kernels in the population is $(0.4537456, 0.5485964)$. We are 95% confident, that between 45.4% and 54.9% of the kernels in the population are purple. The confidence interval could be used to test the hypothesis that half of the kernels in the population are purple. The null statistical hypothesis $H_o : p = 0.5$ would not be rejected working with $\alpha = 0.05$.

11.4.3 Sample size to estimate a proportion

Assume that a survey is being planned with the purpose of estimating a population proportion. In expression (11.31) the margin of error is:

$$m = z^* \sqrt{\frac{p(1-p)}{n}} \tag{11.32}$$

Solving for n (11.32) provides the following expression to calculate the necessary sample size:

$$n = [\frac{z^*}{m}]^2 p(1-p) \tag{11.33}$$

The value of z^* depends on the confidence desired; some values are given in Table 11.2. The margin of error is decided by the person designing the survey according to the precision desired for the estimation. The value of p (proportion in the population) is unknown. There are two options to come up with a value for p to input in (11.33):

- 'to borrow' a value from a previous or similar study

- to assume the 'worst case scenario' and use $p = 0.5$. This produces the largest sample size possible for the specified margin of error and confidence. The reason is that $p(1-p)$ takes its maximum value when $p = 0.5$ as shown in Figure 11.11.

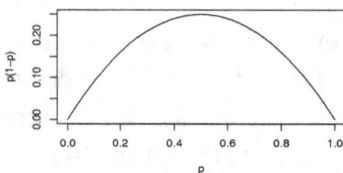

Figure 11.11: Value of $p(1-p)$ for $0 < p < 1$

Example 11.6 How many trees?

Somebody wants to estimate the proportion of the trees in a forest that are affected by a given disease. The confidence is fixed at 95% and the margin of error needs to be no greater than 0.05 (or ±5%). How many trees need to be examined if the researcher does not have any idea about what proportion of trees might be affected?

Applying the formula in (11.33),

$$n = [\tfrac{1.96}{0.05}]^2 0.5(1 - 0.5) = 384.16 \sim 385.$$

Now assume that there is some historical information. It is known that even in the worst seasons, the proportion of trees affected has been at most 20%. If in expression (11.33), p is substituted by 0.2, the necessary sample size becomes

$$n = [\tfrac{1.96}{0.05}]^2 0.2(0.8) = 245.8624 \sim 246.$$

11.4.4   Testing hypotheses about a proportion

The binomial distribution can always be used, as in Chapter 5, to test hypotheses about a population proportion. However, it is still popular to perform tests of hypotheses using the normal approximation described in expression (11.31) when the sample size is very large. Most statistical software give the option of conducting the test with the exact (binomial) or the approximated (normal) distribution.

When the null hypothesis

$$H_o : p = p_o$$

is true, the expression in (11.31) becomes:

$$z = \frac{\hat{p} - p_o}{\sqrt{\frac{p_o(1-p_o)}{n}}} \sim_{approximately} N(0,1) \tag{11.34}$$

Thus, the statistic

$$z = \frac{\hat{p} - p_o}{\sqrt{\frac{p_o(1-p_o)}{n}}} \tag{11.35}$$

can be used to test $H_o : p = p_o$.

The p-value is calculated as an area under the normal curve (to the right, left or at both sides) depending on the nature of the alternative hypothesis, as indicated by Figure 11.6.

Example 11.7 Current smokers

Somebody claims that 1/4 of the adult population of a region are current smokers. Somebody else thinks that it might be a little lower because of all the common knowledge about how bad tobacco is for human health. The null and alternative statistical hypotheses are:

$$H_o : p = 0.25$$
$$H_a : p < 0.25$$

The value of α (maximum value allowed for probability of type I error) is set at 0.05. Assume that a random sample of 2466 individuals is drawn and 574 of them (or 23.28%) are classified as 'current smokers.' The value of the test statistic in (11.34) becomes:

$$z = \frac{0.2328 - 0.25}{\sqrt{\frac{0.25*0.75}{2466}}} = -1.972574$$

The area to the left of $z = -1.972574$ can be found using a standard normal table or statistical software. Using **R**, **pnorm(-1.972574)=0.02427206**. The null hypothesis is rejected because $0.02427206 < \alpha$.

The hypothesis was tested again using the exact test for proportions (using the binomial distribution) from Chapter 5. The exact p-value is 0.02473. Notice that the p-value obtained with the exact test and the normal approximation are very similar (actually equal up to the fourth decimal). The sample size is quite large and the proportion of current smokers in the population is neither close to 0 nor close to 1, therefore the normal approximation works well.

R & MINITAB:
test for
one proportion

Using statistical software:
To perform the test using the normal approximation as in Example 11.7, type in R:
prop.test(x=574,n=2466,p=0.25,alternative=c("less"),correct=FALSE)
The output is

```
1-sample proportions test without continuity correction
data:  574 out of 2466, null probability 0.25
X-squared = 3.9065, df = 1, p-value = 0.02405
alternative hypothesis: true p is less than 0.25
95 percent confidence interval:
 0.0000000 0.2470515
sample estimates:    p      0.2327656
```

To apply the exact test using the binomial distribution, type

$$\text{binom.test}(574, 2466, \text{p} = 0.25, \text{alternative} = \text{c}(\text{ ''less'' }))$$

The output is:

```
 Exact binomial test
data:  574 and 2466
number of successes = 574, number of trials = 2466,
p-value = 0.02473
alternative hypothesis: less than 0.25
95 percent confidence interval:
 0.0000000 0.2471974
sample estimates:  probability of success  0.2327656
```

In MINITAB, the test for proportions is under STAT>Basic statistics> Test for 1 proportion. MINITAB gives the option of working with both the normal approximation and the exact test.

11.4.5   Testing hypotheses about two proportions*

There is a test to compare two populations with regard to a categorical variable. The null hypothesis is:

$$H_o : p_1 = p_2$$

The alternative hypothesis can be

$$H_a : p_1 \neq p_2, \quad H_a : p_1 < p_2 \quad \text{or} \quad H_a : p_1 > p_2$$

When the null hypothesis is true, the proportions are the same in the two populations. Thus, it makes sense to have a single estimated value for the common proportion. The estimator is a 'pooled sample proportion,' meaning that the observations of both samples are combined together:

$$\hat{p}_{pooled} = \frac{x_1 + x_2}{n_1 + n_2} \tag{11.36}$$

where x_1 and x_2 are the number of successes in each sample. The test statistic is:

$$z = \frac{\hat{p}_1 - \hat{p}_2}{SE_{pooled}(\hat{p}_1 - \hat{p}_2)} \tag{11.37}$$

where

$$SE_{pooled}(\hat{p}_1 - \hat{p}_2) = \sqrt{\frac{\hat{p}_{pooled}(1 - \hat{p}_{pooled})}{n_1} + \frac{\hat{p}_{pooled}(1 - \hat{p}_{pooled})}{n_2}} \tag{11.38}$$

Once the test statistic in (11.37) is calculated, the p-value is found using the normal distribution.

Revisiting Example 2.15.1 Poliomyelitis vaccine

The research question in that experiment is whether the vaccine reduces the proportion of children who get polio. The null and alternative statistical hypotheses are:

$$H_o : p_{placebo} = p_{vaccine}$$
$$H_a : p_{placebo} > p_{vaccine}$$

Using the information in Table 2.13, the necessary calculations are done. The 'pooled' estimated value of the common proportion (11.36), under the null hypothesis, is:

$$\frac{142+56}{400000} = 0.000495$$

The standard error in expression (11.38) is:

$$SE_{pooled}(\hat{p}_1 - \hat{p}_2) = \sqrt{\frac{0.000495 \times (0.999505)}{200000} + \frac{0.000495 \times (0.999505)}{200000}} = 0.00007033882$$

The test statistic in expression (11.37) is:

$$z = \frac{142/200000 - 56/200000}{0.00007033882} = 6.113267$$

The p-value is the area to the right of the value $z=6.113267$ under the curve of the standard normal distribution. If using a normal table we notice that the table does not go beyond $z=3.5$, the area up to 3.49 is 0.9998. Thus, the area to the right of 6.11 is practically 0. The area can be calculated with R typing the command **1-pnorm(6.113267)**=4.880593e-10 =0.000000004880583. Thus, the null hypothesis is rejected, and we conclude that the Salk vaccine used in the experiment reduced the proportion of children who got polio.

Only a very small proportion of the population got polio even in those days. The quality of the normal approximation might be considered questionable when p is close to 0 (or to 1). However, the sample size n was very large (200,000) in each group. The restriction to use the normal approximation is usually given in terms of the number of 'successes' and 'failures.' Different authors give different restrictions indicating that the number of successes and failures should be greater than 10 or 15. There were 56 and 142 cases of polio in each group, so there are not serious concerns about applying the normal approximation.

Fisher's exact test, learned in Chapter 3, could also be used to solve this problem. The p-value for the one-sided Fisher's exact test is 3.972e-10=0.0000000003972. This puts our mind at rest with respect to the validity of the normal approximation; the p-values obtained with the two tests (the exact and the approximated) are pretty similar. If small samples had been used, the normal approximation might not have worked so well with a value of the population proportion so close to 0. When the alternative hypothesis is two-sided, the test of homogeneity learned in Chapter 10 can also be used to compare the vaccine and placebo groups.

Using statistical software:
To perform the test for equal proportions for the poliomyelitis example with R, type:

```
prop.test(x=c(142,56), n=c(200000,200000),
alternative = c("greater"),correct=FALSE)
```

Fisher's exact test was applied writing two commands, one to read the data in Table 2.13, the other to perform the test

```
polio<-matrix(c(142,199656,56,199944),nr=2)
fisher.test(polio,alternative='greater')
```

The test for equal proportions is available in MINITAB under STAT > Basic statistics.

11.5   Sample size for testing hypotheses

When planning either an observational study or an experiment, it is important to decide how many individuals to include. In Sections 11.2.2 and 11.4.3, the issue of necessary sample size was discussed in the context of estimation. In that context, the decision of sample size depends on how small we want the margin of error to be (precision), the desired confidence, and the variability existing in the population. More heterogeneous populations require larger samples.

Equally important is the sample size issue in the context of hypothesis testing. In this case, the sample size depends on α (probability of type I error), the power of the test for a given 'effect size,' and the variability in the population. The concepts of power and 'effect size' were introduced in Chapter 5 and the reader is advised to review Section 5.7 before proceeding. Power is the probability of rejecting the null hypothesis when it is FALSE, or in other words, it is the probability of being able to find out that H_o is false when it is false. Naturally, researchers like the power to be high. The effect size is the distance between reality and the value specified by the null hypothesis, that we would care to be able to notice if it existed. The effect size depends on the context, and it is for a chosen effect size that we are specially interested in the test having high power.

Assume that an experiment like the one in Example 3.1 (Chickens and phytate) is being planned. The null statistical hypothesis is $H_o : \mu_1 = \mu_2$. Maybe the researcher would not be worried if phytate increases the mean amount of mucin in the excreta of the chicken by 0.2 grams. However, if phytate increases the mean amount of mucin by 3 grams, the researcher wants for some special reason to be able to notice it and reject the null hypothesis as a result of his/her experiment. In other words, the researcher wants the test to have high power when the difference in means between treatment and control is 3 or more. The variability in the population also plays a role. It is not the same to be able to notice a difference of 3 if the standard deviation of the variable is under 1.5 (as it is in the case of the chickens on a regular diet) than if the standard deviation was 4 or 5. The more variability, the larger we need the sample to be in order to achieve the same power for a given effect size. Sometimes we talk about the effect size in absolute terms (as MINITAB does) and sometimes it is expressed in terms of the standard deviation (as with **R**).

In Chapter 5, the power is calculated by adding entries in the binomial table because the distribution used to calculate the p-values is discrete. Where in the table we start to add the values for the power, depends on the selected value of α. In this chapter, since the normal and the t-distribution are used as distributions of reference to calculate the p-values, the power is an area under the normal or t-distribution because they are continuous distributions. Assume that you are testing the hypothesis $H_o : \mu = 0$ against the alternative $H_a : \mu > 0$ and that you want to work with $\alpha=0.05$ and that σ is known so you are using the z test. The null hypothesis will be rejected when the test statistic based on expression (11.2) is over the threshold of 1.645 (the area to the right of 1.645, when H_o is true, is 0.05). We would like to know the power of the test, for example when the true value of μ is 3 (if we care for an effect size of 3). Figure 11.11 displays:

1. The critical value 1.645 marked by the vertical line; if the value of the statistic is to the right of that line, we reject H_o.

2. The Type I error, α, is the shaded area under the first normal distribution centered at 0, because the null hypothesis is true but we are rejecting it.

3. Power is the area to the right of the vertical line and under the second curve, because the null hypothesis is being rejected AND the true value of the parameter is not 0 but 3.

4. β, the probability of Type II error, is the area to the left of the critical value but under the second normal curve, because the null hypothesis is NOT being rejected BUT it is false. The power and β add up to 1; together they form the total area under the second density curve.

Figure 11.12 illustrates the concepts but is not practical for deciding the sample size, because it is difficult to visually quantify areas and the plot refers only to one effect size. Therefore, in practice, for sample size selection, power curves plots like the one in Figure 11.13 are used. It is useful to learn at least how to interpret such plots.

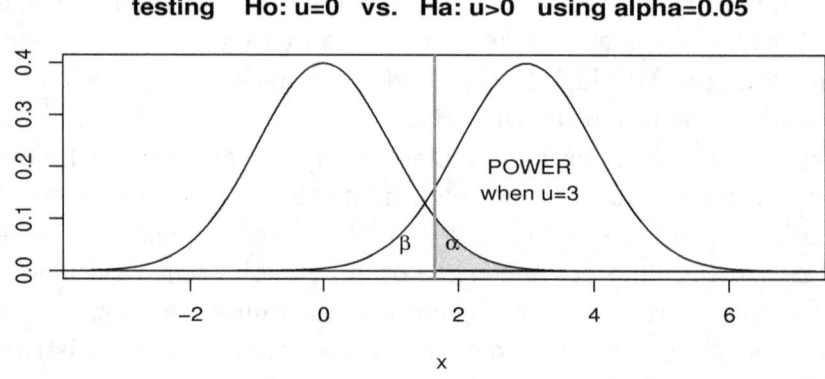

Figure 11.12: Power, α= P(Type I error) and β=P(Type II error)

In Figure 11.13, the power is represented on the vertical axis. The horizontal axis is for the distance between the true value of the parameter and the value specified in H_o, so that one cannot only look at one effect size, but go back and forth in the horizontal axis and observe the power for different effect sizes or differences. Figure 11.13 includes the power for several sample sizes so that the researcher can decide his/her best option for sample size knowing how much power the test can attain. Figure 11.13 offers a good overview of the power that could be attained under different conditions of sample size AND effect size. To prepare such plots, information about the variability in the population is needed as well as the decision of what value of α to use.

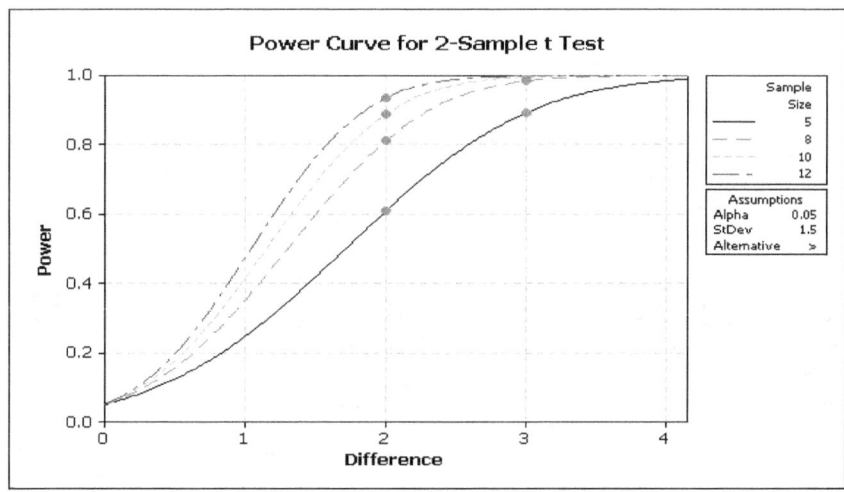

Figure 11.13: Power curves in terms of difference

Figure 11.13 was prepared for a case similar to Example 3.1 (chickens and phytate). A treatment will be compared with a control, and the two-sample t-test will be used with $\alpha = 0.05$. The standard deviation of the variable is thought to be 1.5. Looking at the curves, the researcher realizes that if he/she only cares for a difference of 3, a test using a small sample such as $n = 5$ will have high power. However, if he/she wants to be able to detect an increment of 2 (if the treatment produced that mean increment), then it is better to use a sample size $n = 8$, or even better $n = 10$, in order to have a power close to 0.9. The final decision about sample size usually depends on the desired power and the cost of the experiment.

The effect size can be expressed in a relative way with respect to the standard deviation. In the example, the standard deviation in mucin is assumed to be 1.5. An absolute difference of 3 is equal to a difference of

2 standard deviations ($3 = 2 \times 1.5$). Figure 11.13 displays the power plot, when $n = 8$, in terms of the 'effect size=difference/stdev.' Notice that power for 'distance=1.5' in Figure 11.13 is the power for 'effect size=1' in Figure 11.14.

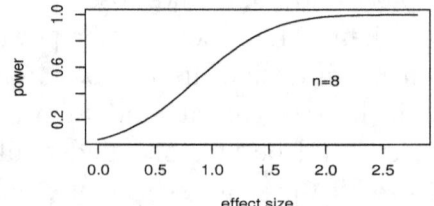

Figure 11.14: Power curve with effect size relative to standard deviation

MINITAB:
Sample size
for tests

Using statistical software:
In MINITAB, choose STAT> Power and Sample size. Choose the type of test you plan to apply, for example 2-sample t. Then specify the value of α and the standard deviation. Enter the values of some effect sizes and sample sizes you are interested in, as shown in Figure 11.15. In 'Options,' indicate the nature of the alternative hypothesis.

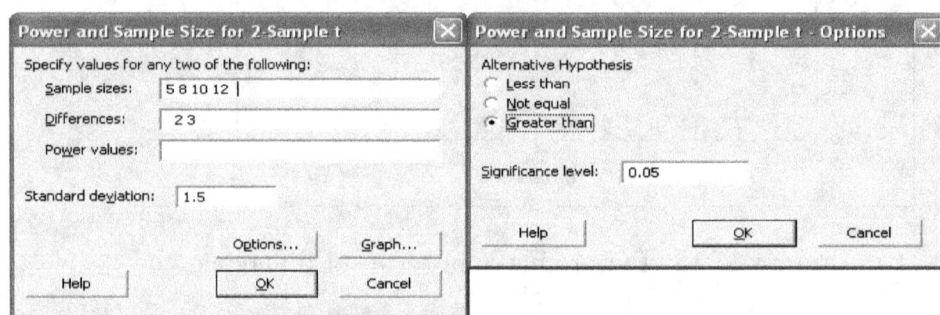

Figure 11.15: Producing a power plot with MINITAB

R:
power
plots

*Power plots can also be produced with **R**. It is necessary to install (once) and upload (at the beginning of each session) the package **pwr** (Select the option 'Packages' from the menu). The following commands were used to produce Figure 11.13.*

```
d<-seq(0,2.8,by=0.05)   ##  effect sizes
pow<-pwr.t.test(n=8,d=d,sig.level=0.05,type=c("two.sample"),
alternative=c("greater"))
pow  ## displays complete output
plot(d,pow$power, 'l',xlab='effect size',ylab='power')
text(2,0.5, 'n=8') ## to include sample size in the plot
```

11.6 Beyond Biology

The methods learned in this chapter are the classical methods of statistical inference generally learned in all introductory statistics courses and applied by researchers from many fields. Test of hypotheses to compare two means or two proportions are used by professionals in psychology, education, medicine, agriculture, among other fields, to compare treatments and methods. Social scientists conduct surveys and need to estimate confidence intervals. Results from opinion polls are frequently reported in the media, including estimated proportions with a corresponding margin of error (check the website of Gallup). The Bureau of Labor Statistics estimates the unemployment rate, which is a population proportion. Calculating the necessary sample size is a necessity for researchers in many fields.

Note: More approaches to statistical inference, and hypothesis testing in particular, are studied in advanced courses on statistical theory.

11.7 Exercises

11.7.1 Review questions

1. What is the sampling distribution of a statistic?

2. What is the mean and standard deviation of the sample means considering all the samples of size n that can be obtained with replacement from a population with mean μ and standard deviation σ ?

3. What is the shape of the sampling distribution of the sample mean for random samples of size 10 drawn from a population where the variable has a normal distribution?

4. What is the shape of the sampling distribution of the sample mean for random samples of size 100 drawn from a population where the variable has a moderately skewed distribution?

5. What is the meaning of '95% confidence' in the context of interval estimation?

6. A random sample of 20 leaves from a sugar maple tree was selected and their mass (in grams) recorded. The 95% confidence interval for

the population mean is (0.43, 0.57). Which of these interpretations of the confidence interval is adequate?

(a) We are 95% confident that the mean weight of all the leaves in the tree is between 0.43 and 0.57 grams.

(b) We are 95% confident that the mean weight of the 20 leaves in the sample is between 0.43 and 0.57 grams.

(c) We are 95% confident that all the leaves in the tree weigh between 0.43 and 0.57 grams.

(d) We are 95% confident that all the leaves in the sample weigh between 0.43 and 0.57 grams.

(e) 95% of the leaves in the tree weigh between 0.43 and 0.57 grams.

7. When is that the t-Student distribution is used instead of the normal distribution to calculate confidence intervals and do tests of hypotheses about the population mean?

8. Why is that, especially if the sample size is small, we perform a test for normality, before applying the t-test or the t-confidence interval? Is normality indispensable?

9. We calculated a 95% confidence interval for the mean of a population based on a sample of size 20 from a population for which the standard deviation is 5. Will the confidence interval become wider or narrower in each one of the following situations?

(a) The confidence is increased from 95% to 99% (keep constant the sample size and standard deviation).

(b) The sample size is increased to 40 (assume confidence and standard deviation remain the same).

(c) We had the wrong standard deviation, it should be 3, not 5 (keep confidence and sample size the same). (Hint: look at any of these formulas (11.6-11.8) and think; you don't need to do calculations).

10. How large should the sample be if we want to estimate the mean canopy volume of black cherry trees? The standard deviation of canopy volume for this type of trees can be considered to be 18 units. The estimation will be done with a 95% confidence interval for which a maximum width of 10 units is wanted (margin of error =5).

11. Write the null and alternative hypotheses for this story: We wonder if the mean canopy volume of two species of trees (species 1 and species 2) can be considered similar or not.

12. We want to test the hypothesis H_o in question 11. A random sample of 20 trees is taken from each one of those species. For each sample we will calculate the mean and the standard deviation. Assume that the variable 'canopy volume' has an approximately normal distribution for each one of the species. What test would you apply?

13. We have studied two versions of the t-test to compare the means of two populations (the regular t-test and the Welch version of the t-test). When is that you apply one or the other?

14. Why is that in the confidence interval formula (11.33), only the value of the sample proportion $\hat{p}$ is used, and in the calculation of the test statistic (11.37) both $\hat{p}$ and p_o are involved? What does p_o represent?

15. A survey is being planned to ask the population of a region some questions about environmental issues. An objective of the survey is to estimate, with a 95% confidence interval, the proportion of the population that agrees with a certain existing environmental policy. The agency in charge of the survey would like to have a margin of error of at most 5% (or 0.05). How large should the sample size be?

16. What distribution can ALWAYS be applied to do inference about a population proportion?

17. What conditions are necessary to use the normal distribution to do inference about a population proportion?

11.7.2   Lab for Chapter 11

Use statistical software to solve exercises 4, 5, 6, 8c, 13, and 15 from the **Lab for** Section 11.7.3 'Exercises for discussion or homework.' In each question write **Chapter 11** a short paragraph summarizing your results and insert the computer output.

11.7.3   Exercises for discussion or homework

1. **A confidence interval for mean reaction time.** Assume these values are the reaction times (measured in milliseconds with a reaction time ruler) of a random sample of individuals from a given population:

130, 143, 150, 150, 150, 150, 155, 160, 165, 170, 170, 180, 190, 190, 190, 200, 235, 250. Check if there is no serious violation to the assumption of normality. Calculate and interpret a 95% confidence interval for the population mean.

2. **How far do crickets jump?** Assume that the data on distance jumped in Exercise 6 in Section 2.17.3 come from a random sample of 20 crickets. Build a 95% confidence interval for the mean distance that the crickets in the population jump. Interpret the confidence interval that you obtain. How comfortable do you feel assuming normality or at least symmetry for the distribution of the variable X: *distance jumped?*

3. **How long is on average the handspan of college female students?** Assume that the data for handspan (inches) reported in Exercise 8 in Section 2.17.3 correspond to a random sample of 23 female college students. Calculate and interpret a 95% confidence interval for the mean handspan in the population from which that sample was drawn.

4. **How heavy on average are the leaves of a sugar maple tree?** Exercise 9 in Section 2.17.3 reports the mass (g) of 20 randomly selected leaves. Use the data to calculate and interpret a 95% confidence interval for the mean mass of sugar maple leaves.

5. **Fly activity.** Read Example 11.4 in Section 11.6.2. The following data set corresponds to the two six-hour periods of light during the 5th day of the experiment. What is your decision about the null statistical hypothesis of no difference in mean activity between the first and second periods based on this new data set?
first:167,37,596,227,0,344,305,428,118,478,383,207,281 ,329,216,240, 358,463,339,123,156,408,373,484, 99,167 ,653,614,205,206,122,209
second:133,112,670,181,1,503,1121,329,559,976,623 66,784,720,583,671, 510,734,1308,300,868,687 607,797,323,418,985,952,640,335,365,288

6. **Are the cells in the onion root smaller in the apical region?** In Chapter 2, the randomization test was applied to test the hypothesis that the mean length of the onion root cells was the same in both regions of the onion root, the 'apical meristem' region and the 'elongation' region. The null and alternative hypothesis are:

$$H_o : \mu_{elongation} = \mu_{apical}$$
$$H_a : \mu_{elongation} > \mu_{apical}$$

where μ represents the mean length of cells.

The observations (in millimeters) are:
apical: 0.033,0.021,0.031,0.033,0.032,0.020,0.025,0.010,0.021,0.026
elongation: 0.107,0.052,0.127,0.043,0.033,0.058,0.045,0.057,0.052,0.038
A picture of the cells is in Figure 2.19 and the lengths are shown in Figure 2.20 in Chapter 2. Discuss how comfortable you would feel applying the two-sample t-test in this case.

7. **Comparing sycamore and sugar maple leaves.** We want to know if the leaves of the sycamore and the sugar maple trees have the same mean length, or if the sycamore leaves are longer on average. Write the null and alternative hypotheses. A random sample of 20 leaves is collected from each species (Example 2.8.2, Figure 2.22). Is this a paired observations case or a two independent samples case? Perform the test and state your conclusions. If the null hypothesis is rejected, calculate a confidence interval for the difference of means and interpret it. For your convenience, here are the data:

```
sycamore<-c(11.2,13,12.5,10,11.3,7,10,10.2,10.6,6.7,9.3
,11,12.8,9.5,12.2,13.5,13.1,11.3,12.2,14.2)
maple<-c(7.6,5.8,7.7,9,8.1,8.6,6.3,7.5,8.6,8.7,7.8,6.5,
6.1,8,7.5,8,7,6.2,7.9,8.2)
```

8. **The bark of spruce trees.** Laasasenaho et al. (2005) studied a large number (1864) of Norway spruce trees. For exercise purposes, the values they report for the bark thickness (cm) at breast height will be considered population values: $\mu = 1.25$ cm and $\sigma = 0.54$. Assume that the variable bark thickness follows a normal distribution.

   (a) If samples of size 10 are taken from that population, what is the distribution of the sample mean?

   (b) Now assume that the distribution of bark thickness is not normal, but slightly skewed. If samples of size 40 are drawn from that population, what is the distribution of the sample mean? Why?

   (c) Assume that you belong to a team that will do a study of the bark of spruce trees in a local forest, and you are the person in charge of deciding how many trees will be selected for the sample.

You talk with the other members of the team and come up with the decision that you want to estimate the mean thickness of bark with a margin of error of at most 0.15 centimeters with 95% of confidence. You are willing to assume that the standard deviation of thickness of bark is 0.54 as reported by Laasasenaho et al. How large does the sample need to be?

(d) You want to learn about the thickness of bark of spruce trees in a local forest. You select a random sample of 25 spruce trees, and the mean thickness of the bark at breast height is 1.15 cm. Calculate a 95% confidence interval for the mean thickness of the bark spruce trees in the region from which you extracted the sample. For now, work under the assumption that you know σ (the standard deviation of the population).

(e) You want to learn about the thickness of bark of spruce trees in a local forest. You select a random sample of 25 spruce trees, and the mean thickness of bark at breast height is 1.15 cm, and the standard deviation in the sample is 0.35 . Calculate a 95% confidence interval for the mean thickness of the bark spruce trees in the region where you extracted the sample. Work with the standard deviation of the sample instead of the value 0.54 given in the original paper.

(f) You wonder if around here spruce trees also have bark with a mean thickness of 1.25 centimeters. Somebody hypothesizes that here the bark might be thinner. The null and alternative statistical hypotheses are: $H_o : \mu = 1.25$ and $H_a : \mu < 1.25$. You select a random sample of 25 spruce trees and find that the mean thickness of bark of the trees at breast height in the sample is 1.15 and the standard deviation is 0.35. Use the standard deviation of the sample. Would you reject the null hypothesis at the 0.05 significance level?

9. **Does staking the potato plant produce higher average yield?**
Exercise 6 in Section 3.4.2 reported the following values for the yield in pounds harvested 131 days after planting for 8 potato plants. Four of them (control group) had no stakes to support the plants, and their yields were 0.4, 1.55, 1.10, and 1.05. (Data produced by V. Burke) Four plants were staked, and their yields were 3.15, 2.85, 1.75, and 2.5. Consider the null hypothesis $H_o : \mu_{staked} = \mu_{non-staked}$ versus the alternative hypothesis that staked plants produce a higher yield

on average. Would you consider a 'paired t-test' or a 'two-independent sample t-test'? Why? Are the data evidence against the null hypothesis at the 0.05 significance level? Apply the test and report the p-value and your conclusion.

10. **Are girls slower when they are distracted?** Assume that the data for girls in Exercise 4 in Section 2.17.3 come from a random sample of 14 female fifth graders. Discuss what type of test learned in Chapter 11 you will apply to answer this question. Write the null and alternative hypotheses and perform the test. Clearly state your conclusions.

11. **Are boys slower when they are distracted?** Assume that the data for boys in Exercise 4 in Section 2.17.3 come from a random sample of 14 male fifth graders. Discuss what type of test learned in Chapter 11 you will apply to answer this question. Write the null and alternative hypotheses and perform the test. Clearly state your conclusions.

12. **Does distraction affect the reaction time of boys and girls in a similar way?** Assume that the 14 boys and 14 girls in Exercise 4 in Section 2.17.3 are two independent random samples from male and female fifth graders in a school district. Discuss which of the tests learned in Chapter 11 you would apply in order to answer the research question. We want to compare the change in reaction time (distraction-quiet) for boys and girls. Apply the test and clearly state your conclusions.

13. **Nesting success of mallards.** Nesting success is defined as the % of nests that successfully hatch at least one egg. Nesting success varies across species and it also depends on environmental conditions. According to the conservationist organization Ducks Unlimited, the nesting success necessary to maintain the mallard population is 15%.

    (a) Somebody is interested in estimating the proportion of successful nests in a northern region. How large should the sample of nests be in order to do the estimation with 95% of confidence and a margin of error no larger than 0.05?

    (b) Consider again the previous question with the added information that p is believed to be no greater than 0.2.

    (c) In a grassland region, a simple random sample of 600 nests was selected, and 210 of them were successful. Calculate a 95% confidence interval for the proportion of successful mallard nests in that region.

(d) In another region they are worried about the mallard population and want to test the hypotheses Ho: p=0.15 against Ha: p<0.15. A random sample of 525 nests is selected and only 50 are successful. Working with alpha=0.05, should we reject the null hypothesis? Is the mallard population endangered in that region?

14. Perform a test of equal proportions, with a two-sided alternative hypothesis, for the proportion of individuals who have tried marijuana among males and females. Use the information given in question 22 of the Review Questions section in Chapter 2 (section 2.17.1).

15. Perform a test of equal proportions, with a one-sided alternative hypothesis ($p_{treatment} > p_{placebo}$) for the two tables (Coronary heart disease and Stroke) in the hormone replacement therapy study described in Exercise 4 in Section 10.7.3. Clearly state your conclusions. Also apply Fisher's exact test and compare the results.

References

1. Delk, K. (2009) *Effects of Cherries on Arthritis Pain.* Undergraduate honors thesis. East Tennessee State University.

2. Efron, B. (1969) Student's *t*-test under symmetry conditions, *Journal of the American Statistical Association* **328**: 1278-1302.

3. Laasasenaho, J, Melkas, T. and Alden, S. (2005) Modelling bark thickness of *Picea abies* with taper curves, *Forest Ecology and Management* **206**: 35-47.

4. Onyango, E.M., Asem, E. and Adeola, O. (2009) Phytic acid increases mucin and endogenous amino acid losses from the gastrointestinal tract of chickens, *British Journal of Nutrition* **101**: 836-842.

5. Student (1908) The Probable Error of a Mean. *Biometrika* **6**: 1-25

Chapter 12

Regression models

In Chapter 2, the strength and direction (positive or negative) of the linear association between two quantitative variables is quantified with the correlation coefficient. In this chapter, we will learn how to write equations that describe the association or relationship existing between the two variables. The focus will be on the simple linear regression model, also called the regression line. However, a non-linear model will also be included because of its importance in the study of relationships between two variables for the same organism in the context of the studies of size and shape. Multiple regression will be briefly discussed.

12.1 Introduction to regression

12.1.1 What is a regression model?

A regression model is an equation that expresses a 'response variable' in terms of 'explanatory variables'

Models in general are succinct representations of some aspect of reality. In mathematics and statistics, models are written using equations. Statistical models include random variables and acknowledge random individual departures from a typical pattern.

Different variables pertaining to the same individuals might be associated; the length of the petiole of a leaf might be related to the mass of the leaf it supports; the length of a foot might be associated with the person's height; the body mass index is likely to be associated with the waist of the person; the canopy volume of a tree might be related to the diameter of the tree trunk or its height; the length of a leaf might be related to its width. The association can be between between a variable of the environment and the response of individuals to the environment, as in the classical example of the number of chirps of crickets and the temperature of the air (Dolbear, 1897).

Revisiting Example 2.1.12

In Section 2.12 it was observed that an association exists between the number of red blood cells and the altitude at which humans live. The correlation coefficient quantifies the strength and direction of a linear association between two variables. In Example 2.12.1, the value r=0.8028 and the scatterplot in Figure 2.29 indicate a positive and strong linear association: the higher the altitude of residence, the higher the number of red blood cells. The association is not perfect, and there is considerable individual variability. An equation or 'model' can be written to describe the association between a response variable Y and an explanatory variable X. We could write the model as:

estimated number of RBC= 4.8365885 + 0.0001303 altitude of residence

Models are used to understand relationships and to predict the value of the response variable

A regression model is an equation that expresses the **response variable** Y as a function of one or more variables X called **explanatory variables**. Sometimes the response variable is called 'dependent' variable and the explanatory variables are called 'independent' variables. Writing a model and estimating its parameters allows us to answer some questions. For example, it is possible to say how much, on average, the number of red blood cells increases with each 1,000 feet of altitude.

Why are models useful?

Models do more than simply satisfy a curiosity; they have a practical application: to estimate the value of a variable that is difficult to measure or observe, based on the value of another variable whose values are easier to obtain. For example, measuring the canopy volume of every tree is not an easy task. However, a random sample of trees can be selected and their canopy volume, height, and diameter at breast height measured. Based on the data collected, a model can be written to explain canopy volume in terms of height and trunk diameter. Later, that model can be used to predict the canopy volume of trees of the same species based on their height and trunk diameter (variables that are much easier to measure).

Two examples (12.1 and 12.3), related to the leaves of sugar maple trees (data file *sugarmaple*) will be used throughout this chapter.

Example 12.1 Width (X) and mass (Y) of the leaf

When considering leaves of a single species, it makes sense to think that larger leaves tend to be heavier. Figure 12.1 displays the width and mass for a random sample of 20 sugar maple leaves. The width of a leaf is very easy to measure, but to quantify the mass you need a special scale. A model to estimate the mass of a sugar maple leaf from the width of the leaf is:

$$\widehat{mass} = -0.4217 + 0.0988 \ width \tag{12.1}$$

The symbolˆon top of the name (mass) of the variable indicates 'estimated' mass, i.e., the value of mass according to the model. The value 0.0988 (slope) indicates that for each additional centimeter of width, the estimated mass increases by 0.0988 grams.

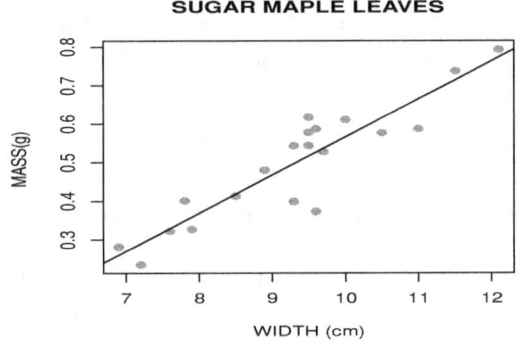

Figure 12.1: Regression model to estimate mass from width in sugar maple leaves

The simple linear regression model

The regression line model studied in this chapter is called 'simple' because there is only one explanatory variable X. Before numerical values are estimated from the data and included in the equation, a model is written in a generic way with symbols representing the unknown values of the constants or **parameters**. Relationships between variables are rarely deterministic, individual variability is almost always present. The symbol ε in equations (12.2) and (12.3) represents that variability, or individual departures from the overall relationship. Those departures can be due just to pure randomness or to the effect of other variables associated with the mass of the leaf that are not included in the model. The expected value of ε is assumed to be 0. For example, before data are available, the simple linear regression model to represent the relationship between mass and width is written as:

Parameters are constants whose numerical values are not known; they are represented by symbols or letters

$$mass = \beta_o + \beta_1 width + \varepsilon \tag{12.2}$$

In general, for two variables: Y (response variable) and X (explanatory variable), the simple linear regression model is:

$$Y = \beta_o + \beta_1 X + \varepsilon \qquad (12.3)$$

Steps in modeling

Stages in the modeling process: IDENTIFICATION ESTIMATION EVALUATION

Modeling is a process considered to have the following three stages:

1. **Identification.** Decide which type of equation to use. Either the type of equation is known (from existing laws in physics, biology, economics, etc.) or a certain type of model is selected from the collection of statistical models available (linear, exponential, etc.) according to the pattern shown by the data (see Section 12.2 on mechanistic and empirical modeling). In Example 12.1, the type of model selected is the equation of a straight line in expression (12.2) because the scatterplot in Figure 12.1 shows a somewhat linear pattern. The **parameters** of the simple linear regression model are the **intercept** and the **slope**.

2. **Estimation.** To estimate the parameters of the model means to assign numerical values to them, based on the data from the sample. In Example 12.1 assume that the estimated value of the intercept is -0.4217 and the estimated value of the slope is 0.0988 (it will be later explained how these values were obtained). The estimated model is written in expression (12.1).

3. **Evaluation.** The first question in the evaluation of the model is: Does the model achieve a good explanation of the response variable? Not all the individuals exhibit the same value for the variable Y. How much of that variability can be explained by the fact that not all individuals have the same value for X?

   The origin of the regression models, as well as the correlation coefficient, is in the work by Francis Galton and Karl Pearson around 1877. Galton was interested in the study of inheritance and conducted several experiments with sweet peas. He classified the peas into seven groups according to their size, measured their diameter, and later measured the diameter of the offspring of those peas. The average sizes of the parents and offspring in each group are displayed next (Weisberg, 2011).

group	1	2	3	4	5	6	7
parent	21	20	19	18	17	16	15
progeny	17.26	17.07	16.37	16.40	16.13	16.17	15.98

The term 'regression' seems to have emerged from the observation that the offspring of large peas were somewhat smaller on average than the parent generation. Conversely, the offspring of dwarf peas were on average a little larger than the parent generation. Galton made the observation that the offspring tended to 'regress' to the median size of the species. See Stanton (2001) for a historical note on regression.

12.1.2 Mechanistic and empirical modeling

There are two-ways to decide what type of model to use: **mechanistic** if the type of relationship between the variables is known, and **empirical** if the data are collected first and, based on the pattern shown by the data, an equation is selected.

Mechanistic modeling

Models are derived from a knowledge of the field of application. Frequently mathematical procedures involving calculus and differential equations are used in the formulation of the model. Once a model is chosen to describe a situation, experiments or observational studies can be conducted in order to obtain data to estimate the numerical value of the parameters. This approach is commonly applied in physics and biochemistry. A simple example is the model known as the exponential growth model to be discussed next. **Using a known law as model**

Example 12.2 Exponential growth

Start with a number a of bacteria, assume the bacteria do not die, suppose that there are unlimited resources, and the hourly reproduction rate is r. After an hour, there will be $a + r \times a$ bacteria. After another hour, there will be the previous $a + r \times a = a(1 + r)$ bacteria plus the new $ra(1 + r)$, that makes a total of $a(1 + r) + ra(1 + r) = a(1 + r)^2$. We can continue and conclude that after t hours there will be $a(1 + r)^t$ bacteria. The parameters of the model are the initial population a and the reproduction rate r, the explanatory variable is time t and the response variable is the size of the current population of bacteria y. The equation of the model is $y = a(1 + r)^t$. Notice that the equation of the model includes parameters, namely a and r. The exponential growth model is NOT a linear model. **Exponential growth**

Empirical modeling

A theoretical relationship between two variables might not be known, but data for those variables still may be available. Plotting the data allows us to observe a pattern that can be represented by a certain type of model. There is a wide variety of statistical models from which to pick. This is like having a bag of tools and grabbing the one that seems most useful for a particular situation. Example 12.1 was such a case, another example is provided next.

Choosing a model by looking at the pattern described by the data

Example 12.3 Mass of the leaf (X) and length of the petiole (Y)

A group of students collected a random sample of 20 sugar maple trees and measured the length (cm) of the petiole and the mass (g) of the leaf. The data are displayed in Figure 12.2a (data file *sugarmaple*). The petiole connects the leaf to the branch. Do heavier leaves have longer petioles? The scatterplot indicates that heavier leaves do tend to have longer petioles, even though the relationship is far from perfect. The pattern does not show any curvature, it seems fairly linear and the correlation coefficient is 0.6529091. The relationship could be represented by a linear model and then we could proceed to estimate the parameters. In this case a model has been selected based on the pattern described by the data; this style of modeling is called **empirical modeling** and is the one frequently applied in statistics.

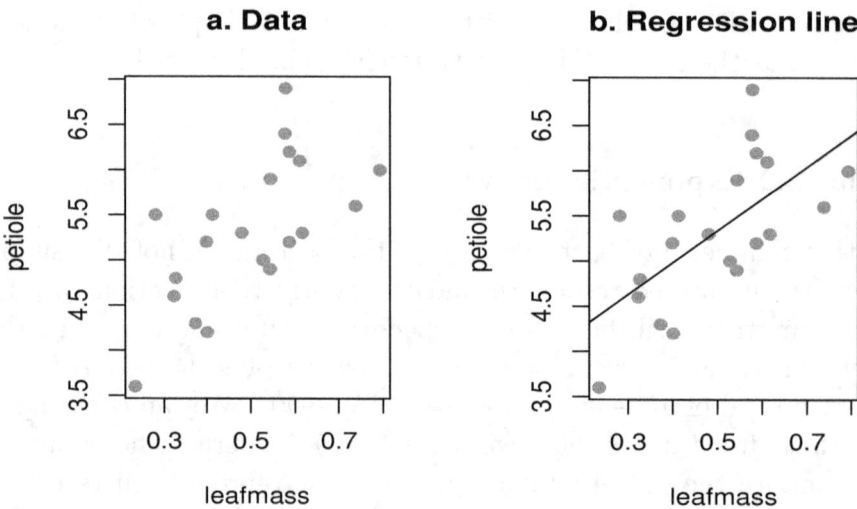

Figure 12.2: Mass of leaf and length of petiole for 20 leaves of sugar maple trees

12.1.3 Understanding a model

The details of the modeling process, using a simple linear regression model, will be described in Section 12.2. In this section, a good understanding of the model will be developed before learning how to produce it. In general, for two random variables Y and X, the simple linear regression model is equation (12.3). The letters Y and X can be replaced with the name of the variables. In Example 12.1, equation (12.3) could be written as equation (12.2). In Example 12.3, equation (12.3) can be written as:

$$petiole = \beta_o + \beta_1 leafmass + \varepsilon \tag{12.4}$$

The error term ε

The symbol ε represents the error term, a random term with an expected value equal to 0. The error represents the departures from what otherwise would look like a deterministic relationship between the variables. Things are hardly deterministic in nature; individual variability around an average pattern is always present. The length of the petiole and the mass of the leaf show an association, but the length of the petiole is not exactly determined by the mass of the leaf. Other variables might be related to the length of the petiole, maybe the length of the leaf (Do longer leaves tend to have longer petioles?), or maybe the distance from the trunk of the tree to the leaf (Does the petiole of leaves at the end of a branch tend to be shorter?). Even after thinking of all the variables that might be related to the length of the petiole, we still might not be able to exactly predict the length of the petiole based on all those characteristics of the leaf. There might be two leaves that have the same measurements and positions, but are supported by petioles of different length. In summary, those departures represented by the symbol ε are a combination of the effect of variables other than the explanatory variable used in the model (mass of the leaf in Example 12.3) and departures due to just pure randomness.

Before data are available to estimate the value of the **parameters** of the model, the parameters are written using symbols such as β_0 and β_1, as in (12.3). Once data are collected, the values of the parameters are estimated, and the error term is replaced by its expected value of 0.

Continuing with Example 12.3

For the mass and petiole example, the model in (12.3) and (12.4) becomes:

$$\widehat{petiole} = 3.575 + 3.524 \ leafmass \tag{12.5}$$

The graph of equation (12.5) is the line plotted over the scatterplot in Figure 12.2.b. Later it will be discussed **how** those numbers (3.575 and 3.524) were calculated. First, we will discuss **what** those numbers represent.

Equation (12.5) is called a regression line or simple linear regression model, 'simple' because it only has one explanatory variable (leaf mass), and linear because the parameters do not appear with exponents or as exponents. The first parameter in the model β_o is called the **intercept**, its estimated value for Example 12.3 is $\hat{\beta}_o = 3.575$. The second parameter in the model is the slope β_1, the one that appears multiplied the explanatory variable. The estimated slope for Example 12.3 is $\hat{\beta}_1 = 3.524$.

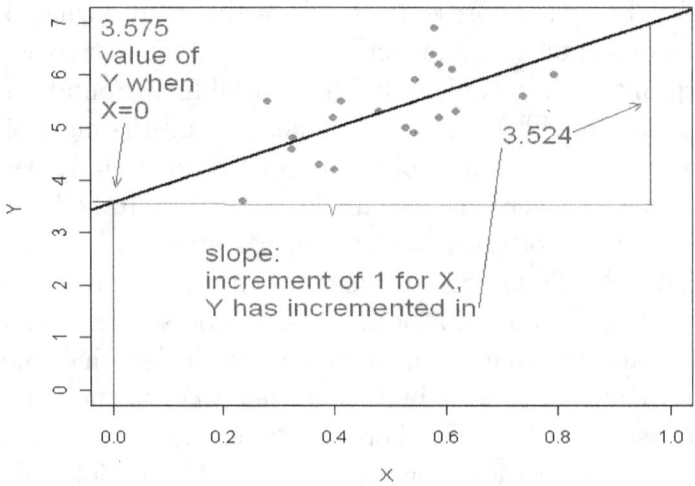

Figure 12.3: Intercept and slope in model (12.5) for Example 12.3

Figure 12.3 helps us to understand the geometrical and statistical meaning of the intercept and the slope. Figure 12.3 is similar to Figure 12.2b only in that the origin $(0,0)$ is included in order to facilitate the interpretation.

Interpreting the intercept

The value 3.575 is the **intercept** of the regression line; that is, the estimated value of the response variable when the explanatory variable takes the value 0. Would it make sense to talk about a leaf with mass 0? Of course not. In this case the intercept has a geometrical meaning, but it does not need to be interpreted from the statistical point of view because it does not make sense to think that the explanatory variable (mass) takes the value 0. If it were reasonable for the explanatory variable to take the value 0, the value of the intercept would be interpreted only if observations near the value $x = 0$

were available. Otherwise, if there are no observations close to $x = 0$, there is no way of knowing if the linear model is valid in the region close to the vertical axis and we would prefer to refrain from interpreting the intercept value.

The slope of the regression line always has a statistical meaning. In the example, the **slope** of the estimated line is 3.524; that means that for each additional gram of mass of the leaf, the estimated length of the petiole increases by 3.524 centimeters. Look at Figure 12.3: for a increment of 1 unit in the horizontal axis, the change with reference to the vertical axis is 3.524 units. One characteristic of the straight line is that the increment in $\hat{Y}$ per unit increment in X is constant, the increment in $\hat{Y}$ is the same when X has low values and when X has high values. The slope of a straight line is the same in any region of the line. Thus, when one decides to use a linear model to represent the relationship between two variables, an implicit assumption is that the change in the response variable per unit increment in the explanatory variable is constant. **Interpreting the slope**

Regression through the origin

A reduced version of the simple linear regression model is a line that passes through the origin. **A line that passes through the origin**

$$Y = \beta_1 X + \varepsilon \qquad (12.6)$$

That model is used when the variables X and Y are such that if X is 0, it only makes sense for Y to take the value 0.

12.2 The modeling process

The modeling process involves the identification of an appropriate type of model, estimation of the parameters, and evaluation of the model. It will be described using the example of the sugar maple leaves.

12.2.1 Identification

The data file *sugarmaple* contains the values of 4 variables (length of petiole, mass, length, and width) for 20 randomly selected sugar maple leaves. Two questions that motivated the collection of the data were: **Looking for a model**

- Is the length of the petiole, which supports the leaf, associated with the mass of the leaf? The objective is to understand the relationship between these two variables.

- What characteristic of the leaf is a good predictor of its mass? A special scale is needed to weigh leaves. However, the dimensions of the leaf such as length or width can be measured with a simple ruler. This example mimics a pretty common situation in real life; variables that are easier or cheaper to measure are used to predict other variables, whose measurement is more difficult to obtain.

Looking for a model that answers the first question, a scatterplot for mass of leaf and length of petiole was prepared (Figure 12.2a). The scatterplot suggests a linear relationship between length of petiole and mass (r=0.6529091).

Looking for a simple model to answer the second question, the correlations between mass and each one of the other variables were calculated: cor(*mass, length*)= 0.88, cor(*mass, width*)= 0.91, cor(*mass, petiole*)=0.65. The highest correlation is the one between mass and width, thus the variable *width* was selected as an explanatory variable to explain the response variable *mass*. Figure 12.1 clearly shows that the relationship between those two variables is linear.

12.2.2   Estimation

It is easy to estimate the values of the parameters of the regression model using software. In MINITAB, the option STAT>Regression>Regression is used. In **R**, after reading the data, the command

lm(petiole∼leafmass)

will produce the equation of the linear model and the command

abline(lm(petiole∼leafmass))

Estimating the parameters
will add the regression line to the scatterplot. For more details, see the section *Using statistical software*. Graphing calculators can also be used to estimate the parameters of a regression model.

But, how does the software calculate those estimated values? In Figure 12.4, some line segments have been added to the scatterplot in Figure 12.2b. Those segments represent the distances from the points that represent the observations to the line that represents the model. We want the regression line to pass as close as possible to the points, but what is understood for 'as close as possible'? One reasonable criterion would be that a line is traced such that the sum of the absolute values of the distances from the points to the line is as small as possible. However, due to mathematical practicality (working with squares is easier than working with absolute values), the criterion that has become widely adopted is that the **sum of squares of the distances from the points to the line is as small as possible**. That criterion is known as the **least squares principle**. Calculus, not a prerequisite for this course, is usually applied in order to derive the formulas for the estimators of the parameters. Thus, the formulas will be provided but, unfortunately, their justification cannot be included. If you have learned calculus and are interested in the derivation of the formula for estimating the slope in the regression through the origin model, see Helfgott & Moore (2011).

There is more than one way of writing the formulas to calculate the least squares estimators of the parameters in a simple linear regression model. A convenient one is:

$$\text{slope}: \quad \hat{\beta}_1 = r \times \frac{s_y}{s_x} \qquad \text{intercept}: \quad \hat{\beta}_o = \bar{y} - \hat{\beta}_1 \bar{x} \qquad (12.7)$$

where r is the correlation coefficient and s_x and s_y are the standard deviations **Estimators** of the variables X and Y, respectively. In many textbooks, the symbols used **for slope** for $\hat{\beta}_o$ and $\hat{\beta}_1$ are a and b. **and intercept**

Continuing with Example 12.3

The basic descriptive statistics for $X : mass$ and $Y : petiole$ are:

```
mean(mass)    = 0.49675        sd(mass)  =  0.1493064
mean(petiole) = 5.325          sd(petiole)= 0.8058177
            cor(petiole,mass)= 0.6529091
```

Thus, using the formulas in (12.7):

$$\hat{\beta}_1 = 0.6529091 \times \tfrac{0.1493064}{0.8058177} = 3.523798 \sim 3.524$$
$$\hat{\beta}_o = 5.325 - 3.523798 \times 0.49675 = 3.574553 \sim 3.575$$

12.2.3 Evaluation

The model in (12.5), with regard to Example 12.3, explains the length of **Evaluating** petiole in terms of mass of the leaf it supports (for the species 'sugar maple'). **the model** Evidently, not all the leaves have the same petiole length. How much of that variability is explained by the fact that not all the leaves are of the same mass? How effective is the model in explaining the variability in *petiole* among leaves in terms of *mass*? In the case of simple linear regression, that question is answered by calculating $r^2 = 0.6529091^2 = 0.4262903$. So, even when the length of the petiole and leaf mass are correlated, the mass of the leaf does not tell the whole story about the length of the petiole using this model. Only 42.62% of the variability in petiole length among leaves is explained by the fact that not all leaves have the same mass, when using this model. The usual threshold for $r^2 \times 100$, for the model to be considered satisfactory, is 75%. The mass of the leaf is far from completely explaining all the variability in length shown by the petioles in the sample.

In Example 12.1, $r^2 \times 100 = (0.9113025^2) \times 100 = 83\%$, not bad at all. The model in (12.1) to predict mass based on width does a fairly good job.

The statistic $R - squared$ or R^2 is calculated, in the case of simple linear regression, as the square of the coefficient of correlation. $R - squared$ is also called the **coefficient of determination**. A more general definition of the coefficient of determination, also applicable to models other than simple linear regression, will be given in Section 12.4.

Using a model

If you are happy with a model: USE it !!

Assume that a scale is not available to weigh a sugar maple leaf, but a ruler is available to measure its width. You measure the leaf and it is 9 cm wide. The variable *width* can be substituted by the value 9 in the model in equation (12.1). The estimated value of the mass is:

$$\hat{mass} = -0.4217 + 0.0988 \times 9 = 0.4675 grams$$

The model can be used to make predictions or estimations for other individuals that are similar to the individuals in the sample. It would not be appropriate to use a model for mass in terms of width, built using data of sugar maple leaves for poplar or oak leaves.

R:
linear
regression

Using statistical software:
The following commands read the data for Examples 12.1 and 12.3, produce Figure 12.1, and estimate the model in (12.1)

```
width<-c(9.6, 7.2, 8.9, 11.5, 11.0, 12.1, 7.8, 10.0, 9.5, 9.5, 9.6,
7.9, 6.9, 9.7, 9.3, 9.3, 8.5, 7.6, 9.5, 10.5)
leafmass<-c(0.587,0.235,0.480,0.738,0.587,0.793,0.401,0.611,0.617,
0.578,0.373,0.327,0.281,0.528,0.399,0.543,0.413,0.323,0.544,0.577)
petiole<-c(6.2, 3.6, 5.3, 5.6, 5.2, 6.0, 4.2, 6.1, 5.3, 6.9,
 4.3, 4.8, 5.5, 5.0, 5.2, 4.9, 5.5, 4.6, 5.9, 6.4)
plot(width,leafmass,pch=19, col= 'red ', main='SUGAR MAPLE LEAVES',
xlab='WIDTH (cm)', ylab='MASS(g) ')
abline(lm(leafmass~ width))
lm(leafmass~width)
```

12.3   Estimated values and residuals

Residuals are the differences between the observed values for the variable Y and the values estimated for Y using a regression model. Figure 12.4 displays the same scatterplot and regression line as Figure 12.1, except the 'residuals' (represented by segments) have been added.

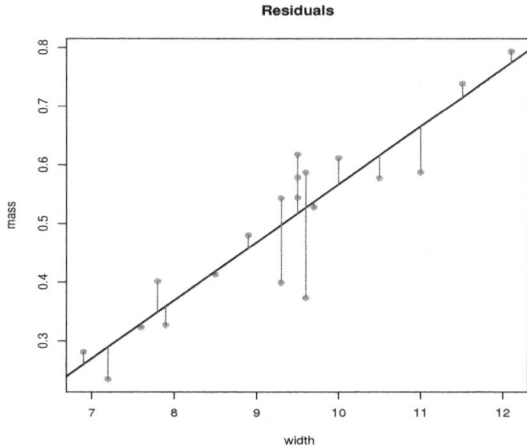

.   Figure 12.4: Residuals in the simple linear regression line

The estimated value of the variable Y for a given value of the variable X is the value obtained from the equation of the model when replacing the X variable by its given value. Using the model in (11.3), for the ith observation,

$$\hat{y}_i = \hat{\beta}_o + \hat{\beta}_1 x_i \qquad (12.8)$$

The difference between the observed value and the value estimated by the model is called the **residual**.

$$e_i = y_i - \hat{y}_i \qquad (12.9)$$

Residual: difference between observed and estimated value

Continuing with Example 12.1

Figure 12.5 and Figure 12.1 are scatterplots for the same data set. Consider the widest leaf in the data set. The widest leaf is observation #6 in the original data file and it is identified in Figure 12.5. The width is $x_6 = 12.1$ centimeters, thus

$$m\hat{a}ss = -0.4217 + 0.0988 \times 12.1 = 0.77378$$

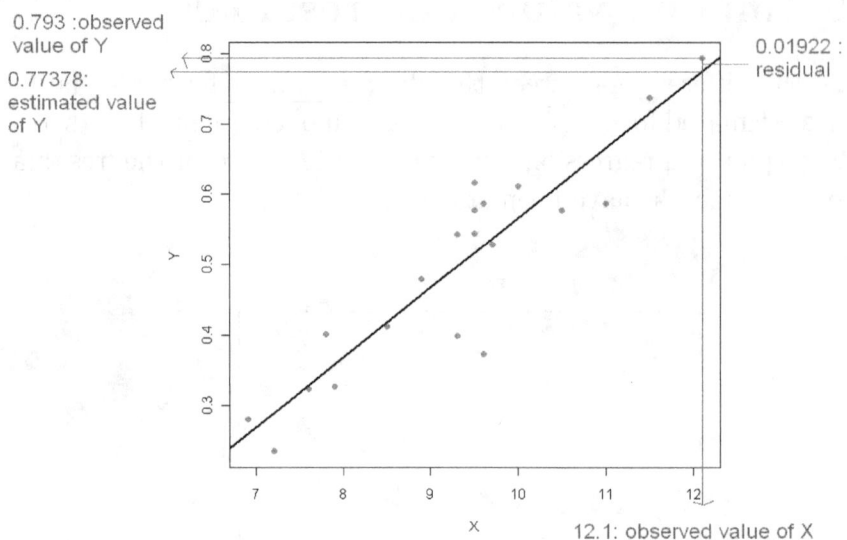

Figure 12.5: Values of x_i, y_i , $\hat{y}_i$, and e_i for one observation

In other words, the model in (12.1) estimates that sugar maple leaves that are 12.1 centimeters wide have on average a mass of 0.77378, based on the observations of the 20 leaves in the sample. Table 12.1 contains the original data and several calculations for Example 12.1. The estimated values of mass $\hat{y}_i$ for all the observations ($i = 1, ..., 12$) are in the third column. The leaf with width 12.1 is in the sixth row. For the sixth observation in the sample (which was 12.1 cm wide), the actual mass is 0.793 g, $y_6 = 0.793$. The residual for that specific observation is:

$$e_6 = y_6 - \hat{y}_6 = 0.793 - 0.77378 = 0.01922$$

Figure 12.5 marks the observed value for width and mass, the estimated value for mass, and the residual for that observation. The residuals for all the 20 observations are shown in column 5 of Table 12.1. The fifth column, (e_i), was calculated by simply subtracting column 4 from column 2.

SSE is the name given to the **sum of square of residuals**. This is an important quantity that summarizes how far the estimated values are from the actually observed values for the response variable. SSE is the sum of the squares of all the residuals in Figure 12.4; those squares are in the sixth column of Table 12.1. In Example 12.1, SSE=0.07180. SSE is used in the next section in the evaluation of the model.

SSE:
Sum of squares
of residuals

$$SSE = \sum_i^n e_i^2 = \sum_i^n (y_i - \hat{y}_i)^2 \qquad (12.10)$$

Table 12.1: Calculation of total and unexplained variability for Example 12.1

Leaf ID	Mass (g)	Width (cm)	Estimated mass	Residual	Squared residual	Squared distance to mean
i	y_i	x_i	$\hat{y}_i$	$e_i = y_i - \hat{y}_i$	e_i^2	$(y_i - \bar{y})^2$
1	0.587	9.6	0.526886	0.060114	0.0036137	0.0081451
2	0.235	7.2	0.289749	-0.054749	0.0029974	0.0685131
3	0.480	8.9	0.457721	0.022279	0.0004963	0.0002806
4	0.738	11.5	0.714620	0.023380	0.0005466	0.0582016
5	0.587	11.0	0.665216	-0.078216	0.0061178	0.0081451
6	0.793	12.1	0.773904	0.019096	0.0003646	0.0877641
7	0.401	7.8	0.349033	0.051967	0.0027005	0.0091681
8	0.611	10.0	0.566409	0.044591	0.0019883	0.0130531
9	0.617	9.5	0.517005	0.099995	0.0099989	0.0144601
10	0.578	9.5	0.517005	0.060995	0.0037203	0.0066016
11	0.373	9.6	0.526886	-0.153886	0.0236810	0.0153141
12	0.327	7.9	0.358914	-0.031914	0.0010185	0.0288151
13	0.281	6.9	0.260107	0.020893	0.0004365	0.0465481
14	0.528	9.7	0.536767	-0.008767	0.0000769	0.0009766
15	0.399	9.3	0.497244	-0.098244	0.0096519	0.0095551
16	0.543	9.3	0.497244	0.045756	0.0020936	0.0021391
17	0.413	8.5	0.418198	-0.005198	0.0000270	0.0070141
18	0.323	7.6	0.329272	-0.006272	0.0000393	0.0301891
19	0.544	9.5	0.517005	0.026995	0.0007287	0.0022326
20	0.577	10.5	0.615813	-0.038813	0.0015064	0.0064401

12.4 Coefficient of determination: R^2

In Section 12.2.3, the model has been evaluated by calculating the **coefficient of determination** (R^2) as r^2, where r is the correlation coefficient. This way of calculating R^2 is correct for the simple regression model. However, a more general definition for R^2 exists; which is applicable to a wider range of models:

$$R^2 = \frac{Explained\ Variability}{Total\ Variability} = \frac{Total\ - Unexplained\ Variability}{Total\ Variability}$$

(12.11)

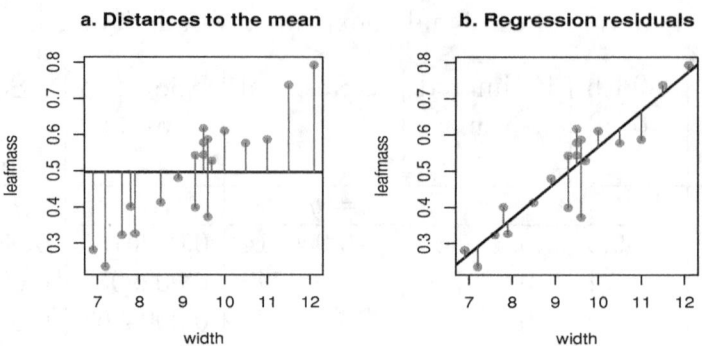

Figure 12.6: Distances to the mean and regression residuals for Example 12.1

Total variability

SST_m **or total variability**

Let us focus on the response variable Y: *leaf mass* in Example 12.1. Not all the leaves have the same mass, so we need to quantify that variability. Figure 12.6a compares the individual values of mass with the mean of mass. For the widest leaf, its mass is well above the mean mass (0.49675) of the 20 leaves in the sample, $y_i - \bar{y}$=0.793-0.49675=0.29625. The last column of Table 12.1 contains the square of the distances of the values of the response variable (mass) to the mean for all the 20 observations in the sample. The total variability in mass is quantified by adding the squares of the distances from the individual values to the mean (this should remind you of the numerator of the expression for the variance in equation (2.4) in Chapter 2). For the data in Example 12.1, $SST_m = 0.42356$, where

$$Total \;\; variability = SST_m = \sum_{i=1}^{n}(y_i - \bar{y})^2 \qquad (12.12)$$

Unexplained variability

SSE or unexplained variability

Compare Figures 12.6a and 12.6b. Focus on the widest leaf. Its mass is well above the mean mass, and in part this is because it is the widest leaf in the sample. However, its mass is above what was predicted by the model (Figure 12.5). That excess mass (0.01922), or residual, cannot be explained in terms of width. The residuals, for all the observations, are displayed in Figure 12.6b and in the second to last column of Table 12.1. The sum of the squares of those residuals (or SSE) is the **unexplained variability**. For the data in Example 12.1, SSE=0.07180

Explained variability

Explained variability is the difference between the total variability and the unexplained variability. In Example 12.1, the explained variability would be the variability in mass that can be explained by the fact that not all the leaves are of the same width. For the data in Example 12.1: Explained variability=Total variability − Unexplained variability=$SST_m - SSE = 0.42356 - 0.07180 = 0.35176$.

Now that all the components have been defined, equation (12.11) can be re-written as:

$$R^2 = \frac{SST_m - SSE}{SST_m} \tag{12.13}$$

The coefficient of determination is the fraction of the total variability in the response variable that is explained by the explanatory variables using a particular model. For the data in Example 12.1:

$$R^2 = \frac{0.42356 - 0.07180}{0.42356} = \frac{0.35176}{0.42356} = 0.8304845 \text{ or } 83\%.$$

The value of the coefficient of determination, 0.83 or 83%, indicates that the model can be used. The usual threshold for considering a model adequate is 0.75 or 75%. The value of *R-squared* is calculated automatically by the statistical software.

12.5   The dangers of extrapolation

In the previous sections, a model has been used to estimate the values of the response variable. It is important to be aware of the observed values for the explanatory variable. It is dangerous to extrapolate and use the model to predict the value of Y for a value of X that is far from the interval in which observations were made. A dramatic example of the dangers of extrapolation is presented next. A linear model to explain stature in terms of age was fitted using data for males between 4 and 8 years of age, a period in which growth is practically linear (first plot in Figure 12.7). The variable X is age in months and the variable Y is the 50th percentile or median of stature in centimeters for the U.S. male population. The data file *statureM4-8* was prepared with data available from the Centers for Disease Control and Prevention website. The linear correlation is almost 1, and the estimated linear regression model is:

$$estimated\ median\ stature = 76.8081 + 0.5345\ months \tag{12.14}$$

or

$$\hat{Y} = 76.8081 + 0.5345\ X. \tag{12.15}$$

Warning: extrapolating can be risky

Stature by age

The model gives a good description of the height of males in the USA from 4 to 8 years of age (48 to 96 months) in relation to age. For each additional month of age, the estimated median stature is 0.5345 inch higher. However, it would be inappropriate to use the model to predict the median height for individuals much younger than 4 years old or much older than 8 years old, because growth is not linear throughout the entire life span. The second plot in Figure 12.7 displays the values of stature for ages 0 to 20 years of age (0 to 240 months) (data file *statureM0-20*). The interval 48-96 months has been marked on the X axis, and the model in equation (12.14) has been added to show how poor the predictions would be outside the interval (X:48-96) for which the model was fitted.

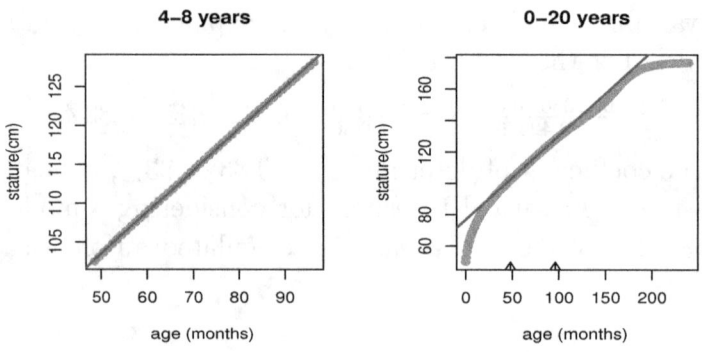

Figure 12.7: Median stature (cm) by age in months

12.6   Outliers, leverage and influential points

Consider the following data set for the variables cigarette consumption per capita in 1930 and deaths (per million habitants) by lung cancer (Freedman et al., 1976).

This is a very interesting data set. The complete data set is displayed in the first plot in Figure 12.8. The second plot displays the data set after removing the observation corresponding to the USA. Compare the regression results in both graphs:

	With all the data in Table 12.2	Without the USA
Regression model	$\hat{y} = 67.6 + 0.228x$	$\hat{y} = 9.1 + 0.369x$
R-Squared	54.4%	88.9%

Table 12.2: Cigarette consumption and male deaths by lung cancer 20 years later

Year:	1930	1950
	Cigarettes	Deaths
Country	per capita(X)	per million (Y)
1 Australia	480	180
2 Canada	500	150
3 Denmark	380	170
4 Finland	1100	350
5 Great Britain	1100	460
6 Iceland	230	60
7 Netherlands	490	240
8 Norway	250	90
9 Sweden	300	110
10 Switzerland	510	250
11 USA	1300	200

Source : Freedman et al. (1976)

Notice that the estimated slope of the simple linear regression model changes with the removal of the USA from the data set. The value of the correlation coefficient r, and consequently of the coefficient of determination R^2, change dramatically. An observation whose presence or absence makes a noticeable difference in the regression results is called an **influential observation**. Now comes the question: Why does the presence of this observation make such a substantial difference? For one thing, the relationship between cigarette smoking and deaths by lung cancer seems to be different for the USA than for the other countries. The dot in the scatterplot corresponding to the USA is outside the pattern described by the other countries. The number of deaths is low for the high consumption of cigarettes. Observations that are clearly outside the pattern described by the rest of the data are called **outliers**. It is not enough to be a clear outlier in order to have a strong impact on the regression results; the position of the observation also matters. The farther the value of the explanatory variable X is from the mean, the more **leverage** the observation has. An observation x close to the mean $\bar{x}$ does not have the same leverage as observations that are at either extreme and far from the mean.

What to do if there are outliers?

What do we do with outliers? First, check if the observation is valid. If it is valid, DO NOT remove it, specially if the data come from a sample. There might be other individuals in the population who show a behavior

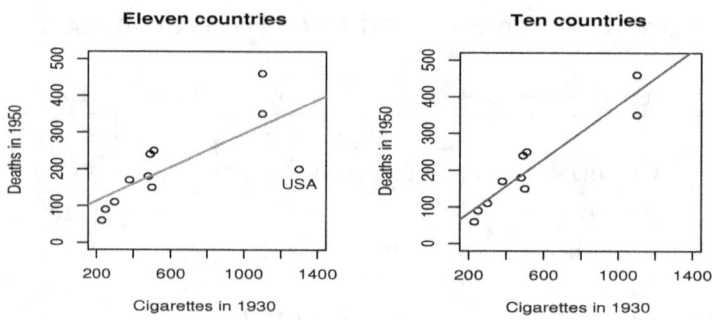

Figure 12.8: Cigarette smoking and deaths by lung cancer

similar to that of the outlier. The situation is a little different if the data set corresponds to all the individuals in a population or subpopulation, for example if the 11 countries in Table 12.2 were the only countries in a subpopulation of industrialized countries where cigarettes were smoked in 1930. Then, if the outlier is removed, one could say 'the model describes the relationship between cigarette smoking and lung cancer for all the countries in the subpopulation except the USA.'

In least squares regression, the leverage is proportional to the square of the distance $(x_i - \bar{x})$. The least squares method is very sensitive to outliers that are in a position of high leverage. If you do not want a single observation to have such impact in the regression results, look for an alternative to the least squares principle. Nowadays, resistant line calculations (that are not as sensitive to outliers as least squares estimators) are performed by most statistical packages.

Outliers should not be considered a nuisance; they might be the starting point for an interesting line of research. For example: Why is the death rate in the USA lower than expected given the high consumption of cigarettes? That is a question for the specialists in nursing, public health, or medicine to answer.

Observations in a position of high leverage can also make the linear relationship look stronger than it really is according to the majority of the data. The scatterplot for simulated data in Figure 12.9a shows a low correlation. In Figure 12.9b, a single point has been added to the previous data set, and the correlation goes up from -0.103 to 0.867. It is always recommended to not only look at the numerical value of the correlation, but also at the scatterplot in order to have a clear picture of the situation.

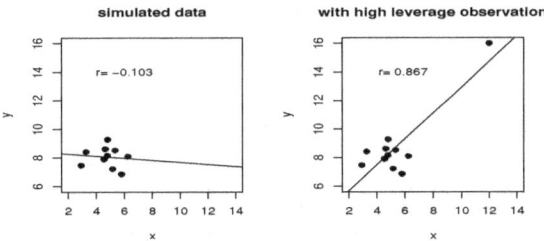

Figure 12.9: Effect of one observation with high leverage

	Data in Figure 12.9a	Data in Figure 12.9b
Regression model	$\hat{y} = 8.39 - 0.072x$	$\hat{y} = 3.87 + 0.904x$
R-Squared	1.1%	75.2%

12.7 Multiple regression

More than one explanatory variable can be used in a model. Employing statistical software, the estimation of the parameters is done in a very similar way to simple linear regression, but with different formulas than those in expression (12.7). Now we will add an additional explanatory variable to Example 12.1.

Example 12.4 Predicting mass with width and petiole

We want to build a model to predict the mass of a sugar maple leaf based on characteristics that are easy to measure. Consider a model to predict the mass of a leaf from its width AND the length of its petiole.

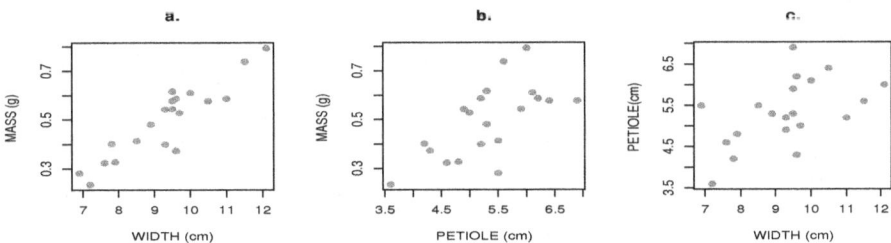

Figure 12.10: Mass, length and petiole of sugar maple leaves

Figure 12.10 indicates a clear linear relationship between width and mass of the leaf. Figure 12.10b indicates that there is also some association between mass and petiole. A regression model with two explanatory variables was fitted:

$$\hat{mass} = -0.53113 + 0.08513 \; width + 0.04444 \; petiole \; .$$

The coefficient for *petiole* indicates that keeping *width* constant, the estimated mean *mass* goes up in 0.044 grams for each additional centimeter of *petiole*. For the simple linear regression model $\hat{mass} = -0.4217 + 0.0988 \; width$, we have SSE=0.07180. For the multiple regression model, SSE= 0.05646. When an explanatory variable is added to a model, the value of SSE always goes down. The total variability $\sum (y_i - \bar{y})^2$ does not change because it is calculated only with the values of the response variable Y. Thus, by adding variables, the explained variability goes up and consequently the coefficient of determination also goes up. However, the increment in R^2 might sometimes be marginal. In the case of Examples 12.1 and 12.4, the value of R^2 for the regression line was 0.83; for the multiple regression model it is 0.8721.

When explanatory variables are added, SSE goes down

One important point to remember when applying multiple regression is that **it is preferable to work with explanatory variables that are not too highly associated with each other**. In Figure 12.10c, notice that there is some association between the two explanatory variables (width of the leaf and length of the petiole). Fortunately, in this case the relationship between width of the leaf and length of the petiole is not a very strong relationship. It is not good that a strong relationship exists between explanatory variables in a multiple regression case. This problem is known as 'collinearity.' One bad consequence of collinearity is that in a further evaluation of the participation of the variables (not included here), we might end up concluding that none of them are important simply because they are more or less telling the same story.

R & MINITAB:
Multiple regression

Using statistical software:
The following commands read the data for the three variables and estimate the parameters of the multiple regression model:

```
width<-c(9.6,7.2,8.9,11.5,11.0,12.1,7.8,10.0,9.5,9.5,9.6,
7.9,6.9,9.7,9.3,9.3,8.5,7.6,9.5,10.5)
leafmass<-c(0.587,0.235,0.480,0.738,0.587,0.793,0.401,0.611,0.617,
0.578,0.373,0.327,0.281,0.528,0.399,0.543,0.413,0.323,0.544,0.577)
petiole<-c(6.2,3.6,5.3,5.6,5.2,6.0,4.2,6.1,5.3,6.9,
4.3,4.8,5.5,5.0,5.2,4.9,5.5,4.6,5.9,6.4)
lm(leafmass ~ width + petiole)
```

12.8 Non-linear models

Example 12.5 Mass and length of wing

Not all relationships are linear. Figure 12.11a displays the typical mass and the typical wing length for females of 28 species of the family Phasianidae (data file *hens* prepared with data from Lislevand et al., 2007). Notice that the pattern shows some curvature as mass increases. In Chapter 2 (section 2.6.2), the logarithmic transformation was introduced. Figure 12.11b displays the scatterplot for the logarithms of mass and wing length; notice that the pattern is closer to being linear than the pattern in Figure 12.11a.

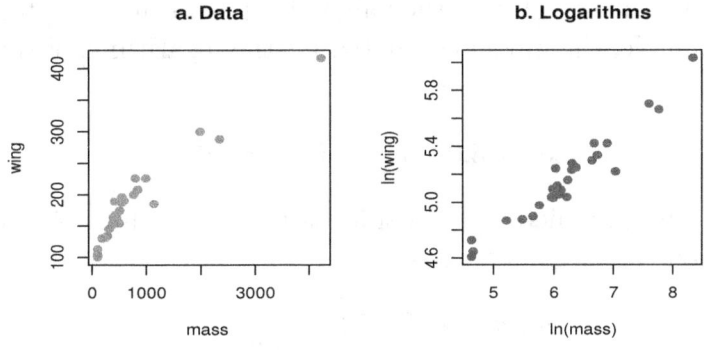

Figure 12.11: Typical mass (g) and length of wing (cm) for females of 28 species of Phasianidae

This is an example in which a non-linear model relating the variables *mass* and *wing* is more appropriate than a regression line. Consider the non-linear model

$$y = c \, x^b \tag{12.16}$$

This particular non-linear model is called a 'power law' or 'allometric' model. It is not only used in biology, but in economics as well. Applying logarithms to both sides of the equation we obtain:

$$ln(y) = ln(c) + b \, ln(x) \qquad or \qquad ln(y) = a + b \, ln(x) \tag{12.17}$$

where $a = ln(c)$ or $c = e^a$. Equation (12.17) indicates that there is a linear relationship between the logarithms of the variables. Thus, when the scatterplot for two variables X and Y shows a pattern similar to the one in Figure 12.11a, i.e., rapid growth at the beginning and slow growth at the end, a model like the one in equation 12.16 is suggested. To check if this type of model is a good choice, a scatterplot with the logarithms of the data is prepared. If the scatterplot for the logarithms shows a linear pattern, then the model in equation (12.16) should be considered. This type of model is suggested for Exercises 12.5 and 12.6. Next, two-ways of estimating the parameters of the nonlinear model will be discussed.

12.8.1 Transforming the variables to make the model linear

The parameters a and b of the linear model in (12.17) can be estimated using the formulas in (12.7) or statistical software for simple linear regression. Then the model in (12.16) can be written using the estimated value for b and c, remembering that $c = e^a$ and the value of a has been estimated. For the mass/wing example, the model estimated for the logarithms of the data is:

$$ln(\widehat{wing}) = 3.0066 \ + \ 0.3478 \ ln(mass)$$

The function exp is applied to both sides of the equation to obtain the estimated model on the original variables:

$$\hat{wing} = 20.21854 \ mass^{0.3478}$$

where $exp(3.0066) = 20.21854$. In this section, the model was 'linearized' by using logarithms in order to apply the estimation process common to linear models. The model, with the parameters estimated using linearization, is represented by the curve that appears at the bottom in Figure 12.12.

12.8.2 Non-linear estimation *

A second optional stage in the estimation of the parameters of a non-linear model is to apply non-linear estimation. Non-linear estimation is a numerical procedure, there are no formulas such as the ones in (12.7). The idea is to start from some initial values for the parameters, make slight changes in the parameters, calculate the SSE and continue in that path until SSE can no longer be reduced. The algorithm used by computer software looks for the shortest path to arrive at the values that minimize the sum of squares of residuals SSE. Algorithms for non-linear estimation require initial values.

The estimated values obtained by using transformations, as in the previous section, are excellent initial values for the non-linear estimation.

The model for Example 12.5, with the parameters estimated using non-linear estimation, is represented by the curve that appears at the top in Figure 12.12. The sum of squares of residuals (SSE) when the model is estimated using linearization (7689.211) can be reduced to 6860.829 by using nonlinear estimation. The estimated model becomes:

$$\hat{wing} = 16.9099140 \ mass^{0.3759}.$$

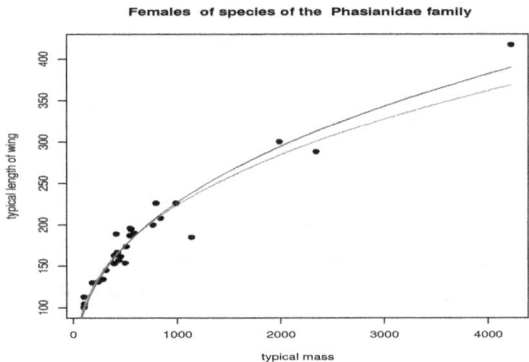

Figure 12.12: Non-linear model $\hat{wing} = c * mass^b$, using two methods of estimating the parameters

In this example, the relationship between the two variables is quite clear and both methods of estimation produce very similar models (the curves in Figure 12.12). In situations where there is a lot of variability, the difference in SSE between the two methods of estimation can be more dramatic.

The following commands in R read the data and fit the model using non-linear estimation. The values 10.21854 and 0.2478 are the initial estimates obtained from the linearization method. The equation of the non-linear model, in this case the power law model, should be included in the command 'nls':

R: *Nonlinear estimation*

```
mass<-c(183.7,2343.8,501,439,425,238.5,284,315,397.5,394,547,455,
545,588,510,1136,386,103,106,103,839,555,414,988,766,1985,795,4222)
wing<-c(130,288,154,157,167,131,134,145,153,163,196,162,187,190,
174,185,154,113,104.3,100.3,208,195,189,226,200,300,226,417)
nls(wing~c*mass^b,start=list(c=20.21854,b=0.3478), trace=TRUE)
```

The output is:

```
7689.211 :   20.21854  0.34780
7153.312 :   16.7007703  0.3752177
6860.833 :   16.9123273  0.3758416
6860.829 :   16.9099140  0.3758534
Nonlinear regression model
  model:  wing ~ c * mass^b
   data:  parent.frame()
       c        b
16.9099  0.3759
 residual sum-of-squares: 6861
Number of iterations to convergence: 3
Achieved convergence tolerance: 4.916e-06
```

Once the parameters of the model were estimated, the commands used to produce Figure 12.12 are listed below. Of course, once the equation of the model is obtained, any software can be used to plot the data and the curves.

```
plot(mass,wing,pch=19,main='Females  of species of the
Phasianidae family',xlab='typical mass',ylab='typical length of wing')
curve(20.21854*x^0.3478,from=0,to=4222,add=TRUE, col='red')
curve(16.9099*x^ 0.3759,from=0,to=4222,add=TRUE, col='blue')
```

MINITAB: Nonlinear regression

Starting with version 16, MINITAB also is capable of non-linear estimation. Open the data file hens. From the MINITAB menu, select STAT > Regression > Non-linear regression. Indicate the response variable (wing), and from the button 'catalog of functions,' pick the 'power concave' option and click OK. In the 'actual predictor' window, enter the variable X (mass), click OK. Press the button 'Parameters' and enter the initial values 20.21854 and 0.3478. Do NOT click the windows 'lock,' and do not enter anything in 'optional constraints.' Click OK. The non-linear model will be automatically plotted over the scatterplot.

12.8.3 Allometry

Some of the examples in this chapter deal with associations between size and shape in biology. The relationship between different parts of the body or between the total mass of the body and the size of one of its parts has an effect on the strength of an organism as it grows, its metabolic rate, its optimal and maximum running or flight speed, and other factors. This field of study is known as Allometry. Many examples of such relationships can be seen in both invertebrates and vertebrates where a change in different aspects of overall body size can result in disproportionate changes in appendages, such as a stag's antlers, insect horns and fish jaws. Two different types of relationships are **isometric** and **allometric** relationships. Models are used to represent those relationships. Isometric relationships are generally represented by linear models, and allometric relationships are generally represented by power models like the one in equation (12.16).

Note: All the variables were quantitative in the regression models studied in this chapter. However, there are other statistical models that allow us to consider categorical and quantitative variables, both as response or as explanatory variables.

12.9 Exercises

12.9.1 Review questions

1. A simple linear regression model was estimated using the data in Table 2.6 (Example 2.7.2). The two variables involved are the typical length of the wing (mm) for males and females of 23 species of Australasian robins. The estimated model is:

   female wing= -0.9842 + 0.9668 male wing

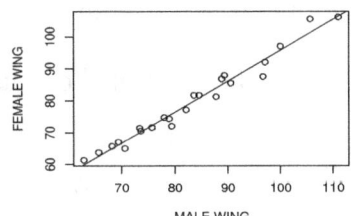

   (a) Which variable is in the role of 'response variable'?

   (b) Which variable is in the role of 'explanatory variable?

   (c) What is the value of the slope? Interpret it.

   (d) What is the value of the intercept? Would it be reasonable to try to interpret it? Why?

   (e) The correlation is $r = 0.986103$. Calculate and INTERPRET the value of R^2.

(f) What is the typical length of the wing for females of the Jacky-winter species according to Table 2.6? What is the estimated typical length of the wing for females of the Jacky-winter species according to the model? Will the residual be positive or negative?

(g) Do you see any really serious outlier?

(h) Mention the name of a species that is in a position of high leverage.

(i) Look only at the observations that are below the regression line. Those are the species with negative residuals. Among those, which species seems to have the largest (in absolute value) residual? Calculate the estimated value $\hat{y}$ and the residual for that species. A large negative residual indicates that the wing for females of that species tends to be shorter from what we would have expected.

2. The relationship between temperature and number of chirps made by crickets appears frequently in popular science and science education websites. Dolbear (1897) reports that during the night, crickets chirp in a very regular way and in relation to the temperature. Based on his observations (he does not report a table of observed values), he proposes the following rule to estimate temperature (F) based on the rate of chirp of the crickets:

$$T = 50 + \frac{N-40}{4}$$

where N is the number of chirps per minute.

(a) Write Dolbear's law in the form of the equation of a simple linear regression model.

$$T = intercept + slope \times chirps$$

(b) Describe the equation you obtained in verbal form, as if you were trying to explain to a friend how to estimate the temperature from the number of cricket chirps per minute.

(c) Interpret the value of the slope.

(d) Dolbear(1897) mentions: 'Below a temperature of 50° the cricket has no energy to waste in music...' Would you consider interpreting the value for the intercept that you calculated? Why?

(e) Bessey & Bessey (1898) include a scatterplot for data they collected in Lincoln, Nebraska, for the chirping of the tree cricket

(*Oecanthus niveus*), under different temperatures, during the summer. They mention that a straight line gives a good approximation when the temperature is between 60 and 80 degrees, but that a curve might be a better fit if lower temperatures were included. From their scatterplot, we selected 6 data points and wrote their approximated values in the table below. Prepare a scatterplot with those six data points. Use statistical software, a calculator, or the formulas in equation (12.7) to estimate the intercept and the slope of a regression model.

Chirps per minute	125	150	135	101	168	181
Temperature (F)	66.8	71.2	69	61.2	76.4	78.3

The basic statistics of those observations are:

	Chirps	Temperature
Mean	143.33	70.48
Standard deviation	29.22	6.30
Correlation	0.9970649	

(f) We do not have the exact numerical values of Bessey & Bessey (1898). Also, we do not know if Dolbear (1897) observed the same species of crickets as Bessey & Bessey. Thus, we do not expect to come up with exactly the same model than Dolbear (1897). However, observe how close your regression model is to Dolbear's law. Working with more than 30 observations, Bessey & Bessey (1898) proposed the model $T = 60 + (N - 92)/4.7$. Write Bessey & Bessey's model in the form of a simple regression model $T = intercept + slope \times chirps$.

3. What is the difference between 'simple' and 'multiple' regression?

12.9.2   Lab for Chapter 12

Work Exercises 1, 3, and 4 in Section 12.9.3 using statistical software. Write a report with your answers; insert the computer output in your report including figures.

12.9.3   Exercises for discussion or homework

1. Using the data on altitude of residence (X) and number of red blood cells (Y) in Table 2.9 and Figure 2.26 in Chapter 2, estimate the pa-

rameters of a simple linear regression model using software (data file *altRBC*). Interpret the values of the estimated slope and R^2.

2. Use the data in Exercise 8 in Chapter 2 to fit a linear regression model for 'foot length' using 'height' as explanatory variables. Plot the data and the regression line. Interpret the slope. Would it be appropriate to interpret the intercept? Why?

3. Use the data in Table 3.3 (Exercise 11 in Chapter 3) to estimate a simple linear regression model to predict the yield (lb) of a potato plant using leaf area as predictor. Interpret the value of the slope. What is the value of R^2? Is this model acceptable? Would you recommend to fit a non-linear model?

4. The data file *falcons* contains information about typical mass, and length of wing and tail, both for males and females of the species of the Falconidae family. It also has information for typical egg mass and clutch size. Prepare scatterplots (for males and females separately) for the following pairs of variables: length of wing vs. mass, length of wing vs. length of tail. Also prepare a scatterplot for clutch size vs. egg mass, and for egg mass vs. female mass.

    (a) Is the number of eggs per nest related to the size of the eggs?

    (b) What type of relationship do you see between the length of the wing and the length of the tail, linear or non-linear? Fit an appropriate model.

    (c) Is the length of the wing related to the mass of the falcons? What type of relationship do you observe, linear or non-linear? Fit a model for males and another for females.

    (d) Is the typical egg mass for the species related to the typical mass of the females of the species? What type of relationship do you observe? Linear or non-linear? Fit an appropriate model.

5. The brain weight (grams) and body weight (kilograms) for 96 species of mammals was analyzed by Sacher & Staffeld (1974).

    A subset of 22 species (Figure 12.13) has been selected for this exercise (data file *brainbody*). Select an appropriate model, estimate the parameters, and write the equation of the model. Use the model to estimate the brain size of a mammal species that has a typical body mass of 600 kg.

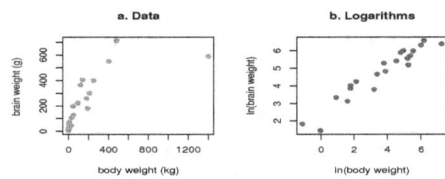

Figure 12.13: Typical body weight (kilograms) and brain weight (grams) for 22 species of mammals

6. Tazawa et al. (2001) report the mean fresh egg mass (grams) and mean embryonic heart beats per minute at 80% of the incubation duration for 9 species of birds. The data are displayed in Table 12.3; the data and their logarithms have been plotted in Figure 12.14. Describe to the best of your ability the relationship between egg mass and heart beat. Fit a model to represent that relationship. Write the equation of the model to explain heart rate in terms of egg mass, using the estimated values for the parameters.

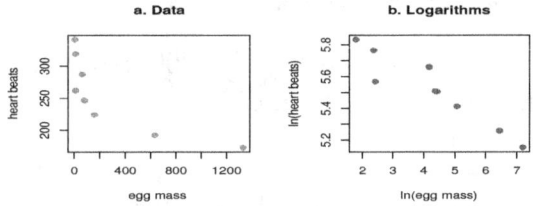

Figure 12.14: Mean egg mass and embryonic heart rate for nine species of birds

References

Bessey, C.A. and Bessey, E.A. (1898) Further notes on thermometer crickets. *The American Naturalist* **32**: 263-264

CDC Growth Charts. Percentile Data Files. Centers for Disease Control and Prevention.

http://www.cdc.gov/growthcharts/percentile_data_files.htm

Dolbear, A.E. (1897) The Cricket as a thermometer. *The American Naturalist* **31**: 970-971

Table 12.3: Mean egg mass and mean embryonic heart rate at 80% of incubation period

Species	egg mass	heart rate
King quail	6.0	341
Japanese quail	10.7	319
Chicken	64.9	287
Duck	79.0	247
Turkey	82.9	246
Peafowl	11.3	262
Goose	158.3	224
Emu	634.0	192
Ostrich	1331.0	173

Source: Tazawa et al. (2001)

Freedman, D., Pisani, R., Purves, R.(1976) *Statistics.* New York: Norton.

Helfgott, M. and Moore, D. (2011) *Introductory Calculus for the Natural Sciences.* ISBN 9781453880838.

Lislevand, T., Figuerola, J. and Szkely, T. (2007) Avian body sizes in relation to fecundity, mating system, display behavior, and resource sharing. *Ecology* **88**: 1605.

Sacher, A. and Staffeld, E.F (1974) Relation of gestation time to brain weight for placental mammals *The American Naturalist* **108**: 593-615.

Stanton, J.M. (2001) Galton, Pearson, and the Peas: A Brief History of Linear Regression for Statistics Instructors. *Journal of Statistics Education* **9**: 3.

Tazawa, H., Pearson, J.T., Komoro, T., and Ar, A. (2001) Allometric relationships between embryonic heart rate and fresh egg mass in birds. *The Journal of Experimental Biology* **204**: 165-174.

Weisberg, S (2011) R-package alr3 http://cran.r-project.org/web/packages/alr3/alr3.pdf

APPENDICES

.1 Preparing a two-way table

An alternative to the use of probability trees for the calculation of a reverse probability is the preparation of a two-way table for a hypothetical population. Rossman and Short (1995) recommend the use of two-way tables to facilitate the understanding of conditional probabilities in an introductory statistics course. Table 6.2 was prepared using the information given in Fan and Levine (2007). The same information was used to prepare the probability tree in Figure 6.4:

1. Among the pregnancies with high-risk of Trisomy 21 (due to the age of the mother and the results from a blood test) the probability of the fetus having Down syndrome is 1/80, thus the probability of NOT having Down syndrome is 79/80.

2. For fetuses with Down syndrome, half show a 'normal marker' in the ultrasound and half show the 'abnormal marker.'

3. All fetuses without Down syndrome show the 'normal marker' in the ultrasound and none shows the 'abnormal marker.'

The details for the preparation of a table like Table 6.2 are described next. How many hypothetical individuals do we need to consider? Since the probabilities of Down and No Down are 1/80 and 79/80, a multiple of 80 is suggested. A large number gives more flexibility in the splitting of the row totals into two columns. For the example, N could be 160 or 160,000 as in Table 6.3.

You are reminded of the warning given at the end of Section 6.3: tables prepared for hypothetical populations should not be used to test hypotheses or calculate confidence intervals. In those two procedures, the sample size makes a difference and here we are working with a hypothetical and arbitrary sample size.

Based on the information given, the totals of the rows are:

	Normal marker	Abnormal marker	Row total
Down syndrome			2
No Down syndrome			158
Column Total			160

Now consider the information given in 2), the total of the first row (2) is split into equal parts: 1 and 1 because, for fetuses with Down syndrome, half of them show the normal marker and half of them show the abnormal marker.

	Normal marker	Abnormal marker	Row total
Down syndrome	1	1	2
No Down syndrome			158
Column Total			160

All fetuses without Down syndrome show normal marker in the ultrasound; thus, there is nobody in the cell 'No Down AND abnormal marker'

	Normal marker	Abnormal marker	Row total
Down syndrome	1	1	2
No Down syndrome	158	0	158
Column Total			160

Now calculate the column totals and the table will be ready to use.

	Normal marker	Abnormal marker	Row total
Down syndrome	1	1	2
No Down syndrome	158	0	158
Column Total	159	1	160

To calculate $P(Down\ syndrome/normal\ marker)$ look at the total of the column 'Normal marker' and notice that there are 159 of those individuals. Only one of them is in the Down syndrome row. Thus,

$$P(Down\ syndrome/normal\ marker) = 1/159.$$

For fetuses in the high-risk group, the probability of Down syndrome is $1/80$. However, if the ultrasound test is negative, the probability of Down syndrome is reduced to $1/159$.

Notice that in the creation of the table for 160 hypothetical individuals, the information given was written into two rows. To find the answers, a column of the table was read. This makes sense because the conditional probabilities in one direction are given and the conditional probabilities in the other direction are calculated. The construction of a table in the general case can be described as follows:

1. There is a partition of the population in two or more categories or classes $(A_1, A_2, A_3, ... A_k)$. Create a table with as many rows as classes.

2. Fix a value for N, the total number of individuals in the hypothetical population.

3. The percent of the population in each class is known, or, equivalently $P(A_1)$, $P(A_2)$,...,$P(A_k)$ are known. Those percentages or probabilities are used to determine the totals at the end of each row.

4. There is an event B to consider, so two columns are needed, one for 'B' and one for 'no B.'

5. The probabilities $P(B/A_i)$ for each one of the classes is known, thus the frequencies in the 'B' column can be written for each row.

6. The frequencies of the second column are calculated by just subtracting the frequencies in the first column for the row totals.

The next figures explain the process of building a two-way table:

	B	No B	Total
A1			
A2			
A3			
Total			N

PREPARE A TABLE WITH THE CLASSES OR GROUPS
IN THE ROWS AND THE EVENTS B AND NO B IN THE
COLUMS
DECIDE A NUMBER OF INDIVIDUALS N SUCH THAT ALL
THE COUNTS WILL BE INTEGERS

	B	No B	Total
A1			N1.
A2			N2.
A3			N3.
Total			N

FILL THE CELLS OF THE ROW TOTALS
ACCORDING TO HOW FREQUENT THE CLASSES
OR CATEGORIES ARE IN THE POPULATION.

N1. = N P(A1)
N2. = N P(A2)
 etc.

	B	No B	Total
A1	N1b		N1.
A2	N2b		N2.
A3	N3b		N3.
Total			N

FILL THE CELLS OF EACH ROW BASED ON THE
KNOWN PROBABILITIES P(B/A1)
FOR EXAMPLE :
N1b = N1   P(B/A1)
THE CELLS IN THE COLUMN No B ARE FILLED JUST
BY CALCULATING THE DIFFERENCES.

	B	No B	Total
A1	N1b		N1.
A2	N2b		N2.
A3	N3b		N3.
Total	Nb		N

CALCULATE THE TOTAL OF THE COLUMN FOR
THE EVENT B (we are calling it Nb)

READ THE COLUMN FOR B IN ORDER TO CALCULATE
THE DESIRED CONDITIONAL PROBABILITY,
FOR EXAMPLE
P(A1/B)= N1b/Nb

.2 Answers to selected review questions

Chapter 1

4. Population: Pima cotton seeds; sample: the 100 seeds planted; variable: if a seeds germinates or not; parameter: proportion of all seeds in the population that would germinate if planted; statistic: proportion of seeds in the 100 in the sample that germinate, $\hat{p} = 0.78$.

5. Population: potato plants of that variety hypothetically growing under the given conditions; parameter: mean yield of all the potatoes in the population; sample: the 20 potatoes planted; variable: yield in pounds of a potato plant, statistic: mean yield of the 20 plants, $\bar{x}=4.75$ lb.

7. A   **8.** C   **9.** A

Chapter 2

1. Quantitative **5.** A)0.028 mm   B)0.025 mm   C)the mean   D) The distribution is skewed left   E) No F)0.006204837 mm   G) Because the 5 cells are not all of the same length. **7.** Check if it is not a typo or incorrect observation. **8.** That the population is actually a mixture of two population. The onion root has an apical region, where cells divide and are smaller, and an elongation region, where cells are longer. When a sample is taken from the whole root, there will be cells from the two regions and the distribution of the variable length might be bimodal. **9.** 12, 9, 2, 4. **11.** $(42 - 32)5/9$, $10(5/9)$ **13.** Coefficient of variation **15.** Side by side boxplots or side by side dotplots **17.** There is a strong negative association, when X increases, Y tends to decrease. B) No **19.** A matrix of scatterplots. **21.** $(188/8506)/(147/8102)=1.2182$, post menopausal women who are treated with estrogen and progestin have risk of coronary heart disease 21% higher than women who do not receive this treatment. **23.** Because coverage is intrinsically a quantitative variable, the intervals of coverage are naturally presented in order. Blood type is a nominal categorical variable, the order in which the blood types (A,B,AB,O) are presented is arbitrary and the 'shape' will depend of the order.

Chapter 3

1. D   **3.** $2/100=0.02$ Reject H_o.   **5.** C   **7.** A   **9.** (162.225, 188.14) Each dot is the mean of a sample of size 10 obtained with replacement from the original sample.   **11.** A

Chapter 4

3. $P(A \cup B)$ is the probability that at least one of the two events happens (A, B, or both). $P(A \cap B)$ is that probability that both of them happen. $P(A)$ is the probability that A happens. **5.** The basic Bernoulli experiment is to generate a letter with the roulette. That experiment is repeated 20 times, independently. The probability of getting an A is 1/4 in each trial or replicate of the experiment. The variable of interest is # of successes in the 20 trials. $P(X = 12) = \binom{20}{12}0.25^{12}(0.75)^8 = 0.0007517$. In **R** that probability can be calculated with the command **dbinom(12,20,0.25).** **7.** $n = 20$, $p = 0.3$, a) $P(x = 7) = \binom{20}{7}0.3^7(0.7)^{13} = 0.164262$, use the command dbinom(7,20,0.3) b) $P(X \leq 7) = 0.7722718$, use the command pbinom(7,20,0.3)

Chapter 5

1. C **3.** A **5.** Working with $\alpha = 0.0431745$ means that the null hypothesis is rejected if 5 or more individuals in the sample of 20 have the trait. The power is then $P(X \geq 5)$. When $p = 0.3$, $P(X \geq 5) = 0.7624922$. When $p = 0.5$, $P(X \geq 5) = 0.994091$. Yes, it makes sense because $p = 0.5$ is farther away than $p = 0.3$ from the value of $p = 0.1$ specified in the null statistical hypothesis. **7.** B

Chapter 6

1.a) 1929/4304 **1.b)** 722/4304 **1.c)** 722/1929
3.b) 53/72 **3.c)** 1 **3.d)** 0 **3.e)** 19/72
5) PPV = P(melanoma/test +) = (0.04*0.93)/(0.04*0.93+0.96*0.0725)
NPV = P(no melanoma/test−) = (0.96*0.9275)/(0.04*0.07+0.96*0.9275)

Chapter 7

8.a) 0.3985 **8.b** 0.6015 **8.c)** 0.0662 **9.** $((0.97)^9)*0.03 = 0.0228$ **10.a)** Binomial $n = 20$, $p = 0.2$,
$P(X = 5) = 0.1746$ **10.b)** Poisson with $\lambda = 5$ **10.c)** Geometric, $p = 0.10$, $P(X = 5) = 0.06561$

Chapter 8

1. P(Type A)=3/4 , P(no Type A)= 1/4

M	F	Genotype	Phenotype	Probability
	A	AA	Type A	1/4
A	a	Aa	Type A	1/4
a	A	Aa	Type A	1/4
	a	aa	no Type A	1/4

3. Aa, AA, aa **5.** a) Binomial $n=3$, $p=0.5$, $P(X = 0)= 0.125$, c)Binomial $n=3$, $p=0.25$, $P(X = 0)=0.421875$. **7.** $[\frac{1}{4}]^6=0.0002441406$ **11.** $4^4=256$, $4^2=16$. **13.** a) $n=6$, $p=1/4$, $P(X \geq 5=0.004638672$ b) $n=12$, $p=1/4$, $P(X \geq 10)= 3.761053e\text{-}05$.

Chapter 9

3. The parameters are a and b, we see the line of the density function starting at a and ending at b. **7.** $1-pnorm(-0.36)=0.6405764$ **9.** $pnorm(60, 76, 12)= 0.09121122$, $1 - pnorm(80, 76, 12)= 0.3694413$. **11.** $qnorm(0.2, 76, 12)= 65.9$ **13.** The number of degrees of freedom k, the mean is also k.

Chapter 10

1. $obs < -c(651, 207)$; $prob < -c(3/4, 1/4)$; $chisq.test(obs, p = prob)$. Output: X-squared = 0.3497, df = 1, p-value = 0.5543. No, the null hypothesis is not rejected.
3. The expected values are :
 108.7541   105.2459
 108.2459   104.7541
To perform the test in R, type: $x < -matrix(c(102, 115, 112, 98), nc = 2)$; chisq.test(x,correct=FALSE). Output: X-squared = 1.7098, df = 1, p-value = 0.191. The hypothesis that color and shape are independent is not rejected.
5. Reject H_o, the probability of getting venous thromboembolism is not the same if a woman takes estrogen + progesting than if she doesn't. **7.** a) 0.25,0.25,0.5 b)30, 30, 60. c) Homogeneity test d) Do not reject the null hypothesis, the data do not constitute evidence against the model that says that when parents are heterozygous the probability that the offspring will be

heterozygous is 1/2, and the probability that it will be homozygous of the dominant allele is 1/4, and the probability that it will be homozygous of the recessive allele is also 1/4. **9.** Do not reject the hypothesis of normality.

Chapter 11

3. Normal **5.** If many samples were drawn from a population, 95% of those samples would produce confidence intervals for the population parameter that actually contain the true value of the parameter.

7. We use the t when the value of the population standard deviation (σ) is not known, regardless of sample size.

9.a) wider **9.b)** narrower **9.c)** narrower.

11. $H_o : \mu_1 = \mu_2$, $H_a : \mu_1 \neq \mu_2$

13. We apply the regular t-test when the variances are assumed to be equal in both populations, and the Welch test when that assumption can not be done. However, if we apply the unequal variances option (Welch) and the standard deviations of the samples happen to be similar, both versions of the test give similar p-values and degrees of freedom.

15. $n = [(1.96/0.05)^2]0.5 \times 0.5 \sim 385$.

16. Binomial.

Chapter 12

1.a) female wing **1.c)** 0.9668, for each additional mm in the typical wing length of the males, the estimated typical wing length of the females goes up 0.9668 mm **1.d)** -0.9842, we do not interpret that value because it does not make sense to think of a robin species with typical wing length of 0. **1.f)** 87.9, -0.9842+0.9668*89.4=85.44772, residual= 87.9-85.44772=2.45228 is positive because the observed value is higher than the estimated value, i.e. the dot is above the regression line. **1.h)** Ashy robin, because the typical length wing of males is far from the mean value.

Introduction to Statistics in a Biological Context

Index

www.ingramcontent.com/pod-product-compliance
Lightning Source LLC
Chambersburg PA
CBHW081103170526
45165CB00008B/2308